AF443661

METHODS IN MOLECULAR BIOLOGY™

Series Editor
John M. Walker
School of Life Sciences
University of Hertfordshire
Hatfield, Hertfordshire, AL10 9AB, UK

For other titles published in this series, go to
www.springer.com/series/7651

Apoptosis

Methods and Protocols

Second Edition

Edited by

Peter Erhardt and Ambrus Toth

Boston Biomedical Research Institute, Watertown, MA, USA

Editors
Peter Erhardt
Boston Biomedical Research Institute
Watertown, MA
USA

Ambrus Toth
Boston Biomedical Research Institute
Watertown, MA
USA

ISSN 1064-3745   e-ISSN 1940-6029
ISBN 978-1-60327-016-8   e-ISBN 978-1-60327-017-5
DOI 10.1007/978-1-60327-017-5
Springer Dordrecht Heidelberg London New York

Library of Congress Control Number: 2009931045

© Humana Press, a part of Springer Science+Business Media, LLC 2004, 2009
All rights reserved. This work may not be translated or copied in whole or in part without the written permission of the publisher (Humana Press, c/o Springer Science+Business Media, LLC, 233 Spring Street, New York, NY 10013, USA), except for brief excerpts in connection with reviews or scholarly analysis. Use in connection with any form of information storage and retrieval, electronic adaptation, computer software, or by similar or dissimilar methodology now known or hereafter developed is forbidden.
The use in this publication of trade names, trademarks, service marks, and similar terms, even if they are not identified as such, is not to be taken as an expression of opinion as to whether or not they are subject to proprietary rights. While the advice and information in this book are believed to be true and accurate at the date of going to press, neither the authors nor the editors nor the publisher can accept any legal responsibility for any errors or omissions that may be made. The publisher makes no warranty, express or implied, with respect to the material contained herein.

Cover illustration: Apoptotic cells display the same distinct morphology detectable by DIC microscopy in different embryonic development stages (Fig. 25-2).

Printed on acid-free paper

Springer is part of Springer Science+Business Media (www.springer.com)

Preface

The ability to detect and quantify apoptosis, to understand its biochemistry, and to identify its regulatory genes and proteins is crucial to biomedical research. In this second edition of *Apoptosis* in *Methods in Molecular Biology*, expert researchers describe the techniques to best investigate the critical steps involved in the apoptotic process.

These readily reproducible step-by-step instructions are presented from several different research perspectives. The first part of the book provides an overview on the general techniques to detect apoptotic cell death, including methods such as caspase activity measurement, flow cytometry, live cell imaging, histopathology, and apoptosis detection in cell-free systems. In contrast, the second part lists methods to assess two forms of non-apoptotic cell death, necroptosis and autophagy.

Apoptotic proteins often undergo posttranslational modifications that alter their activity toward their downstream substrates. Techniques are described to analyze transglutamination, *S*-nitrosylation, and redox modifications of apoptotic proteins. Subsequently, several chapters are devoted to techniques that help dissect the major regulatory pathways of cell death and survival, including p53-dependent and independent and cell cycle regulatory proteins; the role of mitochondrial membrane permeabilization, unfolded protein response and ER stress, uncoupling protein-2, and microRNAs in programmed cell death; as well as the mechanism of phagocytosis by macrophages.

The fifth part of the book contains specific methodology required to evaluate apoptosis in various organs such as central nervous and cardiovascular system, myeloid progenitor cells as well as skeletal muscle. Techniques to detect apoptotic cell death during mammalian development are also described here. The final part of the book summarizes the approaches to study apoptosis in nonmammalian model organisms such as yeast, *Drosophila*, and *Caenorhabditis elegans*.

The protocols follow the *Methods in Molecular Biology* series format, each of them offering detailed laboratory instructions, an introduction outlining the principle behind the technique, lists of equipment and reagents, and tips on troubleshooting on how to avoid common pitfalls.

Apoptosis: Methods and Protocols, Second Edition, constitutes a key technical reference to the significant methodologies used in the field, and offers beginners and experienced researchers powerful tools to illuminate the phenomena of programmed cell death.

Watertown, MA   *Peter Erhardt*
August 2008   *Ambrus Toth*

Contents

Contributors

NEELAM AZAD • *Department of Pharmaceutical Sciences, West Virginia University, Morgantown, WV, USA*

GYORGY BAFFY • *Division of Gastroenterology & Liver Research Center, Brown Medical School and Rhode Island Hospital, Providence, RI, USA; Brigham and Women's Hospital, Harvard Medical School and VA Boston Healthcare System, Boston, MA, USA*

JOHN A. BLAHO • *Department of Microbiology, Mount Sinai School of Medicine, New York, NY, USA*

LISA BOUCHIER-HAYES • *Department of Immunology, St. Jude Children's Research Hospital, Memphis, TN, USA*

CHRISTINE BROWN • *Biology Department, University of Massachusetts, Amherst, MA, USA*

GEORGE A. CALIN • *Department of Experimental Therapeutics, The University of Texas M.D. Anderson Cancer Center, Houston, TX, USA*

DHYAN CHANDRA • *Department of Pharmacology and Therapeutics, Roswell Park Cancer Institute, Buffalo, NY, USA*

SAMUEL CONNELL • *Department of Immunology, St. Jude Children's Research Hospital, Memphis, TN, USA*

CHRISTOPHER R. COTTER • *Department of Microbiology, Mount Sinai School of Medicine, New York, NY, USA*

EVA CSIZMADIA • *Center for Vascular Biology, Beth Israel Deaconess Medical Center, Boston, MA, USA*

VILMOS CSIZMADIA • *Millennium Pharmaceuticals, Inc., Cambridge, MA, USA*

ZBIGNIEW DARZYNKIEWICZ • *Brander Cancer Research Institute, New York Medical College, Valhalla, NY, USA*

DIPAK K. DAS • *Cardiovascular Research Center, University of Connecticut, School of Medicine, Farmington, CT, USA*

ALEXEI DEGTEREV • *Department of Biochemistry, Tufts University School of Medicine, Boston, MA, USA*

ZOLTAN DERDAK • *Division of Gastroenterology & Liver Research Center, Brown Medical School and Rhode Island Hospital, Providence, RI, USA*

LORETTA DORSTYN • *Hanson Institute, Adelaide, Australia*

PETER ERHARDT • *Boston Biomedical Research Institute, Watertown, MA, USA*

LARS-PETER ERWIG • *Department of Medicine and Therapeutics, Institute of Medical Sciences, University of Aberdeen, Foresterhill, Aberdeen, UK*

LASZLO FESUS • *Departments of Biochemistry and Molecular Biology and Signaling and Apoptosis Research Group, Hungarian Academy of Sciences, Research Center for Molecular Medicine, University of Debrecen Medical and Health Sciences Center, Debrecen, Hungary*

ANDREW FRIBLEY • *Department of Biological Chemistry, The University of Michigan Medical Center, Ann Arbor, MI, USA*

ZHENGLIANG GAO • *Molecular and Cellular Biology Program, University of Massachusetts, Amherst, MA, USA*

TAMAKO A. GARCIA • *Division of Gastroenterology & Liver Research Center, Brown Medical School and Rhode Island Hospital, Providence, RI, USA*

SEAN P. GARRISON • *Department of Biochemistry, St. Jude Children's Research Hospital, Memphis, TN, USA*

GABRIEL GIL-GOMEZ • *Institut Municipal d'Investigació Mèdica (IMIM), Barcelona, Spain*

HONOR GLENN • *Pioneer Valley Life Sciences Institute, Springfield, MA, USA*

DOUGLAS R. GREEN • *Department of Immunology, St. Jude Children's Research Hospital, Memphis, TN, USA*

NARASIMMAN GURUSAMY • *Cardiovascular Research Center, University of Connecticut, School of Medicine, Farmington, CT, USA*

GYÖRGY HAJNÓCZKY • *Department of Pathology, Anatomy and Cell Biology, Thomas Jefferson University, Philadelphia, PA, USA*

J. MARIE HARDWICK • *Department of Pharmacology and Molecular Sciences, Johns Hopkins School of Medicine, Baltimore, MD, USA; W. Harry Feinstone Department of Molecular Microbiology and Immunology, Johns Hopkins University Bloomberg School and Public Health, Baltimore, MD, USA*

XIANGWEI HE • *Department of Human and Molecular Genetics, Baylor College of Medicine, Houston, TX, USA*

ANAND KRISHNAN V. IYER • *Department of Pharmaceutical Sciences, West Virginia University, Morgantown, WV, USA*

JOHN R. JEFFERS • *Department of Biochemistry, St. Jude Children's Research Hospital, Memphis, TN, USA*

RANDAL J. KAUFMAN • *Departments of Biological Chemistry, Internal Medicine and the Howard Hughes Medical Institute, The University of Michigan Medical Center, Ann Arbor, MI, USA*

NADEEM KHAN • *EPR Center for Viable Systems, Dartmouth Medical School, Hanover, NH, USA*

SHARAD KUMAR • *Hanson Institute, Adelaide, Australia*

YOUNGSOO LEE • *Department of Genetics and Tumor Cell Biology, St. Jude Children's Research Hospital, Memphis, TN, USA*

LIN LIN • *Medarex Inc., Bloomsbury, NY, USA*

DAN LIU • *Verna and Marrs McLean Department of Biochemistry and Molecular Biology, Baylor College of Medicine, Houston, TX, USA*

RICHARD A. LOCKSHIN • *Department of Biological Sciences, St. John's University, Queens, NY, USA*

NAN LU • *Verna and Marrs McLean Department of Biochemistry and Molecular Biology, Baylor College of Medicine, Houston, TX, USA*

ADEL MANDL • *Boston Biomedical Research Institute, Watertown, MA, USA*

KIMBERLY MCCALL • *Department of Biology, Boston University, Boston, MA, USA*

PETER J. MCKINNON • *Department of Genetics and Tumor Cell Biology, St. Jude Children's Research Hospital, Memphis, TN, USA*

KATHLEEN A. MCPHILLIPS • *Department of Pediatrics, National Jewish Medical and Research Center, Denver, CO, USA*

BENCHUN MIAO • *Department of Biochemistry, Tufts University School of Medicine, Boston, MA, USA*

ZOLTÁN NEMES • *Departments of Psychiatry and Signaling and Apoptosis Research Group, Hungarian Academy of Sciences, Research Center for Molecular Medicine, University of Debrecen Medical and Health Sciences Center, Debrecen, Hungary*

ANDREW OBERST • *Department of Immunology, St. Jude Children's Research Hospital, Memphis, TN, USA*

SANGRAM S. PARELKAR • *Molecular and Cellular Biology Program, University of Massachusetts, Amherst, MA, USA*

CARLOS PENALOZA • *Department of Biology, Queens College, Flushing, NY, USA*

JEANNE S. PETERSON • *Department of Biology, Boston University, Boston, MA, USA*

DARREN C. PHILLIPS • *Department of Biochemistry, St. Jude Children's Research Hospital, Memphis, TN, USA*

TRACY L. PRITCHETT • *Department of Biology, Boston University, Boston, MA, USA*

YON ROJANASAKUL • *Department of Pharmaceutical Sciences, West Virginia University, Morgantown, WV, USA*

RAMON ROSET • *Institut Municipal d'Investigació Mèdica (IMIM), Barcelona, Spain*

SOUMYA SINHA ROY • *Department of Pathology, Anatomy and Cell Biology, Thomas Jefferson University, Philadelphia, PA, USA*

LAWRENCE M. SCHWARTZ • *Pioneer Valley Life Sciences Institute, Springfield, MA, USA; Biology Department, University of Massachusetts, Amherst, MA, USA*

JOANNA SKOMMER • *Queen's Medical Research Institute, Edinburgh, UK*

ZHOU SONGYANG • *Verna and Marrs McLean Department of Biochemistry and Molecular Biology, Baylor College of Medicine, Houston, TX, USA*

RICCARDO SPIZZO • *Department of Experimental Therapeutics, The University of Texas M.D. Anderson Cancer Center, Houston, TX, USA*

STEPHEN W.G. TAIT • *Department of Immunology, St. Jude Children's Research Hospital, Memphis, TN, USA*

DEAN G. TANG • *Department of Carcinogenesis, The University of Texas M.D. Anderson Cancer Center, Smithville, TX, USA*

XINCHEN TENG • *Department of Pharmacology and Molecular Sciences, Johns Hopkins School of Medicine, Baltimore, MD, USA*

AMBRUS TOTH • *Charles River Laboratories, Wilmington, MA, USA*

DONALD WLODKOWIC • *The Bioelectronics Research Center, University of Glasgow, Glasgow, UK*

YIXIA YE • *Department of Biology, Queens College, Flushing, NY, USA*

XIAOMENG YU • *Verna and Marrs McLean Department of Biochemistry and Molecular Biology, Baylor College of Medicine, Houston, TX, USA*

ZAHRA ZAKERI • *Department of Biology, Queens College, Flushing, NY, USA*

GERARD P. ZAMBETTI • *Department of Biochemistry, St. Jude Children's Research Hospital, Memphis, TN, USA*

KEZHONG ZHANG • *Department of Biological Chemistry, The University of Michigan Medical Center, Ann Arbor, MI, USA; Center for Molecular Medicine and Genetics, Wayne State University School of Medicine, Detroit, MI, USA*

ZHENG ZHOU • *Verna and Marrs McLean Department of Biochemistry and Molecular Biology, Baylor College of Medicine, Houston, TX, USA*

Part I

Detection of Apoptosis

Chapter 1

Analysing Caspase Activation and Caspase Activity in Apoptotic Cells

Sharad Kumar and Loretta Dorstyn

Summary

Apoptotic cell death is characterised by various morphological and biochemical changes. Cysteine proteases of the caspase family play key roles in the execution of apoptosis and in the maturation of proinflammatory cytokines. During apoptosis signalling, caspase precursors undergo rapid proteolytic processing and activation. Activated caspases then function to cleave various vital cellular proteins, resulting in the death of the cell. Thus, the measurement of caspase activation and caspase activity provides a quick and convenient method to assess apoptosis. This chapter outlines various commonly used assays for measuring caspase activity and detecting active caspases in cultured cells or tissue extracts.

Key words: Apoptosis, Caspase activation, Synthetic peptides, Electrophoresis, Immunoblotting

1. Introduction

Apoptosis, or programmed cell death, is an active cellular signalling process triggered by a variety of stimuli such as deprivation of growth/survival factors, exposure to cytotoxic drugs or DNA damaging agents, activation of death receptors and action of cytotoxic cells. The process of apoptosis serves a crucial role in controlling cell number and eliminating harmful or virus-infected cells to maintain cell homeostasis throughout development. Apoptosis is tightly regulated by the family of cysteine aspartic proteases, termed caspases (*cysteine aspases*), which function by cleaving their substrates following an aspartate residue *(1–5)*. These proteases are the mammalian homologues of the *Caenorhabditis elegans*

Peter Erhardt and Ambrus Toth (eds.), *Apoptosis*, Methods in Molecular Biology, vol. 559
DOI 10.1007/978-1-60327-017-5_1
© Humana Press, a part of Springer Science + Business Media, LLC 2004, 2009

death protease CED-3 and include 13 mammalian and 7 *Drosophila* members *(6)*.

There are two major functions assigned to caspases. While caspase-1, -4, -5 and -11 are primarily involved in the processing and activation of proinflammatory cytokines, several others, including caspase-2, -3, -6, -7, -8 and -9 have been implicated in the execution phase of apoptosis *(1, 7, 8)*. All caspases exist as inactive precursor molecules or zymogens, which are activated by dimerization and/or proteolytic processing to generate active enzyme *(3–5, 9)*. The structural studies on active caspases predict that the mature enzymes have a heterotetrameric configuration composed of two heterodimers derived from two precursor molecules (*see* **Fig. 1**) *(10–13)*. In addition to the regions that give rise to two subunits, procaspases contain amino terminal prodomains of varying lengths. Caspases can be divided into two classes based on the length of their prodomain *(5, 14)*. Initiator caspases have long prodomains and include mammalian caspase-2, -9, -8, -10 and *Drosophila* DRONC. Effector or downstream caspases have short or absent prodomains and include mammalian caspase-3, -6, -7 and *Drosophila* Drice, Dcp-1. The long prodomains comprise protein–protein interaction motifs such as the caspase recruitment domain (CARD) in caspase-2, -9 and DRONC or a pair of death effector domains (DED) in caspase-8 and -10 which play a crucial role in caspase activation. These protein-interaction domains facilitate caspase recruitment to specific death adaptor complexes. Once activated, initiator caspases process and activate effector caspases, which then mediate the cleavage of a wide range of vital cellular proteins, resulting in the characteristic cellular morphological changes including membrane blebbing, nuclear condensation, fragmentation of DNA and ultimately the demise

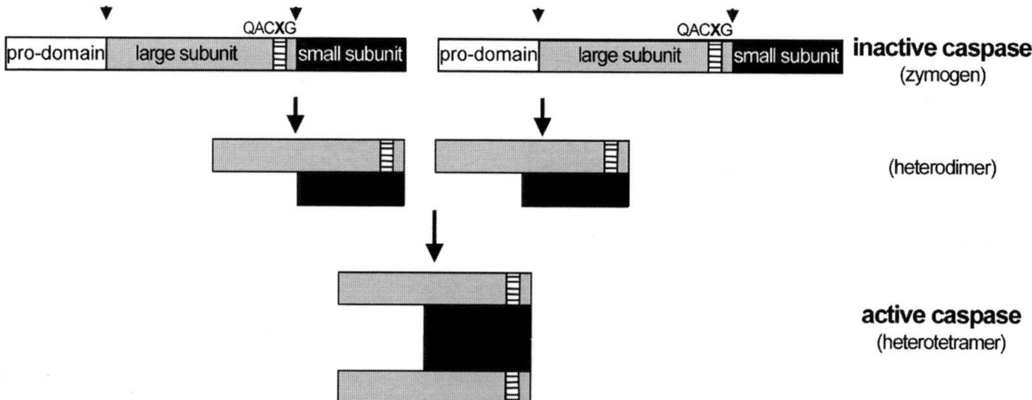

Fig. 1. Schematic representation of activated caspases. Cleavage sites following the prodomain and large subunits are indicated by *small arrowheads*. The active site "QACXG" is indicated by a *hatched box* in the large subunit.

of the cell *(4)*. The activation of caspases is of fundamental importance in cell death commitment and hence substantial efforts have been devoted to the understanding of mechanisms that underlie their activation *(5, 14–16)*.

Induction of apoptosis is almost always associated with the activation of caspases; therefore, measurement of caspase activity is a convenient way to assess whether the cells are undergoing apoptosis. There are several ways to measure caspase activation. Most common ones involve use of chromogenic or fluorogenic peptide substrates that release the chromogen or fluorescent tag upon cleavage by a caspase. Activated caspases can also be labelled in cells using fluorescent-conjugated antibodies specific for active caspases or by affinity labelling using biotin-conjugated peptide substrates. Other qualitative methods include monitoring the cleavage of in vitro synthesised ^{35}S-labelled caspase substrates, or measuring the cleavage of endogenous caspase substrates by immunoblotting using specific antibodies. In this chapter all these techniques are described.

The most direct and quantitative method for measuring caspase activity is by using synthetic peptide substrates. There are 13 mammalian caspases and optimal substrate specificities for many of these have been determined using peptide combinatorial libraries *(17, 18)*. The minimum substrate required for a caspase is usually a tetrapeptide sequence with an aspartate residue in P_1 position, a glutamate residue in the P_3 position and variable P_2 to P_4 residues based on cleavage specificity of individual caspases. With some caspases, such as caspase-2 and the *Drosophila* caspase DRONC, the presence of a P_5 residue greatly enhances substrate cleavage *(17, 19, 20)*. The most commonly used and commercially available substrates are listed in **Table 1**. While most caspases exhibit cleavage specificity for certain peptide substrates, it is important to note that most of the commonly used caspase substrates can be cleaved by several caspases, albeit at different efficiencies *(17, 18)*. Therefore, when assaying for caspase activity in crude cell extracts containing many active caspases, it is not possible to distinguish which caspases are contributing to activity by using substrates listed in **Table 1**. Furthermore, the abundance of individual caspases in a cell type can vary greatly, therefore the relative contribution of a single caspase to substrate cleavage is always difficult to assess.

While cleavage of effector caspases is required for activation, some initiator caspases, such as caspase-9 and DRONC are activated by dimerization and may not require activation by proteolytic cleavage *(3, 21)*. Therefore, cleavage is not necessarily a definitive measure of whether these initiator caspases are active. Methods for detecting active initiator and effector caspases in cells include the use of biotin-tagged, irreversible peptide inhibitors which mimic caspase substrates and block apoptosis *(22, 23)*.

Table 1
A list of commonly used synthetic peptide substrates for caspases

Caspase	Optimal substrate	Other substrates
Caspase-1	WEHD	YVAD
Caspase-2	VDVAD	DEVD
Caspase-3	DEVD	VDVAD
Caspase-4	WEHD	YVAD
Caspase-5	WEHD	YVAD
Caspase-6	VEID, VEHD	
Caspase-7	DEVD	VDVAD
Caspase-8	LETD	VEID, DEVD
Caspase-9	LEHD	
Caspase-10	LETD	VEID, DEVD

The peptide substrates usually have an Ac- or z- amino terminal blocking group and either AFC, AMC or pNA reporter at the carboxyl terminus. The optimal substrates are based on in vitro cleavage specificities determined by screening peptide combinatorial libraries using recombinant caspases expressed in *E. coli* (*4, 17, 18, 20*). Alternative substrates that can also be used for caspase assays are listed in the third column

Active caspases can then be isolated using immobilised streptavidin and in conjunction with immunoblotting for specific caspases, this method provides an indication of the specific caspases that are activated following specific death stimuli.

2. Materials

2.1. Cells Culture and Lysis

1. Humidified incubator at 37°C with 5% CO_2 for cell culture.

2. Cell culture media such as Dulbecco's Modified Eagle's Medium (DMEM) (SAFC Biosciences) supplemented with 10 mM 4-(2-Hydroxyethyl)-1-piperazineethanesulfonic acid (HEPES), 100 mM penicillin/streptomycin (CSL Biosciences) and 10% foetal bovine serum.

3. Trypsin (0.25%) is used for dissociation of adherant cells and is diluted in Hank's buffered salt solution (HBSS) (SAFC Biosciences) and stored in aliquots at –20°C.

4. Cell extraction buffer: 50 mM HEPES, pH 7.5, 50 mM NaCl, 10 mM dithiothreitol (DTT), 0.5 mM ethylenediaminetetraaceticacid (EDTA), 0.1% (3-[3-Cholamidopropyl)-dimethylammonio]-1-propanesulfonate (CHAPS), 10% sucrose, pH 7.0 and protease inhibitor cocktail such as Complete™ (Roche) (*see* **Notes 1–4**).

2.2. Caspase Substrates

1. The fluorogenic substrates with *N*-acetyl- (Ac) and C-terminal conjugated 7-amino-4-trifluoromethylcoumarin (AFC) or 7-amino-4-methylcoumarin (AMC) reporters, and colorimetric substrates with *p*-nitroanilide (pNA) reporter are available from various commercial sources.

2. Cell permeable caspase substrates are synthesized with a benzyloxycarbonyl group (z) at the N-terminus and O-methyl side chains to enhance cellular permeability thus facilitating their use in both in vitro cell culture as well as in vivo animal studies.

3. Caspase inhibitors are commonly conjugated to chloromethylketone (cmk), fluoromethylketone (fmk) or CHO-aldehyde, which act as effective irreversible inhibitors with no added cytotoxic effects. There are various biotin-conjugated caspase inhibitors available; the most commonly used are biotin-valine–alanine–aspartate-fluoromethyl ketone (bVAD-fmk) and biotin-aspartate–glutamate–valine–aspartate-CHO (bDEVD-CHO). Two of the earliest known suppliers are Enzyme Systems Products, Inc. (USA) and Bachem (Switzerland), but many commonly used caspase substrates and inhibitors can now be bought from numerous different suppliers. AMC/AFC calibration standards are available from various suppliers (e.g. BioMol International) (*see* **Note 5**).

2.3. Spectrometers

1. For the measurement of fluorescence, a luminescence spectrometer, such as Perkin-Elmer LS50B fluorimeter or a FLUOstar Optima Luminescence Spectrometer (BMG LabTech), preferably equipped with a thermostated plate reader is required.

2. If using pNA colorimetric substrates, a spectrophotometer, preferably equipped with a thermostated cuvette or plate holder is required.

2.4. Caspase Assay Buffers

1. Caspase assay buffer: 100 mM HEPES pH 7.0, 10% sucrose, 0.1% CHAPS, 0.5 mM EDTA and 10 mM DTT and store in aliquots at –20°C. Alternatively, assay buffer without DTT can be stored at room temperature for several months and DTT added to 10 mM from a fresh 1 M stock as required (*see* **Note 1**).

2. 2× Protein Loading Buffer: 100 mM Tris–HCl, pH 6.8, 200 mM DTT, 20% glycerol, 4% SDS, 0.2% bromophenol blue.

2.5. In Vitro Translated Proteins

1. A convenient kit for in vitro coupled transcription/translation using rabbit reticulocyte lysate is commercially available from Promega Corporation. Alternatively, reagents for in vitro transcription and in vitro translation can be purchased separately.

2. For the synthesis of radiolabelled proteins, we commonly use ^{35}S-Methionine (ICN Biochemicals) and follow the protein transcription/translation instructions provided by the manufacturer. Translated proteins can be stored for up to 2 weeks at $-70°C$.

2.6. Protein Electrophoresis and Transfer

1. 8–15% acrylamide gels: 375 mM Tris–HCl, pH 8.8, 0.1% SDS, 0.1% APS.

2. Stacking gels: 5% acrylamide, 125 mM Tris–HCl, pH 6.8, 0.1% SDS, 0.1% APS. 0.1% N,N,N,N'-Tetramethyl-ethylenediamine (TEMED) is added to gels to polymerise acrylamide.

3. 40% acrylamide/bis solution (37.5:1 with 2.6% C) can be purchased from Bio-Rad (note that acrylamide is a neurotoxin when unpolymerised and so care should be taken to avoid exposure).

4. Ammonium persulfate is prepared as a 10% solution in water and frozen in aliquots at $-20°C$. Repeated freeze-thaw cycles can greatly reduce product stability.

5. Running buffer (1×): 25 mM Tris, 192 mM glycine, 0.1% (w/v) SDS. A 5× buffer can be pre-made and stored at room temperature.

6. Prestained molecular weight markers can be purchased from various suppliers and include Kaleidoscope markers (BioRad), prestained protein markers (Invitrogen) or unstained protein markers (Invitrogen).

7. A standard protein electrophoresis apparatus and a semi-dry protein transfer apparatus (such as Hoefer™ SemiPhor) are required. Details of protein electrophoresis and transfer protocols can be found in various protocol books such as *Molecular Cloning (24)*.

2.7. Immunoblotting

1. Antibodies against many caspase substrates and secondary conjugates are commercially available. Most commonly used caspase substrate is poly (ADP)ribose polymerase (PARP). The anti-PARP antibody supplied by Roche Molecular Biology cleanly detects the 115-kDa PARP precursor and the 89-kDa cleavage product *(25)*. Other common sources of antibodies include BD Biosciences Pharmingen (USA) and SantaCruz (USA). Antibodies are commonly used at concentrations of 0.5–1 µg/mL (as specified by the manufacturer) and are diluted in 1–5% skim milk in phosphate buffered saline (PBS) containing 0.05% Tween 20 (PBST).

2. Secondary antibodies used are commonly conjugated to alkaline phosphatase (AP) or horse radish peroxidase (HRP) and proteins detected by enhanced chemifluorescence (ECF) and enhanced chemiluminescence (ECL), respectively, according to manufacturers' protocols (GE Healthcare/Amersham).

2.8. Affinity Capture of Active Caspase

1. Affinity labels, such as biotin-VAD-fmk or biotin-DEVD-CHO, are diluted to 50 µM working solution in Buffer A: 50 mM NaCl, 2 mM MgCl$_2$, 5 mM ethylene glycol-bis[β-aminoethyl ether]-N',N',N',N'-tetraacetic acid (EGTA), 10 mM HEPES, 1 mM DTT, pH 7.

2. Cells are resuspended in Buffer B: 50 mM KCl, 50 mM piperazine-N,N'-bis[2-ethanesulphonic acid] (PIPES), 10 mM EGTA, 2 mM MgCl$_2$, 1 mM DTT, 0.1 mM phenylmethanesulfonyl fluoride (PMSF) and containing protease inhibitor cocktail.

3. Immobilised streptavidin, Streptavidin-sepharose, is available from GE Healthcare/Amersham.

---

3. Methods

3.1. Measurement of Caspase Activity Using Synthetic Peptide Substrates

3.1.1. Preparation of Cell Extracts

1. Grow cells in culture using standard methods. Prepare protein extracts from untreated cells and cells treated with the appropriate apoptosis stimuli.

2. For preparation of cell extracts from animal tissue samples, homogenise frozen tissue cut into small pieces in Extraction Buffer using a tissue homogeniser prior to cell lysis.

3. For cultured cells in suspension, spin down cells at $200 \times g$ for 10 min and wash once in ice cold PBS.

4. For adherent cells, gently scrape cells into medium, spin down cell pellet at $200 \times g$ for 10 min and wash once in cold PBS.

5. Resuspend cells at approximately 10^7 cells/mL in extraction buffer.

6. Freeze/thaw cells three times in liquid nitrogen/ice cold water.

7. Centrifuge extracts at $15,000 \times g$ for 10 min at 4°C and carefully transfer supernatant (cytosolic extract) to a clean tube, leaving the pellet undisturbed.

8. After determining protein concentration, using a standard BCA assay, the extracts can be stored on ice until use, or for long-term storage extracts can be frozen at -70°C in small aliquots for several months without any significant loss of enzyme activity.

*3.1.2. Measurement of
Caspase Activity*

Caspase assays should be performed continuously if the spectro-photometer is equipped with a regulated temperature chamber that can accommodate cuvettes or 96-well plates, otherwise the release of AFC, AMC or pNA can be monitored after a fixed period of incubation (usually 1 h at 37°C). To save reagents, carry out assays in a final volume of 50–100 µL. If the fluorimeter is equipped with a plate reader, several assays can be carried out simultaneously using a 96-well plate. If this is not possible, reactions can be carried out in microfuge tubes or cuvettes. Appropriate controls may include untreated cells or cells treated with a caspase inhibitor such as z-VAD-fmk.

1. Add varying concentrations of the cell lysates (10–50 µg) to caspase assay buffer supplemented with 0.1 mM of an appropriate caspase substrate and monitor the release of fluorochrome or chromogen at 37°C in the thermostat fitted spectrophotometer (*see* **Notes 1–3**). For AMC fluorescence detection adjust the excitation and emission wavelengths to 360 nm and 460 nm, respectively. For AFC, excitation and emission wavelengths are 400 nm and 505 nm, respectively. pNA absorbance should be monitored at 405–410 nm.

2. Monitor the release of the fluorochrome or chromogen every 10–15 min over 1–2 h. Plot data as fluorescence (for AFC or AMC) or absorbance (for pNA) vs. time (min) for each sample (*see* **Fig. 2a**). Calculate the slope of the line from the linear portion of the progress curves. Highly active samples can deplete the substrate rapidly, so in the case that substrate depletion occurs too quickly, dilute cell extracts to get a more linear response.

   If continuous monitoring of fluorochrome or chromogen release is not possible, assays can be carried out for various lengths of time, up to 1 h at 37°C. At the end of the incubation, stop the reactions by adding 0.4 mL of ice cold water and storing tubes on ice. Transfer reactions to a cuvette and measure release of AMC/AFC on a fluorimeter.

3. The above data will give a qualitative indication of caspase activity. To quantify caspase activity in each sample, express as pmol substrate hydrolysed/min. This will require setting up an AMC/AFC fluorescence calibration curve. Prepare serial dilutions of the AMC/AFC Calibration Standards in assay buffer to a final volume of 100 µl, such that concentration ranges from 0 to 50 µM, and measure the fluorescence of each standard dilution.

   Plot relative fluorescence units (RFU) against AMC/AFC concentration (µM). The graph should be linear and the slope of the line can be calculated (*see* **Fig. 2b**).

4. Caspase activity can now be calculated as pmol substrate hydrolysed per minute (*see* **Fig. 2c**).

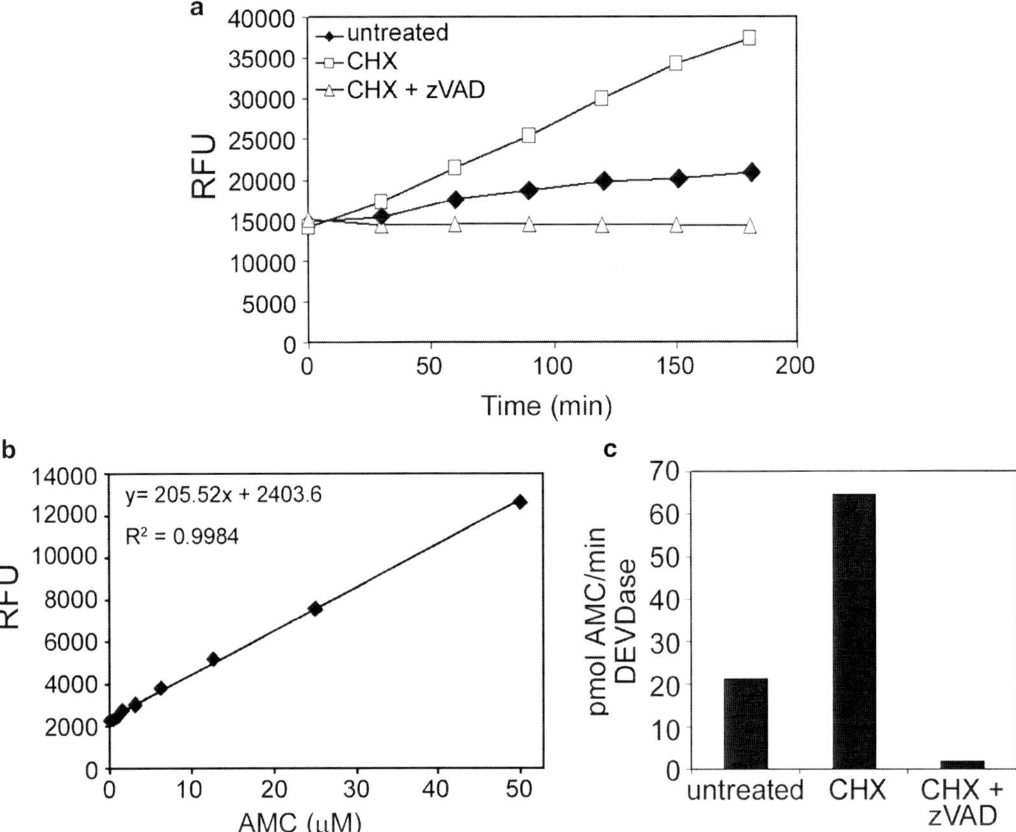

Fig. 2. An example of cleavage of the synthetic fluorogenic substrate Ac-DEVD-AMC by cell extracts. *Drosophila* BG2 neuronal cells were left untreated or were treated with 15 µg/mL cycloheximide (CHX) for 6 h to induce apoptosis. Thirty micrograms of cell extracts were assayed for Ac-DEVD-AMC activity at 30°C over 2 h (Ac-DEVD-AMC final concentration 50 µM). The rate of hydrolysis was measured by release of AMC using a FLUOstar Optima Luminescence Spectrometer (BMG Labtech, excitation 360 nm, emission 460 nm). (**a**) DEVDase activity of BG2 cell extracts expressed as relative fluorescence units (RFU) over time. (**b**) The standard curve of AMC was plotted as RFU over time and the slope of the line calculated. (**c**) The DEVD cleavage activity of the BG2 cell extracts expressed in pmol/min.

$$\text{Activity} = \text{slope of sample } (\text{RFU/min}) \times [1/\text{slope of calibration standard}](\mu M/\text{RFU}) \times \text{assay volume}$$

3.2. Assay of Caspase Activity by Cleavage of ^{35}S-Met Labelled Caspase Substrates

This is a qualitative assay that is suitable for confirming the presence of active caspases in cell extracts. Clone the cDNAs containing caspase cleavage sites, such as PARP *(26, 27)*, DNA-PK catalytic subunit *(28)*, ICAD *(29)*, in plasmid vectors that carry either SP6, T3 or T7 promoters (pBluescript, pGEM and pcDNA3 vectors are all appropriate for this purpose). It is not necessary to clone the entire protein, truncated coding regions containing the caspase site(s) which give rise to easily discernible cleavage products work well *(30)*.

1. Purify plasmids using CsCl centrifugation or by alkaline lysis using a QIAGEN Plasmid kit and carry out in vitro transcription/translation using Promega TNT-rabbit reticulocyte translation system according to the instructions provided by the manufacturer. Typical 50 µL reactions contain 25 µL TNT lysates, 2 µL TNT reaction buffer, 1 µL T3, T7 or SP6 RNA polymerase, 1 µL amino acid mixture lacking Met, 3 µl ³⁵S-Methionine, 1 µL RNase inhibitor, 1 µg plasmid DNA and sterile RNase-free water.

2. Incubate reaction tubes at 30°C for 1.5–2 h, centrifuge at 10,000 × *g* in a microfuge for 5 min and transfer supernatant to fresh tube. In vitro translated proteins can be stored at –70°C for up to 2 weeks.

3. For cleavage assays, incubate 5 µL of labelled protein at 37°C for 2 h with varying amounts of cell extracts (10–50 µg total protein) in caspase assay buffer in a total volume of 20 µL. In control experiments, cell extracts can be preincubated with caspase inhibitors, such as 50 µM zVAD-fmk for 30 min prior to the addition of labelled protein substrate.

4. At the end of incubation period, add 20 µL of 2× protein loading buffer to each tube, boil for 5 min and centrifuge at 10, 000 × *g* in a microfuge for 5 min.

5. Remove supernatant to fresh tube and resolve cleavage products by electrophoresis on 10–15% polyacrylamide/SDS gel.

6. Following fixation, gels can be dried. Alternatively, proteins can be transferred to polyvinylidine difluoride (PVDF) membranes using a semi-dry transfer apparatus for 90 min at 130 mA, and ³⁵S-labelled protein bands visualized by autoradiography. This avoids the possibility of gels cracking during the drying process. In most cases, freshly labelled ³⁵S-proteins and their cleavage products can be detected following an overnight exposure to X-ray film or phosphor screen.

3.3. Assessing Caspase-mediated Substrate Cleavage by Immunoblotting

1. Since caspase activation results in the cleavage of the caspase precursor into subunits, caspase activation can be indirectly observed by immunoblotting using specific antibodies (**Fig. 3**) (*see* **Note 6**). However, a more direct measure of caspase activity, in particular that contributed by the downstream or effector caspases such as caspase-3 and caspase-7, is to determine whether endogenous caspase targets are being cleaved. This can be easily achieved by immunoblotting of cell extracts using a specific antibody against a known endogenous caspase substrates. There are hundreds of proteins now known to be cleaved by caspases *(31)*. The most common one, for which good antibodies are

available from many commercial suppliers, is PARP, a caspase-3 substrate. Prepare samples for electrophoresis by mixing equal volume of protein extract, prepared as described in **Subheading 3.1.1** and 2× protein loading buffer.

2. Cell pellets or small pieces of tissues can also be directly lysed by boiling in 2× protein loading buffer. However, often the lysates prepared in such a way will be very viscous due to the release of DNA. To reduce viscosity, the samples can be passed through a 22 gauge needle 3–4 times or sonicated for 30 s to shear DNA.

3. Boil samples for 5–10 min and centrifuge lysates for 5 min at 10,000 × g in a microfuge to remove any insoluble material. At this stage, if required, the samples can be stored at –70°C indefinitely.

4. Electrophorese 30–50 μg of the protein samples on a 10% polyacrylamide/SDS gel.

5. Transfer proteins to PVDF membrane using a semi-dry protein transfer apparatus.

6. Block membrane in 5% skim milk-PBST for 1 h at room temperature or overnight at 4°C.

7. Dilute primary antibody as suggested by the manufacturer in 1–5% skim milk/PBST and incubate the membrane with the antibody solution for 1 h at room temperature.

8. Wash membrane three times for 10 min each and incubate with the appropriate secondary antibody diluted in 1–5% skim milk/PBST.

9. For detection of signals by ECL or ECF, follow instruction supplied by manufacturer (e.g. GE Healthcare/Amersham). As an example, in healthy cells PARP will appear as a single band of approximately 115 kDa, whereas in cells undergoing apoptosis a gradual decrease in 115-kDa band and appearance of 89-kDa cleavage product should be clearly visible.

3.4. Affinity Labelling of Active Caspases

Active caspases in cells or cell extracts can be labelled using biotin-conjugated peptide inhibitors such as bVAD-fmk or bDEVD-CHO, which bind covalently to the active cysteine site of most caspases. While bVAD-fmk is a general caspase inhibitor and used commonly to isolate both active initiator and effector caspase species (*32–34*), bDEVD-CHO inhibitor is commonly used to specifically capture active effector caspase species such as caspase-3, -6 and -7 (*33*) (*see* **Note 7**).

1. Affinity labels are diluted to 50 μM in Buffer A.

2. Cells are left untreated or are treated with an apoptotic stimulus for the required time and then harvested and resuspended at 1 × 10⁷ cells/mL in Buffer B.

3. Centrifuge cells at 1,000 × g and remove supernatant. Retain cell pellet.

4. Snap freeze cell pellet in liquid nitrogen.

5. Add an equal volume of the 50 µM bVAD-fmk or bDEVD-CHO to the cell pellet and lyse cells by three cycles of freeze-thawing in liquid nitrogen/ice cold water.

6. Incubate lysates at 37°C for 30 min and pellet debris by centrifugation at 100,000 × *g* for 20 min at 4°C.

7. Retain supernatant and transfer to a clean microfuge tube. Add 0.05 volumes of streptavidin sepharose and incubate at 4°C with rotation, overnight.

8. Wash streptavidin sepharose with 10 volumes of Buffer B.

9. Add an equal volume of 2× Protein Loading Buffer to the sepharose and boil for 5 min.

10. Separate proteins on 10–15% polyacrylamide-SDS gels.

11. Transfer proteins to PVDF membrane and immunoblot with the caspase antibodies of interest as described in **Subheading 3.3**.

4. Notes

1. Although most caspases are active at pH 7.0, some have different pH optima. For example, caspase-2 and caspase-9 favour slightly acidic pH *(35)*. If necessary, the assay buffer containing 0.1 M (2-[*N*-Morpholino]ethanesulfonic acid) (MES), pH 6.5 can be used instead of 0.1 M HEPES, pH 7.0.

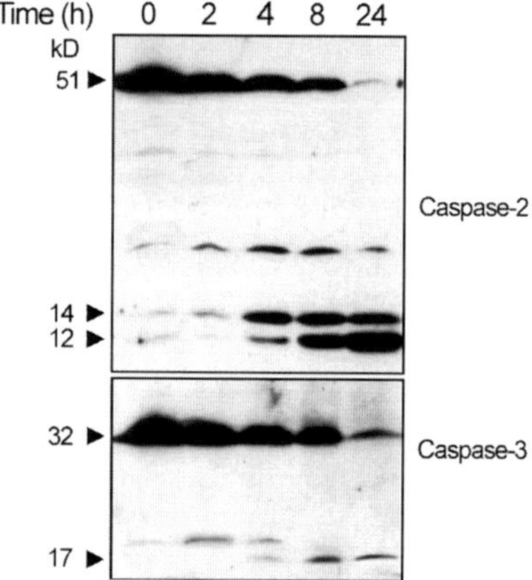

Fig. 3. Detection of processing of caspases in apoptotic cells. Jurkat cells were treated with etoposide (40 µM) over 24 h. Cell extracts were prepared and protein electrophoresed through SDS-polyacrylamide gels and transferred to PVDF membrane. Membranes were immunoblotted with anti-caspase-2L rabbit polyclonal antibody (C20, Santa Cruz Biotechnology) or anti-caspase-3 mouse monoclonal antibody (BD Biosciences Pharmingen), followed by immunoblotting with a HRP-conjugated secondary antibody (GE Healthcare, Amersham), and signals detected by ECL.

2. If necessary, recombinant caspases expressed in *Escherichia coli* can be used for positive controls. A number of publications describe the preparation of recombinant caspases *(28, 35, 36)*. Some commercial suppliers, such as Alexis Biochemicals (Switzerland), also provide a number of recombinant purified caspases.

3. To avoid non-specific hydrolysis of caspase substrates, it is useful to include protease inhibitor cocktail in the cell lysis buffer. Many commercially available protease inhibitor sets can be used provided they do not contain caspase inhibitors.

4. For a positive control, mammalian cell lines treated with apoptosis inducing agents can be used. As a guide, extracts prepared from Jurkat cells treated for 2 h with 200 ng/mL of an anti-Fas antibody (e.g. from Upstate Biotechnology), or for 4 h with 40 μM etoposide, will show significant levels of caspase activity on IETD, DEVD and VDVAD substrates. Extracts from treated cells can be prepared as described in **Subheading 3.1.1**.

5. In our experience, fluorogenic assays are far more (50–100-fold) sensitive than the colorimetric assays. This may be an important consideration when there is a limited availability of starting material (cells or tissue sample). AMC and AFC-conjugated substrates can be stored at –20° C as 5–10 mM stock solution in dimethyl formamide for 1–2 years. Dissolve pNA substrates at 20 mM in dimethyl formamide and store at –20°C. Caspase inhibitors are made up at 10 mM stock in DMSO and stored in aliquots at –20°C. Avoid repeated freeze-thaw cycles, which greatly reduce product stability.

6. To test whether individual caspases are being activated, immunoblot analysis of cell extracts using specific caspase antibodies can be performed. To do this, prepare cell extract blots as described in Subheading 3.3 and probe them with caspase antibodies to determine whether a specific caspase precursor is being cleaved into active subunits. There are numerous commercial sources of caspase antibodies; however, many antibodies on the market are of poor quality. If using a new antibody for the first time, especially when the same antibody has not been used in the published literature, specificity and affinity of the antibody should be empirically established using recombinant caspases. Some antibodies will detect both the precursor and one or more subunits/intermediates, while others are specific for either the precursor or the subunits. In some cell types, the half-life of some active caspase subunits is often very short. In such cases a clear decrease in zymogen signal can be seen but not a corresponding increase in the subunit signal.

7. Affinity labelling of active caspases can prove to be technically difficult and we have tested various published methods. Our

protocol is a modification of the method described in Faleiro et al. *(33)* and works well when detecting effector caspases. bVAD and bDEVD can also be added directly to cell extracts (containing approximately 1 mg protein) and following the procedure (*see* **Subheading 3.4, steps 5–11**). Intiator caspases are more difficult to detect as their activation appears to be more transient. If the above method does not detect active initiator caspases, bVAD-fmk can be directly added to cells in culture at the same time of treatment with apoptotic stimuli. Cells are then incubated in the presence of bVAD-fmk for several hours prior to harvesting and cell lysis.

Acknowledgements

The financial support of the National Health and Medical Research Council is gratefully acknowledged. LD is supported by a Royal Adelaide Hospital Florey Research Fellowship.

References

1. Kumar, S. (2007). Caspase function in programmed cell death. *Cell Death Differ.* **14**, 32–43.

2. Salvesen, G. S., and Abrams, J. M. (2004). Caspase activation – stepping on the gas or releasing the brakes? Lessons from humans and flies. *Oncogene* **23**, 2774–2784.

3. Shi, Y. (2004). Caspase activation, inhibition, and reactivation: a mechanistic view. *Protein Sci.* **13**, 1979–1987.

4. Nicholson, D. W. (1999). Caspase structure, proteolytic substrates, and function during apoptotic cell death. *Cell Death Differ.* **6**, 1028–1042.

5. Kumar, S. (1999). Mechanisms mediating caspase activation in cell death. *Cell Death Differ.* **6**, 1060–1066.

6. Hengartner, M. O. (2000). The biochemistry of apoptosis. *Nature* **407**, 770–776.

7. Siegel, R. M. (2006). Caspases at the crossroads of immune-cell life and death. *Nat. Rev. Immunol.* **6**, 308–317.

8. Ranger, A. M., Malynn, B. A., and Korsmeyer, S. J. (2001). Mouse models of cell death. *Nat. Genet.* **28**, 113–118.

9. Riedl, S. J., and Shi, Y. (2004). Molecular mechanisms of caspase regulation during apoptosis. *Nat. Rev. Mol. Cell Biol.* **5**, 897–907.

10. Wilson, K. P., Black, J. A., Thomson, J. A., Kim, E. E., Griffith, J. P., Navia, M. A., et al. (1994). Structure and mechanism of interleukin-1 beta converting enzyme. *Nature* **370**, 270–275.

11. Walker, N. P., Talanian, R. V., Brady, K. D., Dang, L. C., Bump, N. J., Ferenz, C. R., et al. (1994). Crystal structure of the cysteine protease interleukin-1 beta-converting enzyme: a (p20/p10)2 homodimer. *Cell* **78**, 343–352.

12. Rotonda, J., Nicholson, D. W., Fazil, K. M., Gallant, M., Gareau, Y., Labelle, M., et al. (1996). The three-dimensional structure of apopain/CPP32, a key mediator of apoptosis. *Nat. Struct. Biol.* **3**, 619–625.

13. Mittl, P. R., Di Marco, S., Krebs, J. F., Bai, X., Karanewsky, D. S., Priestle, J. P., et al. (1997). Structure of recombinant human CPP32 in complex with the tetrapeptide acetyl-Asp-Val-Ala-Asp fluoromethyl ketone. *J. Biol. Chem.* **272**, 6539–6547.

14. Kumar, S., and Colussi, P. A. (1999). Prodomains–adaptors–oligomerization: the

pursuit of caspase activation in apoptosis. *Trends Biochem. Sci.* **24**, 1–4.

15. Boatright, K. M., and Salvesen, G. S. (2003). Mechanisms of caspase activation. *Curr. Opin. Cell Biol.* **15**, 725–731.

16. Danial, N. N., and Korsmeyer, S. J. (2004). Cell death: critical control points. *Cell* **116**, 205–219.

17. Talanian, R. V., Quinlan, C., Trautz, S., Hackett, M. C., Mankovich, J. A., Banach, D., et al. (1997). Substrate specificities of caspase family proteases. *J. Biol. Chem.* **272**, 9677–9682.

18. Thornberry, N. A., Rano, T. A., Peterson, E. P., Rasper, D. M., Timkey, T., Garcia-Calvo, M., et al. (1997). A combinatorial approach defines specificities of members of the caspase family and granzyme B. Functional relationships established for key mediators of apoptosis. *J. Biol. Chem.* **272**, 17907–17911.

19. Hawkins, C. J., Yoo, S. J., Peterson, E. P., Wang, S. L., Vernooy, S. Y., and Hay, B. A. (2000). The Drosophila caspase DRONC cleaves following glutamate or aspartate and is regulated by DIAP1, HID, and GRIM. *J. Biol. Chem.* **275**, 27084–27093.

20. Dorstyn, L., Colussi, P. A., Quinn, L. M., Richardson, H., and Kumar, S. (1999). DRONC, an ecdysone-inducible Drosophila caspase. *Proc. Natl Acad. Sci. USA* **96**, 4307–4312.

21. Stennicke, H. R., Deveraux, Q. L., Humke, E. W., Reed, J. C., Dixit, V. M., and Salvesen, G. S. (1999). Caspase-9 can be activated without proteolytic processing. *J. Biol. Chem.* **274**, 8359–8362.

22. Nicholson, D. W., Ali, A., Thornberry, N. A., Vaillancourt, J. P., Ding, C. K., Gallant, M., et al. (1995). Identification and inhibition of the ICE/CED-3 protease necessary for mammalian apoptosis. *Nature* **376**, 37–43.

23. Thornberry, N. A., Peterson, E. P., Zhao, J. J., Howard, A. D., Griffin, P. R., and Chapman, K. T. (1994). Inactivation of interleukin-1 beta converting enzyme by peptide (acyloxy)methyl ketones. *Biochemistry* **33**, 3934–3940.

24. Sambrook, J., and Russell, D. (2001). *Molecular Cloning: A Laboratory Manual*, Vols. 1–3, 3rd ed., Cold Spring Harbor Laboratory, Cold Spring Harbor, NY.

25. Harvey, K. F., Harvey, N. L., Michael, J. M., Parasivam, G., Waterhouse, N., Alnemri, E. S., et al. (1998). Caspase-mediated cleavage of the ubiquitin-protein ligase Nedd4 during apoptosis. *J. Biol. Chem.* **273**, 13524–13530.

26. Kaufmann, S. H., Desnoyers, S., Ottaviano, Y., Davidson, N. E., and Poirier, G. G. (1993). Specific proteolytic cleavage of poly(ADP-ribose) polymerase: an early marker of chemotherapy-induced apoptosis. *Cancer Res.* **53**, 3976–3985.

27. Lazebnik, Y. A., Kaufmann, S. H., Desnoyers, S., Poirier, G. G., and Earnshaw, W. C. (1994). Cleavage of poly(ADP-ribose) polymerase by a proteinase with properties like ICE. *Nature* **371**, 346–347.

28. Song, Q., Lees-Miller, S. P., Kumar, S., Zhang, Z., Chan, D. W., Smith, G. C., et al. (1996). DNA-dependent protein kinase catalytic subunit: a target for an ICE-like protease in apoptosis. *EMBO J.* **15**, 3238–3246.

29. Sakahira, H., Enari, M., and Nagata, S. (1998). Cleavage of CAD inhibitor in CAD activation and DNA degradation during apoptosis. *Nature* **391**, 96–99.

30. Harvey, N. L., Butt, A. J., and Kumar, S. (1997). Functional activation of Nedd2/ICH-1 (caspase-2) is an early process in apoptosis. *J. Biol. Chem.* **272**, 13134–13139.

31. Timmer, J. C., and Salvesen, G. S. (2007). Caspase substrates. *Cell Death Differ.* **14**, 66–72.

32. Ekert, P. G., Silke, J., and Vaux, D. L. (1999). Caspase inhibitors. *Cell Death Differ.* **6**, 1081–1086.

33. Faleiro, L., Kobayashi, R., Fearnhead, H., and Lazebnik, Y. (1997). Multiple species of CPP32 and Mch2 are the major active caspases present in apoptotic cells. *EMBO J.* **16**, 2271–2281.

34. Tu, S., McStay, G. P., Boucher, L. M., Mak, T., Beere, H. M., and Green, D. R. (2006). In situ trapping of activated initiator caspases reveals a role for caspase-2 in heat shock-induced apoptosis. *Nat. Cell Biol.* **8**, 72–77.

35. Garcia-Calvo, M., Peterson, E. P., Rasper, D. M., Vaillancourt, J. P., Zamboni, R., Nicholson, D. W., et al. (1999). Purification and catalytic properties of human caspase family members. *Cell Death Differ.* **6**, 362–369.

36. Harvey, N. L., Trapani, J. A., Fernandes-Alnemri, T., Litwack, G., Alnemri, E. S., and Kumar, S. (1996). Processing of the Nedd2 precursor by ICE-like proteases and granzyme B. *Genes Cells* **1**, 673–685.

Chapter 2

Flow Cytometry-Based Apoptosis Detection

Donald Wlodkowic, Joanna Skommer, and Zbigniew Darzynkiewicz

Summary

An apoptosing cell demonstrates multitude of characteristic morphological and biochemical features, which vary depending on the stimuli and the cell type. The gross majority of classical apoptotic hallmarks can be rapidly examined by flow and image cytometry. Cytometry thus became a technology of choice in diverse studies of cellular demise. A large variety of cytometric methods designed to identify apoptotic cells and probe mechanisms associated with this mode of cell demise have been developed during the past two decades. In the present chapter, we outline a handful of commonly used methods that are based on the assessment of: mitochondrial transmembrane potential, activation of caspases, plasma membrane alterations and DNA fragmentation.

Key words: Flow cytometry, Apoptosis, Single cell analysis, Mitochondria, Caspases, Annexin V, DNA fragmentation

1. Introduction

During the past decade, mechanisms underlying cell death have entered into a focus of many researchers in the diverse fields of biomedicine. These mechanisms include a wide range of signaling cascades that regulate initiation, execution, and postmortem cell disposal mechanisms *(1–3)*. **Figure 1** outlines major morphological and molecular changes occurring during classical caspase-dependent apoptosis vs. accidental cell death (herein termed as necrosis). Alterations in parameters presented in **Fig. 1** become a foundation to development of many markers for microscopy, cytometry, and molecular biology techniques *(1, 4)*. It is important to note, however, that the occurrence of specific apoptotic markers can show a profound divergence. Moreover, burgeoning

Peter Erhardt and Ambrus Toth (eds.), *Apoptosis*, Methods in Molecular Biology, vol. 559
DOI 10.1007/978-1-60327-017-5_2
© Humana Press, a part of Springer Science + Business Media, LLC 2004, 2009

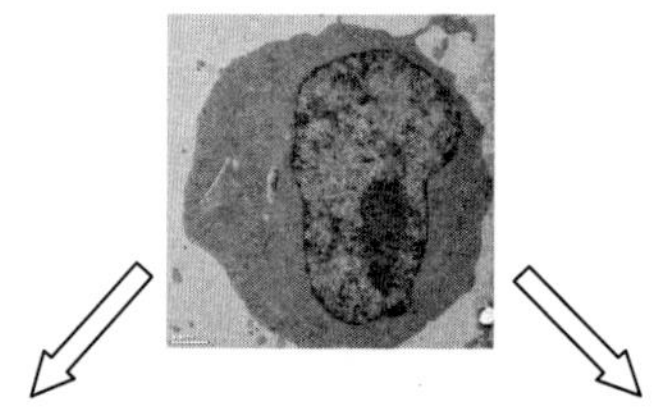

<table>
<tr><td align="center">CLASSICAL APOPTOSIS</td><td align="center">ACCIDENTAL CELL DEATH
(NECROSIS)</td></tr>
</table>

Morphological features

Cell dehydratation & shrinkage Loss of pseudopodia or microvilli Detachment from the surface (anoikis) "Blebbing" of plasma membrane Chromatin condensation Nucleolar segregation Nuclear fragmentation Preservation of mitochondria structure Disassembly of the Golgi apparatus Formation of apoptotic bodies Shedding of apoptotic bodies Engulfment and elimination by phagocytes/neighboring cells	Cell swelling Nuclear and mitochondrial swelling Vacuolization of cytoplasm Rupture of plasma membrane Dissolution of chromatin Dissolution of cell constituents and attraction of inflammatory cells Scar formation

Biochemical & molecular features

Preservation of ATP levels Loss of mitochondrial membrane potential Oxidative stress (ROS generation) Cardiolipin peroxidation Release of cytochrome c from mitochondria Release of AIF, EndoG, Smac/Diablo and HtrA2 from mitochondria Mobilization of intracellular Ca^{2+} (Ca^{2+} flux) Activation of caspases Activation of serine proteases (serpases) Activation of calpains Activation of endonucleases PARP cleavage DNA fragmentation Loss of DNA double helix stability (susceptibility to denaturation) Extensive phosphorylation of histone H2AX Endonucleolytic DNA degradation Separate packaging of DNA and RNA into apoptotic bodies General preservation of plasma membrane integrity (increased permeability only for very small cationic probes e.g. YO-PRO1) Externalization of phosphatidylserine on the outer leaflet of plasma membrane Activation of transglutaminase (TGase2)	Mitochondrial dysfunction Rapid depletion of intracellular ATP Lack of caspase activation Random degradation of DNA Rapid loss of plasma membrane integrity Uncontrolled release of cytoplasmic constituents

Fig. 1. Morphological and biochemical hallmarks of apoptosis and accidental cell death (necrosis). Note that some features characterizing apoptosis may not be present as they heavily depend on particular cell type, stimuli, and cellular microenvironment.

data demonstrate that elimination of many cells may rely on alternative mechanisms (i.e., caspase-independent apoptosis-like PCD [programmed cell death], cornification, autophagy, necrosis-like PCD, mitotic catastrophe, etc.) with critical connotations in both physiological and pathological processes *(5, 6)*. The colloquial term "apoptosis" should be, therefore, restricted only to the demise program featuring all "hallmarks of apoptotic cell death," namely (a) activation of caspases as an absolute marker of cell death; (b) tight (geometric) compaction of chromatin; (c) activation of endonucleases(s) causing internucleosomal DNA cleavage and extensive DNA fragmentation; (d) appearance of distinctive cellular morphology with preservation of organelles; (e) cell shrinkage; (f) plasma membrane blebbing; and (g) nuclear fragmentation followed by formation of apoptotic bodies **(Fig. 1)** *(7, 8)*.

In this context, a gross majority of classical apoptotic attributes can be quantitatively examined by flow cytometry, the preferred platform for rapid assessment of multiple cellular attributes at a single cell level *(1–4, 9)*. The major advantages of flow cytometry include the possibility of multiparameter measurements (correlation of different cellular events at a time), single cell analysis (avoidance of bulk analysis), and rapid analysis time (thousand of cells per second) *(3, 9)*. Flow cytometry overcomes, thus, sensitivity problems of traditional bulk techniques such as fluorimetry, spectrophotometry, or gel techniques (e.g., Western blot). In this chapter, we outline only a handful of commonly used cytometric assays based on the assessment of (a) mitochondrial transmembrane potential ($\Delta\psi_m$ loss), (b) caspase activation, (c) plasma membrane remodeling, and (d) DNA fragmentation *(1–3)*.

2. Materials

2.1. Dissipation of Mitochondrial Transmembrane Potential ($\Delta\psi_m$)

1. Cell suspension (2.5×10^5–2×10^6 ells/mL).
2. 1× PBS.
3. 1.5-mL Eppendorf tubes.
4. 12 × 75 mm Falcon FACS tubes (BD Biosciences).
5. 1 mM tetramethylrhodamine methyl ester perchlorate (TMRM; Invitrogen/Molecular Probes) stock solution in DMSO. Store protected from light at –20°C. Reagent is stable for over 12 months. Caution: although there are no reports on TMRM toxicity, appropriate precautions should always be applied when handling TMRM solutions.
6. 1 μM working solution of TMRM probe in PBS (make fresh as required).
7. TMRM staining mixture (for one sample). Prepare by adding 15 μL of 1 μM TMRM working solution to 85 μL of PBS (make fresh as required).

2.2. Activation of Caspases – FLICA Assay

1. Cell suspension (2.5×10^5–2×10^6 cells/mL).
2. 1× PBS.
3. DMSO.
4. 1.5-mL Eppendorf tubes.
5. 12 × 75 mm Falcon FACS tubes (BD Biosciences)
6. Poly-caspases FLICA reagent (FAM-VAD-FMK; Immunochemistry Technologies LLC) (powder). Store protected from light at –20°C, stable for over 12 months.
7. Reconstituted stock of poly-caspases FLICA reagent. Prepare by adding 50 µL DMSO to the vial and mix by rolling. Store protected from light at –20°C, stable for over 6 months.
8. FLICA working solution. Make fresh as required by 5× dilution of the reconstituted FLICA stock in PBS.
9. 50 µg/mL propidium iodide (PI) stock solution in PBS. Store protected from light at +4°C. Stable for over 12 months. Caution: PI is a DNA binding molecule and thus can be considered as a potential carcinogen. Always handle with care and use protective gloves.
10. Propidium iodide staining mixture. Prepare fresh as required by 10× dilution of PI stock in PBS.

2.3. Apoptotic Changes in the Plasma Membrane – Annexin V assay

1. Cell suspension (2.5×10^5–2×10^6 cells/mL).
2. 1× PBS.
3. Annexin V Binding Buffer (AVBB): 10 mM HEPES/NaOH pH 7.4; 140 mM NaCl, 2.5 mM $CaCl_2$. Store at +4°C as long as no precipitate is visible.
4. 1.5-mL Eppendorf tubes.
5. 12 × 75 mm Falcon FACS tubes (BD Biosciences).
6. Annexin V- FITC or Annexin V-APC conjugate (Invitrogen/Molecular Probes), store protected from light at +4°C. Stable for over 12 months.
7. 50 µg/mL propidium iodide (PI) stock solution in PBS. Store protected from light at +4°C. Reagent is stable for over 12 months. Caution: PI is a DNA binding molecule and thus can be considered as a potential carcinogen. Always handle with care and use protective gloves.
8. Propidium iodide staining mixture. Prepare fresh as required by 10× dilution of PI stock in AVBB.

2.4. Assessment of Fractional DNA Content (sub-G$_1$ fraction)

1. Cell suspension (5×10^5–1×10^6 cells/mL).
2. Cold 70% EtOH (store at –20°C).
3. 1× PBS.

4. 1.5 mL Eppendorf tubes.

5. 12 × 75 mm Falcon FACS tubes (BD Biosciences).

6. 1 mg/mL propidium iodide (PI) stock solution in PBS. Store protected from light at +4°C. Reagent is stable for over 12 months. Caution: PI is a DNA binding molecule and thus can be considered as a potential carcinogen. Always handle with care and use protective gloves.

7. 1 mg/mL RNase A solution in MilliQ water (available from Sigma). Store protected from light at –20°C. Reagent is stable for over 12 months.

8. Staining mixture (for one sample). Prepare fresh as required by adding 954 µL of PBS, 30 µL of RNase A, and 16 µL of PI stock solution.

3. Methods

3.1. Dissipation of Mitochondrial Transmembrane Potential ($\Delta \psi_m$)

The cytometric detection of $\Delta\psi_m$ loss is a sensitive marker of early apoptotic events (*see* **Notes 1–3**). Procedure is based on a tetramethylrhodamine methyl ester perchlorate (TMRM), a fluorescent lipofilic cationic probe readily taken up by live cells and accumulating in energized mitochondria *(10)*. The extent of its uptake, as measured by intensity of cellular fluorescence, is proportional to cellular $\Delta\psi_m$ status (**Fig. 2a**; *see* **Notes 4 and 5**). TMRM probe is particularly useful for multiparameter assays combining diverse apoptotic markers (*see* **Note 6**; **Fig 2b**) *(4, 10, 13)*.

1. Collect cell suspension into 12 × 75 mm Falcon FACS tube and centrifuge for 5 min, 160 ×*g* at room temperature (RT).

2. Resuspend cell pellet in 1–2 mL of PBS and centrifuge for 5 min, 160 ×*g*.

3. Discard supernatant and add 100 µL of TMRM staining mix.

4. Gently agitate to resuspend cell pellet.

5. Incubate for 20 min at +37°C, protected from direct light.

6. Add 500-µL PBS and keep samples on ice.

7. Analyze on a flow cytometer. Use 488-nm excitation line (Argon-ion laser or solid-state laser) and emission collected at 575 nm. Adjust the logarithmic amplification scale to distinguish between viable cells (bright TMRM⁺), apoptotic cells/ necrotic cells with compromised plasma membranes (TMRM⁻) (*see* **Fig. 2a**).

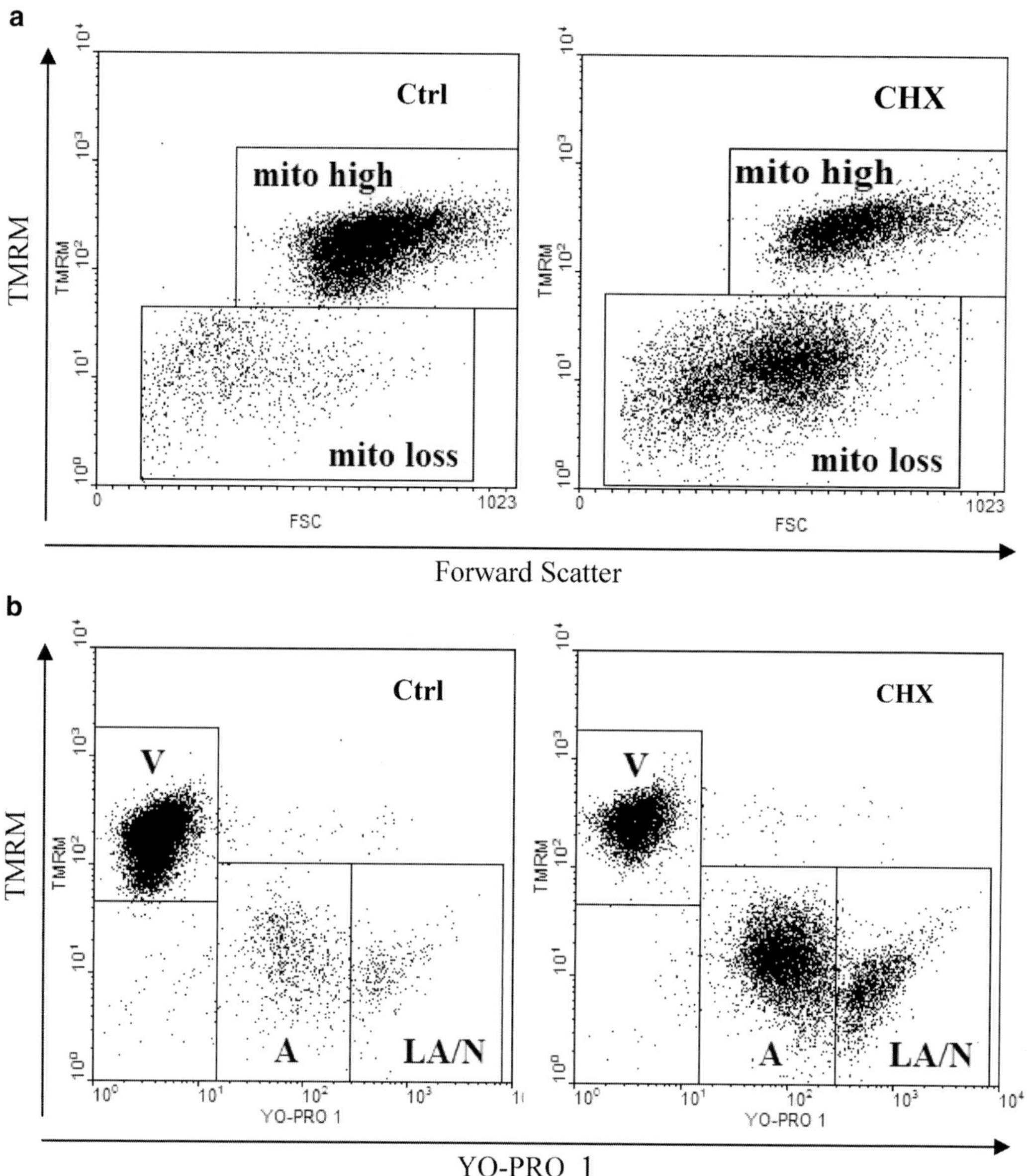

Fig. 2. Dissipation of mitochondrial transmembrane potential ($\Delta\psi_m$). (**a**) Analysis by staining with tetramethylrhodamine methyl ester (TMRM). Human B-cell lymphoma cells were either untreated (Ctrl) or treated with cycloheximide (CHX) to induce apoptosis and supravitally loaded with TMRM as described (*10, 11*). Cells with collapsed mitochondrial transmembrane potential (mito loss) have decreased intensity of orange TMRM fluorescence. Note that by only employing the $\Delta\psi_m$-sensitive probe there is no distinction between early, late apoptotic and necrotic cells. (**b**) Multiparameter analysis employing mitochondrial potential sensitive probes using concurrent analysis of collapse of $\Delta\psi_m$ and early plasma membrane permeability during apoptosis. Cells were treated as in **Fig. 4a** and supravitally stained with both YO-PRO 1 and TMRM probes (*12*). Their green and orange fluorescence was measured by flow cytometry. Live cells (V) are both TMRM[high] and exclude YO-PRO 1. Early apoptotic cells (A) exhibit loss of $\Delta\psi_m$ (TMRM[low]) and moderate uptake of YO-PRO 1. Late apoptotic/secondary necrotic cells (LA/N) are highly permeant to YO-PRO 1 probe. Note that multiparameter analysis of $\Delta\psi_m$-sensitive probe with YO-PRO 1 allows for a lucid distinction between live, early, late apoptotic and necrotic cells.

3.2. Activation of Caspases – FLICA Assay

Use of *fluorochrome-labeled inhibitors of caspases* (FLICA) allows for a convenient estimation of apoptosis by both cytometry and fluorescence microscopy *(13)* (*see* **Notes 1–3**). FLICAs were designed as affinity ligands to active centers of individual caspases and their specificity toward individual caspases is provided by the four amino-acid peptide. Presence of the fluorescent tag (FITC or SR) allows detection of FLICA–caspase complexes inside viable cells *(13, 14)*. When applied together with the plasma membrane permeability marker propidium iodide (PI), several consecutive stages of apoptosis can be distinguished (*see* **Fig. 3**; **Notes 6 and 7**) *(13, 14)*.

1. Collect cell suspension into 12×75 mm Falcon FACS tube and centrifuge for 5 min, $160 \times g$ at room temperature (RT).

2. Resuspend cell pellet in 1–2 mL of PBS and centrifuge for 5 min, $160 \times g$.

3. Discard supernatant and add 100 µL of PBS.

4. Gently agitate to resuspend cell pellet and add 3 µL of FLICA working solution.

5. Incubate for 60 min at +37°C, protected from direct light. Gently agitate cells every 20 min to allow homogenous loading with FLICA probe.

6. Add 2 mL of PBS and centrifuge for 5 min, $160 \times g$ at RT.

7. Discard supernatant and repeat **step 6**.

8. Discard supernatant and add 100 µL of PI staining mix.

9. Incubate for 3–5 min and add 500 µL of PBS. Keep samples on ice.

10. Analyze samples on a flow cytometer. Use 488-nm excitation line (Argon-ion laser or solid-state laser) and emission collected at 530 nm (green, FLICA) and 575–610 nm (orange, PI). Carefully adjust the logarithmic amplification scale and compensation between green and orange channels. Distinguish between viable cells (FLICA$^-$/PI$^-$), early apoptotic cells (FLICA$^+$/PI$^-$), late apoptotic/secondary necrotic cells (FLICA$^+$/PI$^+$) and primary necrotic cells (FLICA$^-$/PI$^+$) (*see* **Fig. 3**).

3.3. Apoptotic Changes in the Plasma Membrane – Annexin V Assay

Under physiological conditions, choline phospholipids (phosphatidylcholine, sphingomyelin) are exposed on the external leaflet while aminophospholipids (phosphatidylserine, phosphatidylethanolamine) are exclusively located on the cytoplasmic surface of the lipid bilayer. This asymmetry is scrambled during apoptosis when phosphatidylserine (PS) becomes exposed on the outside leaflet of the membrane *(15, 16)*. The detection of PS

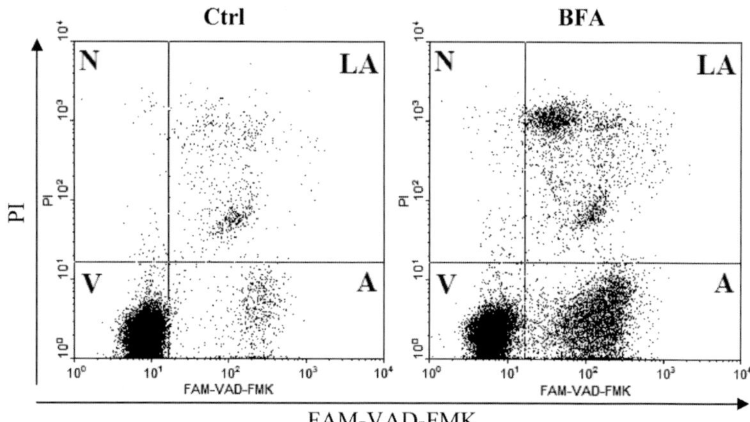

FAM-VAD-FMK

Fig. 3. Detection of activated caspases by fluorescently labeled inhibitors of caspases (FLICA) combined with plasma membrane permeability assessment (propidium iodide; PI). Human B-cell lymphoma cells were either untreated (Ctrl) or treated with Brefeldin A (BFA) to induce apoptosis as described (*12*). Cells were subsequently supravitally stained with FAM-VAD-FMK (pan caspase marker; FLICA) and PI. Their logarithmically amplified green and red fluorescence signals were measured by flow cytometry. Live cells (V) are both FAM-VAD-FMK and PI negative. Early apoptotic cells (A) bind FAM-VAD-FMK but exclude PI. Late apoptotic/secondary necrotic cells (LA) are both FAM-VAD-FMK and PI positive. Primary necrotic and some very late apoptotic cells (N) stain with PI only.

by fluorochrome-tagged 36-kDa anticoagulant protein Annexin V allows for a precise estimation of apoptotic incidence *(16)* (*see* **Fig. 4; Notes 1–3**). This probe reversibly binds to phosphatidylserine residues only in the presence of mM concentration of divalent calcium ions.

1. Collect cell suspension into 12×75 mm Falcon FACS tube and centrifuge for 5 min, $160 \times g$ at room temperature (RT).

2. Resuspend cell pellet in 1–2 mL of Annexin V Binding Buffer (AVBB) and centrifuge as in **step 1**.

3. Discard supernatant and add 100 μL of PI staining mix in AVBB.

4. Add 2–4 μL of Annexin V-FITC or -APC conjugate.

5. Incubate for 15 min at RT.

6. Add 500 μL of AVBB and keep samples on ice.

7. Analyze samples on a flow cytometer. Use 488-nm excitation line (Argon-ion laser or solid-state laser) and emission collected at 530 nm (green, FITC) and 575–610 nm (orange, PI). Alternatively use flow cytometer with 488-nm excitation for PI (emission collected at 530 nm) and 633-nm excitation for Annexin V-APC conjugate (emission collected at 660 nm).

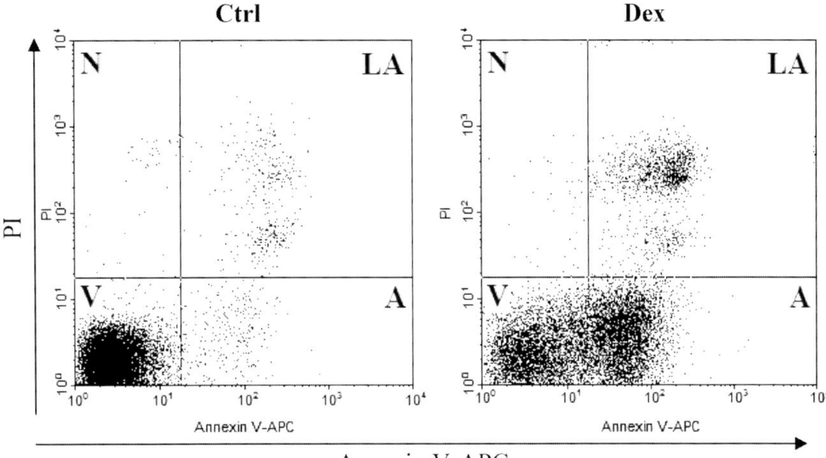

Annexin V-APC

Fig. 4. Apoptotic changes in plasma membrane. Detection of apoptosis by concurrent staining with annexin V-APC and PI. Human B-cell lymphoma cells were untreated (*left panel*) or treated with dexamethasone (*right panel*), as described previously (*11*). Cells were subsequently stained with annexin V – APC conjugate and PI and their far-red and red fluorescence was measured by flow cytometry. Live cells (V) are both annexin V and PI negative. At early stage of apoptosis (A) the cells bind annexin V while still excluding PI. At late stage of apoptosis (N) they bind annexin V-FITC and stain brightly with PI.

Carefully adjust the logarithmic amplification scale and compensation between green and orange channels. No compensation between PI and APC conjugate is needed. Distinguish between viable cells (Annexin V$^-$/PI$^-$), early apoptotic cells (Annexin V$^+$/PI$^-$), late apoptotic/necrotic cells (Annexin V$^+$/PI$^+$) and late necrotic cells (Annexin V$^-$/PI$^+$) as seen in **Fig. 4** (also *see* **Notes 8** and **9**).

3.4. Assessment of Fractional DNA Content (sub-G$_1$ Fraction)

The fragmented, low molecular weight DNA can be extracted from cells during the process of cell staining in aqueous solutions *(17, 18)*. Such extraction takes place when the cells are treated with precipitating fixatives such as ethanol or methanol (*see* **Note 10**). As a result of DNA extraction apoptotic cells exhibit a deficit in DNA content and following staining with a DNA-specific fluorochrome they can be recognized by flow cytometry as cells having fractional DNA content *(18)*. On frequency distribution histograms these events are characterized by a distinctive "sub-G$_1$" peak that represents oligonucleosomal DNA fragments (**Fig. 5**; *see* **Notes 11–13**).

1. Collect 1 mL of cell suspension into Eppendorf tubes and centrifuge for 5 min, 327 ×*g*) at room temperature (RT).

2. Resuspend cell pellet in 60 μL of PBS.

3. While vortexing add drop-by-drop 1 mL of ice-cold 70% EtOH.

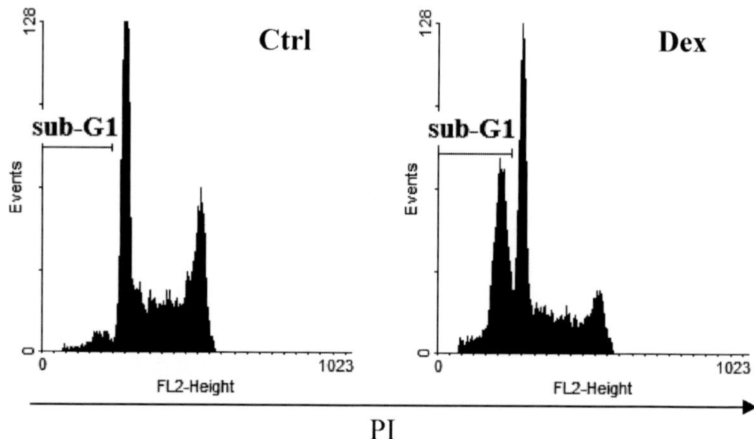

Fig. 5. Detection of fractional DNA content ("sub-G₁ peak"). Apoptosis of human follicular lymphoma cells was induced with dexamethasone (Dex). Ethanol fixed and propidium iodide (PI) stained cells were analyzed on a flow cytometer. Red fluorescence of PI was collected using linear amplification scale. Debris was gated out electronically. Note distinctive sub-G₁ peak. For further details refer to text.

4. Permeabilize cells for at least 1.5 h at –20°C or overnight at +4°C. Samples can be stored for months at –20°C.

5. Centrifuge for 10 min, 392 ×g at room temperature (RT).

6. Gently discard supernatant and add 1 mL staining mixture containing PI and RNase A. Residual EtOH can be left without interference with assay performance.

7. Vortex to resuspend cell pellet and incubate for 60 min at +37°C protected from direct light.

8. Analyze on a flow cytometer. Use 488-nm excitation line (Argon-ion laser or solid-state laser) and emission collected at 575–610 nm. Adjust the linear amplification scale to obtain cell cycle profile and "sub-G1" peak as seen in **Fig. 5** (also *see* **Notes 11** and **12**).

---

4. Notes

1. The universal term "apoptosis," has a propensity to misinterpret the actual phenotype of cell suicide program *(4, 5, 8)*. Thus, the use of the generic term apoptosis should be always accompanied by listing the particular morphological and/or biochemical apoptosis-associated feature(s) that was(were) detected *(4, 7, 8)*.

2. Morphological criteria (examined by the light, fluorescent, and electron microscopy) are still the "gold standard" to define the mode of cell death and confirm the results obtained by flow cytometry *(1–5)*. Lack of microscopic examination may potentially lead to the misclassification and false positive or negative artifacts, and is a common drawback of the experimental design *(1–4)*. The best example of such misclassification is identification of phagocytes that engulfed apoptotic bodies as individual apoptotic cells *(3)*.

3. Cell harvesting by trypsinization, mechanical or enzymatic cell disaggregation from tissues, extensive centrifugation steps, may all lead to preferential loss of apoptotic cells. On the other hand some cell harvesting procedures interfere with apoptotic assays as discussed elsewhere *(1–3)*.

4. Loss (dissipation) of the mitochondrial transmembrane potential appears to be, early and initially, a transient event, followed by permanent collapse later during apoptotic cascade *(3, 4)*. Depolarization of mitochondrial membrane is usually followed by rapid activation of caspases followed by externalization of phosphatidylserine. As a result, loss of staining with TMRM probe precedes binding of fluorescently labeled inhibitors of caspases (FLICA). Our recent studies revealed also that the time-window of apoptosis detected by FLICA binding is much wider than that by the Annexin V binding *(4, 13)*.

5. According to Nernst equation, the intracellular distribution of any cationic mitochondrial probe reflects the differences in the transmembrane potential across both the plasma membrane (i.e., between exterior vs. interior of the cell) and the outer mitochondrial membrane *(2, 3, 10)*. Thus, apart from mitochondria the probes can also accumulate in the cytosol. This is facilitated by both active and passive transport across the plasma membrane. Caution should be also taken, as cationic probes may be targeted to other organelles like endoplasmic reticulum (ER) or lysosomes. Moreover, accumulation of some probes may be influenced by the activity of multidrug efflux pumps (MDR). In each experiment it is advisable to assess probes' specificity by preincubation of cells for 20–30 min with 50–100 μM protonophores CCCP or FCCP. Both agents collapse the mitochondrial transmembrane potential and should be used as positive controls *(2, 3, 10)*.

6. FLICAs are highly permeant to plasma membrane and relatively nontoxic. This provides an unique opportunity to detect caspase activation in living cells where uptake of these reagents is followed by covalent binding to activated caspases. To date, no interference resulting by MDR efflux

pump activity has been reported for FLICA uptake. Extensive multiplexing combinations are compatible with both single- and multilaser instrumentation *(11, 13, 14)*.

7. FLICAs withstand cell fixation (with 4% paraformaldehyde; PFA) and subsequent cell permeabilization with 70% ethanol or methanol. As a result, this assay can be combined with the analysis of cell attributes that can require prior cell permeabilization such as DNA content measurement, DNA fragmentation (TUNEL assay), etc. *(11, 13, 14)*.

8. A range of Annexin V conjugates with organic fluorescent probes is commercially available with the predominant popularity of FITC, PE, and APC conjugates. There is also a considerable progress in inorganic, semiconductor nanocrystals (Quantum Dots; QDs) conjugates *(19)*. Their significant advantages over currently available organic fluorochromes are rapidly attracting attention in both cytometric and imaging applications *(20)*.

9. The interpretation of results from Annexin V assay may be difficult after mechanical disaggregation of tissues to isolate individual cells, enzymatic (e.g., by trypsinization) or mechanic detachment (e.g., by "rubber policeman") of adherent cells from culture flasks, cell electroporation, chemical cell transfection, or high-titer retroviral infections. These conditions reportedly influence phosphatidylserine flipping. A high surface expression of phosphatidylserine has also been detected on some healthy cells such as differentiating monocytes, activated T cells, positively selected B lymphocytes, activated neutrophils, or myoblasts fusing into myotubes *(1–3, 9)*.

10. Fixation with cross-linking fixatives such as formaldehyde, on the other hand, results in the retention of low MW DNA in the cell as they become cross-linked to intercellular proteins. Therefore a formaldehyde fixation is incompatible with the "sub-G_1" assay *(1–3)*.

11. Optimally, the "sub-G_1 peak" representing apoptotic cells should be separated with little or no overlapping from the G_1 peak of the nonapoptotic cell population. The degree of low molecular weight DNA extraction varies, however, markedly depending on the extent of DNA degradation (duration of apoptosis), the number of cell washings, and pH and molarity of the washing/staining buffers. Shedding of apoptotic bodies containing fragments of nuclear chromatin may also contribute to the loss of DNA from apoptotic cells. As a result, the separation of "sub-G_1" is not always satisfactory *(1–3)*.

12. Estimation of the sub-G_1 fraction fails when DNA degradation does not proceed to internucleosomal regions but stops after generating 50–300 kb fragments. Little DNA can be extracted then from the cells and rigid reliance on this method provides false negative results *(1–3)*. If G_2/M or even late S phase cells undergo apoptosis, the loss of DNA from these cells may not produce the sub-G_1 peak. These apoptotic cells often end up with DNA content equivalent to G_1/early S phase and are, thus, indistinguishable *(1–3)*.

13. Some markers (like oligonucleosomal DNA fragmentation) may not be detected in specimens challenged with divergent stimuli or microenvironmental conditions (e.g., cytokines, growth factor deprivation, heterotypic cell culture, etc.). It is always advisable to simultaneously study several markers to provide a multidimensional view of advancing apoptotic cascade *(1–3)*. Multiparameter assays detecting several cell attributes are the most desirable solution for flow cytometric quantification of apoptosis *(3, 4)*.

Acknowledgments

Supported by NCI CA RO1 28 704 (ZD). JS received the L'Oreal Poland-UNESCO "For Women In Science" 2007 Award and MRC Career Development fellowship. Views and opinions described in this chapter were not influenced by any conflicting commercial interests.

References

1. Darzynkiewicz, Z., Juan, G., Li, X., Gorczyca, W., Murakami, T. and Traganos, F. (1997). Cytometry in cell necrobiology: analysis of apoptosis and accidental cell death (necrosis). *Cytometry* **27**, 1–20.

2. Darzynkiewicz, Z., Li, X. and Bedner, E. (2001). Use of flow and laser-scanning cytometry in analysis of cell death. *Methods Cell Biol.* **66**, 69–109.

3. Darzynkiewicz, Z., Huang, X., Okafuji, M. and King, M.A. (2004). Cytometric methods to detect apoptosis. *Methods Cell Biol.* **75**, 307–41.

4. Wlodkowic, D., Skommer, J. and Darzynkiewicz, Z. (2008). SYTO probes in the cytometry of tumor cell death. *Cytometry A.* **73**, 496–507.

5. Leist, M. and Jaattela, M. (2001). Four deaths and a funeral: from caspases to alternative mechanisms. *Nat. Rev. Mol. Cell Biol.* **2**, 589–98.

6. Kroemer, G. and Martin, S.J. (2005). Caspase-independent cell death. *Nat. Med.* **11**, 725–30.

7. Blagosklonny, M.V. (2000). Cell death beyond apoptosis. *Leukemia* **14**, 1502–8.

8. Zhivotovsky, B. (2004). Apoptosis, necrosis and between. *Cell Cycle* **3**, 64–6.

9. Telford, W.G., Komoriya, A. and Packard, B.Z. (2004). Multiparametric analysis of apoptosis by flow and image cytometry. *Methods Mol. Biol.* **263**, 141–60.

10. Castedo, M., Ferri, K., Roumier, T., Metivier, D., Zamzami, N. and Kroemer, G. (2002). Quantitation of mitochondrial alterations associated with apoptosis. *J. Immunol. Methods* **265**, 39–47.

11. Wlodkowic, D., Skommer, J. and Pelkonen, J. (2006). Multiparametric analysis of HA14-1-induced apoptosis in follicular lymphoma cells. *Leukemia Res.* **30**, 1187–92.

12. Wlodkowic, D., Skommer, J. and Pelkonen, J. (2007). Brefeldin A triggers apoptosis associated with mitochondrial breach and enhances HA14-1- and anti-Fas-mediated cell killing in follicular lymphoma cells. *Leukemia Res.* **31**, 1687–700.

13. Pozarowski, P., Huang, X., Halicka, D.H., Lee, B., Johnson, G. and Darzynkiewicz, Z. (2003). Interactions of fluorochrome-labeled caspase inhibitors with apoptotic cells: a caution in data interpretation. *Cytometry A* **55**, 50–60.

14. Smolewski, P., Grabarek, J., Lee, B.W., Johnson, G.L. and Darzynkiewicz, Z. (2002). Kinetics of HL-60 cell entry to apoptosis during treatment with TNF-α or camptothecin assayed by stathmo-apoptosis method. *Cytometry* **47**, 143–9.

15. Koopman, G., Reutelingsperger, C.P.M., Kuijten, G.A.M., Keehnen, R.M.J., Pals, S.T. and van Oers, M.H.J. (1994). Annexin V for flow cytometric detection of phosphatidylserine expression of B cells undergoing apoptosis. *Blood* **84**, 1415–20.

16. van Engeland, M., Nieland, L.J.W., Ramaekers, F.C.S., Schutte, B. and Reutelingsperger, P.M. (1998). Annexin V-affinity assay: a review on an apoptosis detection system based on phosphatidylserine exposure. *Cytometry.* **31**, 1–9.

17. Nicoletti, I., Migliorati, G., Pagliacci, M.C., Grignani, F. and Riccardi, C. (1991). A rapid and simple method for measuring thymocyte apoptosis by propidium iodide staining and flow cytometry. *J. Immunol. Methods* **139**, 271–80.

18. Gong, J., Traganos, F. and Darzynkiewicz, Z. (1994). A selective procedure for DNA extraction from apoptotic cells applicable for gel electrophoresis and flow cytometry. *Anal. Biochem.* **218**, 314–9.

19. Le Gac, S., Vermes, I. and van den Berg, A. (2006). Quantum dots based probes conjugated to annexin V for photostable apoptosis detection and imaging. *Nano Lett.* **6**, 1863–9.

20. Chattopadhyay, P.K., Price, D.A., Harper, T.F., Betts, M.R., Yu, J., Gostick, E., Perfetto, S.P., Goepfert, P., Koup, R.A., de Rosa, S.C., Bruchez, M.P. and Roederer, M. (2006). Quantum dot semiconductor nanocrystals for immunophenotyping by polychromatic flow cytometry. *Nat. Med.* **12**, 972–7.

Chapter 3

Live to Dead Cell Imaging

Stephen W.G. Tait, Lisa Bouchier-Hayes, Andrew Oberst, Samuel Connell, and Douglas R. Green

Summary

Live cell imaging allows several key apoptotic events to be visualized in a single cell over time. These include mitochondrial outer membrane permeabilization (MOMP), mitochondrial dysfunction, phosphatidylserine exposure, and membrane permeabilization. Here we describe a protocol for imaging multiple apoptotic processes in the same cell over time. Initially, this involves generating a cell line stably expressing a fluorescent fusion protein that can act as an apoptotic marker, such as cytochrome c-GFP. By combining various fluorescent fusion proteins and probes, several apoptotic events can be imaged in the same cell. Next, the cells are induced to undergo apoptosis and continuously imaged. Finally, quantitative kinetic analysis of various apoptotic processes is performed postimaging.

Key words: Apoptosis, Mitochondria, Live cell imaging, Confocal microscopy, Mitochondrial outer membrane permeabilisation, Fluorescent fusion protein, Microinjection, Cytochrome c, Smac, Omi, Bax

1. Introduction

Mitochondrial outer membrane permeabilization is a critical step for apoptosis induction by many stimuli (1). Bcl-2 family members such as Bid and Bax regulate MOMP resulting in the cytosolic release of proteins such as cytochrome c, Smac, and Omi that normally reside in the mitochondrial intermembrane space. This leads to caspase activation and cell death. Live cell imaging has greatly advanced the field of apoptosis particularly with respect to studying the mitochondrial pathway enabling visualization of the different events that occur during the process (2–4). Here we describe a general protocol for imaging such events

Peter Erhardt and Ambrus Toth (eds.), *Apoptosis*, Methods in Molecular Biology, vol. 559
DOI 10.1007/978-1-60327-017-5_3
© Humana Press, a part of Springer Science + Business Media, LLC 2004, 2009

including MOMP, mitochondrial dysfunction, plasma membrane reorganization (phosphatidylserine flip), and membrane permeabilization. Imaging these distinct apoptotic events simultaneously relies mainly upon the use of different fluorescent proteins or chromophores with nonoverlapping excitation/emission spectra. Multiparameter imaging is further facilitated by the distinct temporal and spatial differences inherent to different stages of the apoptotic program.

Proteins such as Bax or BH3-only family members (such as Bid) redistribute from a cytoplasmic to mitochondrial localization during apoptosis. By using fluorescent fusions of such proteins, events proximal to or associated with MOMP can be imaged *(5, 6)*. MOMP can by visualized by studying the movement of mitochondrial intermembrane space proteins such as cytochrome *c*, Smac, or Omi fused to a fluorescent protein from the mitochondria to the cytoplasm *(2–4)*. This relocalization is visualized as a change in the subcellular distribution of the fusion protein from a punctate pattern (when in the mitochondria) to a diffuse, cytosolic distribution. Simultaneous analysis of mitochondrial morphology can be achieved by targeting a fluorescent fusion protein to the mitochondrial matrix *(7)*. Such a protein will not be released upon MOMP, thus the mitochondria can still be visualized during post-MOMP events.

The methods described here also permit analysis of the caspase-dependent execution phase of apoptosis that lies downstream of MOMP. Specifically, mitochondrial dysfunction, phosphatidylserine (PS) exposure, and plasma membrane permeabilization can be visualized using fluorescent potentiometric dyes, fluorescent-conjugated Annexin V, and plasma membrane impermeable dyes, respectively *(2)*. Quantitative postimage analysis can be carried out to determine, amongst other parameters, the time of onset and duration of MOMP.

2. Materials

2.1. Generation of Stable Cell Lines

1. Complete medium: DMEM (Invitrogen) containing 10% FCS (Omega).
2. Opti-MEM cell culture medium (Invitrogen).
3. Lipofectamine 2000 transfection reagent (Invitrogen).
4. Polybrene (Sigma): dissolve in PBS to 5 mg/mL (1,000× stock), store at –20°C.
5. Geneticin (Invitrogen) (also known as G418) 50 mg/mL.
6. Hygromycin (Roche) 50 mg/mL.

7. Puromycin (Sigma): dissolve in PBS to 1 mg/mL stock solution.

8. Zeocin (Invitrogen).

2.2. Apoptosis Induction

2.2.1. General Apoptosis Inducers

1. Staurosporine (Sigma): dissolve in DMSO to 1 mM, store at –20°C.

2. Actinomycin D (Calbiochem): dissolve in DMSO to 1 mM, store at –20°C.

3. TNFα (Calbiochem): aliquot and store at –20°C, avoid multiple freeze-thaw cycles.

4. Cycloheximide (Sigma): dissolve in ethanol to 10 mg/mL and store at –20°C.

5. Stratalinker UV crosslinker (Stratagene).

6. qVD-OPH (MP Biomedicals): dissolve in DMSO to 20 mM (1,000× stock), aliquot and store at –20°C, avoid freeze-thaw.

7. z-VAD-fmk (MP Biomedicals): dissolve in DMSO to 100 mM (1,000× stock), aliquot and store at –20°C, avoid freeze-thaw.

2.2.2. Microinjection

1. Microinjector (Eppendorf InjectMan NI 2 or equivalent)

2. Micromanipulator (Eppendorf FemtoJet or equivalent)

3. HeLa cells stably expressing cytochrome *c*-GFP.

4. 3-cm dish with embedded coverslip (Mattek Corp).

5. Complete medium: DMEM, 10% FBS, 1% L-glutamine, 1% pen-strep, 20 mM Hepes.

6. HE buffer: 10 mM Hepes, 1 mM EDTA.

7. 10-kD dextran conjugated to AlexaFluor568 (Invitrogen): light sensitive, store at 4°C.

8. Caspase 8 cleaved recombinant Bid protein (R&D Systems): aliquot and store at –70°C, avoid freeze thaw.

9. Microinjection needles (Femtotips from Eppendorf).

10. Microloaders (Eppendorf).

2.2.3. Protein/peptide Transfection

`1. HeLa cytochrome *c*-GFP cells.

2. 8-well Labtek II chambered coverglass #1.5 (Mattek Corp).

3. Chariot protein transfection reagent (Active Motif).

4. 10-kD dextran conjugated to AlexaFluor568 (Invitrogen): light sensitive, store at 4°C.

5. Peptide corresponding to the BH3 domain of Bid or Bim *(8)*.

6. Opti-MEM cell culture medium (Invitrogen).

7. Complete medium.

8. Complete medium with 20% FBS (Omega).

1. 4-well Labtek II chambered coverglass #1.5 (Mattek Corp)

2. Fibronectin solution 1 mg/mL in PBS: dilute 10-mg/mL fibronectin solution (Chemicon) in PBS. Store at 4°C.

3. Phosphate buffered saline (PBS).

4. Trypsin EDTA (Mediatech). Store at 4°C.

5. Complete medium: DMEM (Invitrogen) containing 10% FCS (Omega).

6. Imaging medium: prepare fresh prior to imaging. 1 M Hepes (50× stock), 55 mM β-mercaptoethanol (1,000× stock), Annexin V AlexFluor647 (200× stock) and tetramethylrhodamine methyl ester (TMRE) are all from Invitrogen. TMRE and Annexin V AlexFluor647 are light sensitive. Make 50 µM stock solution of TMRE in DMSO (1,000×). Make propidium iodide (Sigma) stock solution in PBS (100 µg/mL, 250× stock) Store all at 4°C. Store $CaCl_2$ (Fisher) solution (1 M in water, 400× stock) at room temperature.

7. Appropriate excitation laser lines and emission filters are critical for live-cell imaging. With more laboratories utilizing solid-state lasers, the exact laser lines and emission filters utilized may differ. Examining the excitation and emission spectra of your chosen fluorophores in relation to your available confocal microscopy equipment is essential. In this protocol typical laser choices are as follows: GFP may be excited with 488 nm, propidium iodide and TMRE with 568 nm, and AlexaFluor 647 with 647 nm.

8. Either a spinning disk confocal head or a laser scanning confocal head is advised, with the former being preferred for its reduction in phototoxicity.

9. A high numerical aperture objective enabling the highest light gathering capabilities of your microscope, preferably either 40× 1.3NA or 63× 1.4NA. In this multiparameter protocol, one is utilizing a significant portion of the light spectra; therefore a highly corrected Plan-Apochromat objective is also recommended.

10. An incubator enclosure for maintaining samples and microscope components at physiological temperature.

11. Within the enclosure, a smaller workhead is necessary for providing samples with humidified 5% CO_2 either from a regulator or a tank of premixed 5% CO_2.

1. Metamorph, ImagePro, SlideBook, Imaris or equivalent software.

3. Methods

Generation of stable cell lines expressing a fluorescent fusion protein of interest greatly facilitates apoptosis imaging. It reduces the likelihood of artifacts resulting from transient overexpression, enables one to select cell lines in which the fusion protein is properly localized, allows for consistency between experiments and, in some cases, is a necessity when transient high level overexpression of a fusion protein is toxic (e.g., GFP-Bax). We routinely generate stable cell lines by retroviral transduction/drug selection or drug selection/cell sorting following transient transfection (**Subheading 3.1**). Selection of the apoptosis inducer for time-lapse imaging depends upon various factors. Many chemical inducers of apoptosis take several hours to induce MOMP and apoptosis. This can limit the number of images obtained due to phototoxicity inherent to live cell imaging. However, the individual events that occur once a cell has committed to apoptosis occur quite rapidly and it is often more informative to reduce the delay between images. Accordingly, introduction of certain BH3-only proteins or peptides corresponding to the BH3 domain of such proteins directly into cells either by microinjection or transfection can induce MOMP within 1 h (*see* **Subheading 3.2**). The general imaging protocol described here involves detection of MOMP by analysis of mitochondrial cytochrome *c*-GFP release. Analysis of other apoptotic events (Bax translocation, BH3 only translocation, and mitochondrial fragmentation) can be achieved either by using the appropriate fluorescent fusion protein expressing cell line (e.g., GFP-Bax expressing cells). Simultaneous imaging of some processes, e.g., MOMP and mitochondrial fragmentation, requires the use of a spectrally distinct fluorescent fusion protein, such as matrix-targeted mCherry, in addition to cytochrome *c*-GFP (*see* **Subheading 3.3**). An overview of fluorescent fusion proteins and their utility is shown in **Table 1**. Quantitative analysis with the appropriate software can be done postimage capture (*see* **Subheading 3.4**).

3.1. Generation of Cell Lines Stably Expressing Fluorescent Fusion Protein

3.1.1. Retroviral Transduction

1. Day 1: Plate 3 million Phoenix producer cells per 10-cm plate in complete medium (*see* **Note 1**).

2. Day 2: Transfect Phoenix cells with retroviral vector. Dilute retroviral vector (5 µg) in 400-µL Opti-MEM, without FCS or antibiotics. In a separate tube, dilute 10-µL Lipofectamine 2000 transfection reagent in 400-µL Opti-MEM. Incubate both tubes at room temperature for 5 min, then mix the two together and incubate for a further 20 min. Remove media from Phoenix cells, and replace with 8-mL DMEM without FCS or antibiotics. Add DNA mixture dropwise to the plate

Table 1
Fluorescent fusion proteins used for apoptosis imaging

Fluorescent fusion protein	Usage	Comments
Cytochrome *c*-GFP	Visualizing MOMP	Requires generation of stably expressing cell line and selection of clones displaying correct localization
Smac GFP	Visualizing MOMP	Dependent on cell type, may require generation of stably expressing cell line and selection of clones displaying correct localization
Omi mCherry	Visualizing MOMP	Works well by transient transfection, can easily be combined with GFP fusions for two-color imaging
GFP Bax	Visualizing Bax mitochondrial translocation/activation	Requires generation of stably expressing cell line
Bid GFP	Visualizing Bid mitochondrial translocation/activation	Dependent on cell type may require generation of stably expressing cell line and selection of clones displaying correct localization, utility limited to apoptotic stimuli that engage Bid activity
CoxVIII mts dsRed	Monitoring of mitochondrial morphology throughout apoptosis	Mitochondrial targeting sequence of CoxVIII directs dsRed to the matrix. Works well by transient transfection. Green emission during maturation of dsRed can cause problems when carrying out two color imaging with GFP
Histone 2B GFP	Monitoring nuclear changes	Works well by transient transfection

and swirl to mix, taking care not to detach the Phoenix cells. Incubate at 37°C, 5% CO_2 for 4–6 h.

3. Remove DNA–media mixture from Phoenix cells, replace with complete medium (*see* **Note 2**).

4. Day 3: Plate 1×10^5 target cells in each well of a 6-well plate. Include a nontransduced well to verify the selection procedure.

5. Day 4: Carefully remove the virus-containing media from Phoenix producer cells, and replace with fresh complete medium. Spin media 10 min at $300 \times g$ to remove any Phoenix cells. Add polybrene to the viral supernatant to a final concentration of 5 μg/mL (*see* **Note 3**).

6. Remove media from target cells and replace with viral supernatant. Incubate for 8 h at 37°C, 5% CO_2.

7. Eight hours after incubation repeat **steps 5** and **6**; remove and centrifuge media from Phoenix cells and use it to replace virus-containing media on target cells.

8. Day 5: Replace media on target cells with complete medium.

9. Day 6: Place target cells under selection. Remove media and replace with complete medium containing the required selection agent (*see* **Note 4**).

3.1.2. Generation of Stable Cell Lines by Stable Transfection

1. Day 1: Plate 1 million target cells in complete medium in a 10-cm dish for each line to be generated (*see* **Note 5**).

2. Day 2: Transfect cells with vector. Dilute vector (5 μg) in 400-μL Opti-MEM, without FCS or antibiotics. In a separate tube, dilute 10-μL Lipofectamine 2000 transfection reagent. Incubate both tubes at room temperature for 5 min, then mix the two together and incubate a further 20 min (*see* **Notes 6** and **7**). Remove media from cells, and replace with 8-mL DMEM without FCS or antibiotics. Add DNA mixture dropwise to the plate and swirl to mix. Incubate at 37°C, 5% CO_2 for 4–6 h.

3. Replace cell media with complete medium, continue incubating at 37°C, 5% CO_2.

4. Day 4: Replace cell media with complete medium containing the required selection agent (*see* **Note 4**).

5. Select cells for fluorescent protein expression by flow-cytometry-based cell sorting. Alternatively, use limiting dilution to derive clonal cell lines by diluting cells to less than 1 per 100 μL and adding 100 μL of the cell mix to each well of a 96-well plate (*see* **Note 8**).

3.2. Apoptosis Induction

3.2.1. General Inducers of Apoptosis

1. Add inducers to imaging medium and gently mix prior to adding to cells and imaging (*see* **Notes 9** and **10**)

2. For UV irradiation, remove media, wash once in PBS, remove, UV irradiate, and add sufficient imaging medium to cover cells.

3.2.2 Microinjection of Proapoptotic Proteins

1. Plate HeLa cells expressing cytochrome *c*-GFP on fibronectin-coated glass coverslips in a 3-cm dish for 24 h prior to microinjection such that the cells achieve a density of 50–70%

(roughly 2×10^5 cells per plate) in complete medium (*see* **Notes 11** and **12**).

2. Prepare the solution to be injected. Dilute caspase-8 cleaved Bid (C8-Bid) in HE buffer to 0.1–1 mg/mL (*see* **Notes 13** and **14**). Add a fluorescent dextran that emits at the red end of the spectrum such as AlexaFluor568 dextran to inject at a concentration of 0.08% (w/v) in order to identify injected cells.

3. Prior to injection, spin the C8-Bid solution at top speed for 10 min in a bench top centrifuge to remove any aggregates or other particulates that may clog the needle (*see* **Note 15**). Transfer the supernatant to a new tube.

4. Load the microinjection needle with the C8-Bid solution. Load approximately 2-μl solution into the needle using a microloader.

5. Inject C8-Bid solution into the cytoplasm of each cell (time, 0.2 s; pressure, 200 hPa) (*see* **Note 16**).

6. After the injection has been completed replace the media with fresh media and incubate at 37°C.

7. Analyze cells for cytochrome *c* release by confocal microscopy 1–2 h later. Cytochrome *c*-GFP release in cells injected with tBid should be evident in most of the cells 1 h after injection (*see* **Note 17**).

3.2.3. Transfection of Proapoptotic Peptides

1. Plate HeLa cells stably expressing cytochrome *c*-GFP at 2.5 $\times 10^4$ per well of an 8-chamber cover slide 24 h prior to the experiment (scale up as required for each experiment).

2. If required, preincubate cells with 20 μM qVD-OPH for 2 h before loading, to prevent cells detaching after they undergo apoptosis.

3. Resuspend Chariot protein transfection reagent in 150-μL ddH$_2$O and sonicate in water bath sonicator for 5 min prior to transfection to disrupt any aggregates.

4. Mix peptide (10 μg in DMSO) with 10-kD dextran conjugated to AlexaFluor568 (0.5 μg) in 25-μL PBS.

5. Add Chariot protein transfection reagent (1 μL/reaction) to 25 μL of ddH$_2$O per reaction, add to the peptide solution and incubate for 30 min at room temperature.

6. Wash cells with PBS and add 50 μL of chariot/peptide solution plus 50-μL Opti-Mem to the cells and incubate at 37°C for 1 h.

7. Add an equal volume of medium containing 20% FBS, incubate for a further 2 h and then remove the peptide complexes and replace with regular medium.

8. Analyze cells for cytochrome *c* release by confocal microscopy 1 h later. Cytochrome *c*-GFP release in cells loaded with Bid or Bim BH3 peptide should be evident (*see* **Note 18**).

3.3. Multiparameter Apoptosis Imaging

This protocol enables simultaneous detection of MOMP (determined by cytochrome *c*-GFP release from the mitochondria), loss $\Delta\psi_m$ (detected by loss of TMRE from mitochondria), PS exposure (detected by Annexin V AlexaFluor647 binding to plasma membrane exposed PS), and plasma membrane permeabilization (detected by uptake and nuclear retention of propidium iodide) (*see* **Note 19**).

1. Coat 4-well Labtek chamber slide with 1 mg/mL fibronectin (1 mL per well) for at least 5 min at room temperature. Remove fibronectin (keep and reuse), wash wells once in PBS and remove PBS (*see* **Note 20**).

2. Remove media from stock flask of HeLa cytochrome *c*-GFP expressing cell line, wash once in PBS and trypsinize cells. Count cells, spin down, and resuspend well in complete medium. Make up cell suspension to 4×10^4 cells per mL and add 1 mL to each well of the chamber slide and incubate overnight (*see* **Note 21**).

3. The following day set microscope incubator to 37°C at least 1 h prior to imaging the cells (*see* **Note 22**).

4. Make up imaging media (4.5 mL per 4-well chamber slide). Complete medium contains 10% FCS, 50 nM TMRE, 20 mM Hepes, pH 7.4, 55 μM β-mercaptoethanol, 0.5% (w/v) Annexin V AlexaFluor647, 2.5 mM $CaCl_2$, and 0.4-μg/mL propidium iodide. Warm medium to 37°C (*see* **Note 23**).

5. Remove media from cells and add imaging media to chamber slide (1 mL per well). Add appropriate apoptosis inducing agent to a given well and mix by pipetting (*see* **Note 24**).

6. Turn on 5% CO_2 source, and place chamber slide on microscope stage inside of an incubator enclosure at 37°C and allow at least 15 min for thermal equilibration (*see* **Note 22**).

7. Focus on cells using 40× or 63× objective. Empirically determine the least amount of laser light for a given channel that provides the required signal/noise ratio (*see* **Note 25**).

8. If using a motorized XY stage and microscope with multifield capabilities, set different field positions.

9. Set time interval between image capture (*see* **Note 26**).

10. Start imaging, typically for a 16-h period (*see* **Note 27**).

3.4. Analysis of Confocal Time-Lapse Data

3.4.1. Cytochrome c Release – The Punctate/ Diffuse Index

The duration or extent of the release of cytochrome c and other intermembrane space proteins from the mitochondria can be expressed graphically by the punctate/diffuse index. The punctate/diffuse index is the standard deviation of the average brightness of all the pixels in an individual cell and can be measured using Metamorph or a similar software program. A high standard deviation value represents a cell with a punctate distribution of GFP, because there are many bright green pixels adjacent to black (nonfluorescent) pixels. Conversely a low standard deviation value represents a cell with diffuse or released cytochrome c-GFP since the brightness of the pixels in the cell is evenly distributed *(2)*.

3.4.2. Loss of $\Delta\psi_m$ – Average Intensity

Loss of $\Delta\psi_m$ can be similarly measured and displayed graphically. The signal due to TMRE is lost rather than redistributed. Therefore, the cells are measured for changes in the average intensity of the cell, loss of which is representative of the loss of $\Delta\psi_m$.

3.4.3. Data Analysis Using Metamorph Software

1. Using the appropriate tool draw a region around each of the cells that are to be analyzed (*see* **Note 28**).

2. Use the software to measure each region in each frame of the movie for standard deviation (cytochrome c release) or average intensity (loss of $\Delta\psi_m$). Export the results to an Excel spreadsheet.

3. For each cell, identify the frame just prior to when cytochrome c release (or other event) occurs and identify the corresponding number representing the standard deviation of that frame in the data set for that cell. Label this point as time zero and line up all the cells so that each time zero is in one row of the Excel spread sheet.

4. Correct each value (x) for each cell according to this formula $(x-\text{min})/(\text{max}-\text{min})$ (*see* **Note 29**).

5. Obtain the average standard deviation or punctate/diffuse index of all the cells for each time point. This average can be represented graphically and the release of cytochrome c-GFP is seen as a sudden drop in the punctate/diffuse index *(2)*.

6. Calculate the duration of release as the time it takes between the maximum point in the graph and the lowest point when release is complete.

7. Add error bars to each data point by calculating the standard error of the mean (SEM) of the cells for each time point (*see* **Note 30**).

4. Notes

1. Ecotropic and amphotropic Phoenix cell lines exist. Virus produced from ecotropic Phoenix cells will infect most cell lines besides human cell lines. Virus from amphotropic cell lines will infect most cell lines including human cell lines. Appropriate biosafety procedures must be carried out when carrying out retroviral work especially when using amphotropic Phoenix cells.

2. While the aim of Phoenix cell transfection is to produce high titer virus encoding the gene of interest, some expression of the gene of interest occurs in the packaging cell line. Since the gene of interest is fluorescent, this allows for ready observation under a microscope, enabling easy means of assessing transfection efficiency.

3. Virus-containing media from Phoenix cells may be frozen at −80°C for later use, though viral titer will drop by approximately half with each freeze-thaw cycle.

4. Working stocks of common selection drugs (*see also* **Note 9**): Geneticin: 1 mg/mL, puromycin: 1 µg/mL, Zeocin: 200 µg/mL, hygromycin: 200 µg/mL. Cell lines may vary in their susceptibility to these agents. Selection agents differ in the speed with which they work. Puromycin will kill nontransduced/transfected cells in ~48 h, while G418 can take up to a week; hygromycin and Zeocin work with intermediate speed. Once selection takes place, it is common to observe many small clusters of adherent cells; these represent single surviving clones that are growing out. It may be necessary to split cells before they reach full confluence, so these clusters do not overgrow.

5. This protocol makes use of plasmid vectors that encode a drug resistance cassette (such as eGFPN1). If the plasmid does not contain a drug resistance cassette, then stable cell lines expressing the fusion protein of interest must be selected by flow-cytometry-based cell sorting. Plasmid linearization with an appropriate restriction enzyme prior to transfection may improve genomic integration.

6. If the gene of interest is a strongly proapoptotic molecule, such as GFP-Bax, gene expression in the target cells may lead to apoptosis. It may, therefore, be necessary to reduce the amount of DNA used in the transfection; quantities as low as 20 ng per 10-cm plate may be used (*see also* **Note 7**).

7. Some forethought is required to produce and maintain stable cell lines; because most vectors (retroviral or transient) contain drug resistance cassettes that are expressed from a

promoter separate from that of the gene of interest, it is possible for stable cells to stably integrate the resistance cassette but not the gene of interest. This problem is exacerbated when the gene of interest is strongly proapoptotic; in such cases, cells that lose the gene of interest but maintain drug resistance have an obvious growth advantage. IRES or 2A peptide strategies *(9)*, in which both the gene of interest and the drug selection marker are expressed off the same mRNA, may aid generation of a stable cell line.

8. If a given fluorescence protein localizes incorrectly (e.g., cytosolic rather than mitochondrial), then it may be necessary to use limiting dilution to select clonal cell lines displaying the correct localization.

9. Suggested concentration ranges for common proapoptotic stimuli are 0.5–2 μM for staurosporine, 0.5–2 μM for actinomycin D, and 10–100 ng/mL for TNFα. For TNFα to induce apoptosis, cycloheximide must be added at 10 μg/mL. For UV irradiation a suitable dose range is between 2 and 40 mJ/cm². Different cell types may differ greatly in their sensitivity to apoptotic stimuli.

10. Many apoptotic stimuli induce necrosis at higher doses. Inhibiting caspase-dependent cell death using a caspase inhibitor such as qVD-OPH confirms that a given dose is inducing apoptosis.

11. Cells must be plated on glass to enable the cells to be visualized by confocal or fluorescence microscopy postinjection and to provide an even, flat surface, decreasing the chance of breaking the microinjection needle. Cells tend to adhere less well to glass surfaces than to plastic so we recommend coating the glass with fibronectin prior to plating the cells.

12. The cells will be exposed to the laboratory atmosphere for the duration of the microinjection procedure necessitating the addition of antibiotics and Hepes to prevent contamination and to maintain the pH of the media, respectively.

13. Buffer selection for microinjection: the buffer which is closest to the physiological milieu of the cell contains: 48 mM K_2HPO_4, 4.5 mM KH_2PO_4, 14 mM NaH_2PO_4, pH 7.2 and is generally recommended as injection buffer. However, several other buffers have been used without any obvious effect on cell function such as HE buffer or PBS. The use of DMSO as a solvent should be avoided because it disrupts the integrity of the glass needle thus BH3 peptides that are soluble in DMSO should be introduced into cells by alternate means such as protein transfection.

14. For injection of proteins such as C8-Bid, a concentration of 0.1–1 mg/mL is sufficient to induce cytochrome *c*-GFP

release in 2 h (we use C8-Bid as an example here but this protocol is sufficient for other proteins that activate Bax, including p53).

15. Prior to injection, if required, add a caspase inhibitor such as qVD-OPH (20 μM) or zVAD-fmk (100 μM) to the cells and incubate for 1 h at 37°C to prevent the detachment of cells during apoptosis.

16. The volumes injected are usually reported as being within 5–20% of the cell volume. It is thus estimated that the solution is diluted 10–100-fold upon injection and the volume of a HeLa cell is approximately 4–5 pL *(10)*. Therefore, if the concentration of the injected protein is 0.1 mg/mL, it is estimated that 5–50 fg of each protein is delivered to each cell.

17. If time-lapse experiments are required after microinjection of C8-Bid, then it is recommended that the cells should be immediately placed on the microscope stage and brought to the experimental temperature due to the relatively short time to cytochrome *c* release.

18. At 1 h after completion of the peptide transfection protocol approximately 90–100% of cells will have released cytochrome *c* if 10 μg of Bid or Bim BH3 peptide is loaded. If time-lapse experiments are required, it is recommended that the cells should be placed on the microscope stage and brought to the experimental temperature immediately after adding the 20% FCS containing medium and omitting the step where the complexes are removed. Since the process of cytochrome *c* release occurs so quickly after loading the peptides in the cells, the prolonged presence of the peptide complexes should not adversely affect the experiment.

19. The general protocol for detection of Bax/BH3-only protein mitochondrial translocation and monitoring of mitochondrial morphology does not vary from the above besides the cell line/fluorescent fusion protein being used. See **Table 1** for a list of fusion proteins and their utility in live cell imaging of apoptosis.

20. Fibronectin enhances the adhesion of many cell types to glass (we routinely use MCF7 and HeLa cells); however, for some cell types it may be ineffective or dispensable. Alternatives to fibronectin coating include collagen or poly-l-lysine.

21. It is important that when imaging, the cells are not too confluent. 4×10^4 cells are sufficient for imaging the following day. If imaging is to be carried out 2 days after plating the cells (e.g., because the cells are transfected the following day) then scale down to 2×10^4 cells initially. Scale up and

down cell numbers appropriately according to surface area of chamber slide used.

22. It is critical that if one is using a microscope set up with an enclosed incubator that the incubator is given sufficient time to reach thermal equilibrium. Significant focal drift can occur as the chamber slide thermally equilibrates with the incubator. It is essential that equilibration occurs prior to imaging. If the incubator is not humidified, then one must overlay the medium with mineral oil to prevent evaporation.

23. β-mercaptoethanol mitigates the potentially harmful effects of reactive oxygen species produced during imaging. It can be substituted with DTT (0.5 mM). TMRE is taken into mitochondria dependent upon $\Delta\psi_m$. Loss of $\Delta\psi_m$ (as a result of caspase activity upon mitochondrial function) leads to loss of TMRE staining *(11, 12)*. Annexin V binds to PS exposed on the plasma membrane during apoptosis. Propidium iodide is a cell impermeable dye that enters cells and binds DNA following membrane permeabilization. Annexin V is available conjugated to different fluorophores and alternative cell impermeable dyes exist that can be used (e.g., Sytox Red) should spectral restraints be an issue.

24. Add 2-mL imaging media per 3-cm dish (microinjection) and 0.5 mL per well of an 8-well chamber slide (protein transfection).

25. Imaging cells with low levels of laser light and for short time periods minimizes phototoxicity. It is worth noting that when correctly minimizing phototoxicity, the signal to noise, and the resultant image quality is lower than one would choose when imaging fixed cells or at single time-points. Selection of cells expressing high amounts of a given fluorescent protein reduces the need for high laser powers or long exposure times. Dependent upon the experimental requirement it may not be necessary to use higher magnification (i.e., where possible use 40× 1.3NA rather than 63× 1.4NA). This also reduces the amount of phototoxicity and has the advantage of allowing more cells to be imaged in a given field. Laser levels/exposure lengths for Annexin V AlexaFluor647 and propidium iodide should be determined and noted by staining apoptotic cells with Annexin V AlexaFluor647 and propidium iodide and using these settings for live cell imaging.

26. A good initial starting point for time intervals is 10 min. If cells have been microinjected or transfected with a BH3 only protein/peptide, the time interval can be significantly shortened since the time to MOMP induction will be rapid.

27. The typical order of events during apoptosis is MOMP, loss of TMRE staining, PS exposure, and, finally, plasma membrane

permeabilization. It is essential to control for potential phototoxicity during live cell imaging. Suitable controls include imaging under the same condition without inducing apoptosis (the cells should not die and should enter mitosis). Moreover, inducing apoptosis and imaging at fixed time points poststimulation should reveal a kinetic pattern of events similar as if the cells are imaged continuously.

28. The cells inevitably move during the course of the movie, so it is best to ensure that the chosen region encompasses the cell at the time of the event (cytochrome c release, TMRE loss) and does not include too much extra/negative space, which will lead to spurious results.

29. Since the measurements made by the software are arbitrary numbers and not specific units, it is possible to correct each value with respect to the maximum and minimum values for each cell. The maximum value is the average of the standard deviations for all the frames prior to time zero (they should be approximately equal since cytochrome c remains in the mitochondria) while the minimum value is the lowest number (the point when cytochrome c has been released completely). In this way the standard deviation value of every frame for each cell should fall between 1 (when cytochrome c is in the mitochondria hence punctate) and 0 (when cytochrome c is released hence diffuse).

30. These analyses can give very accurate representations of the changes that occur at a single cell level during apoptosis. However, any nonspecific microscopic aberrations such as focal drift or photobleaching can lead to spurious results when calculating the statistics. Such problems must be taken into account and controlled for in each separate experiment.

References

1. Green, D. R., and Kroemer, G. (2004). The pathophysiology of mitochondrial cell death, *Science* **305**, 626–629.

2. Goldstein, J. C., Waterhouse, N. J., Juin, P., Evan, G. I., and Green, D. R. (2000). The coordinate release of cytochrome c during apoptosis is rapid, complete and kinetically invariant, *Nat Cell Biol* **2**, 156–162.

3. Rehm, M., Dussmann, H., and Prehn, J. H. (2003). Real-time single cell analysis of Smac/DIABLO release during apoptosis, *J Cell Biol* **162**, 1031–1043.

4. Munoz-Pinedo, C., Guio-Carrion, A., Goldstein, J. C., Fitzgerald, P., Newmeyer, D. D., and Green, D. R. (2006). Different mitochondrial intermembrane space proteins are released during apoptosis in a manner that is coordinately initiated but can vary in duration, *Proc Natl Acad Sci USA* **103**, 11573–11578.

5. Wolter, K. G., Hsu, Y. T., Smith, C. L., Nechushtan, A., Xi, X. G., and Youle, R. J. (1997). Movement of Bax from the cytosol to mitochondria during apoptosis, *J Cell Biol* **139**, 1281–1292.

6. Zha, J., Weiler, S., Oh, K. J., Wei, M. C., and Korsmeyer, S. J. (2000). Posttranslational N-myristoylation of BID as a molecular switch for targeting mitochondria and apoptosis, *Science* **290**, 1761–1765.

7. Frank, S., Gaume, B., Bergmann-Leitner, E. S., Leitner, W. W., Robert, E. G., Catez, F., Smith, C. L., and Youle, R. J. (2001). The

role of dynamin-related protein 1, a mediator of mitochondrial fission, in apoptosis, *Dev Cell* **1**, 515–525.

8. Kuwana, T., Bouchier-Hayes, L., Chipuk, J. E., Bonzon, C., Sullivan, B. A., Green, D. R., and Newmeyer, D. D. (2005) BH3 domains of BH3-only proteins differentially regulate Bax-mediated mitochondrial membrane permeabilization both directly and indirectly, *Mol Cell* **183**, 434–442.

9. de Felipe, P. (2002) *Curr Gene Ther* **2**, 355–378.

10. Minaschek, G., Bereiter-Hahn, J., and Bertholdt, G. (1989) *Exp Cell Res* **183**, 434–442.

11. Waterhouse, N. J., Goldstein, J. C., von Ahsen, O., Schuler, M., Newmeyer, D. D., and Green, D. R. (2001). Cytochrome *c* maintains mitochondrial transmembrane potential and ATP generation after outer mitochondrial membrane permeabilization during the apoptotic process, *J Cell Biol* **153**, 319–328.

12. Ricci, J. E., Munoz-Pinedo, C., Fitzgerald, P., Bailly-Maitre, B., Perkins, G. A., Yadava, N., Scheffler, I. E., Ellisman, M. H., and Green, D. R. (2004). Disruption of mitochondrial function during apoptosis is mediated by caspase cleavage of the p75 subunit of complex I of the electron transport chain, *Cell* **117**, 773–786.

Chapter 4

Detection of Apoptosis in Tissue Sections

Eva Csizmadia and Vilmos Csizmadia

Summary

TUNEL-based assays were used to demonstrate the presence of apoptotic cells in tissue sections derived from target tissues of animal models of different diseases. Emphasis was placed on tissue preparation and fixation, as these are crucial to successful histological staining. The protocol suggested here facilitates not only the reliable detection of TUNEL-positive cells but the immunodetection of different proteins in these cells and the surrounding tissues by DAB or fluorescence-based immunostaining.

Key words: Tissue fixation, Zn-fixative, Apoptosis, TUNEL, Caspase-3, Immunohistochemistry, Animal models

1. Introduction

Apoptosis is a form of programmed cell death used in multicellular organisms to dispose of cells in many biological processes, including embryonic development, pathogenesis, and response to therapeutic agents *(1)*. Apoptosis involves a series of cellular perturbations leading to a variety of specific morphologic changes and eventually cell death and the dismantling and removal of dead cells from the location where apoptosis occurred. These phenomena clearly distinguish apoptosis from other types of cell death, such as necrosis and necroptosis and form the basis of apoptosis assays *(2)*. DNA degradation into nucleosomal units is considered one of the hallmarks of apoptotic cell death. Apoptotic DNA degradation in most cases results in the formation of nicks in the cleaved DNA molecule, allowing for terminal uridine deoxynucleotydil transferase dUTP nick end labeling (TUNEL), a common method of identifying dying cells in the last phase of

Peter Erhardt and Ambrus Toth (eds.), *Apoptosis*, Methods in Molecular Biology, vol. 559
DOI 10.1007/978-1-60327-017-5_4
© Humana Press, a part of Springer Science+Business Media, LLC 2004, 2009

apoptosis *(3, 4)*. Because of its simplicity and the commercial availability of assay kits, TUNEL assays are central to demonstrating the occurrence of apoptosis, provided they are performed correctly *(5)*. There are a few controversies about the specificity of the TUNEL assay because apoptosis reportedly can occur without DNA nick formation and appear as a negative result in TUNEL; in addition, not only apoptotic but nonapoptotic cells can possess cleaved DNA and appear TUNEL-positive *(6, 7)*. Because of these and other similar apoptosis-related issues, it is advisable and occasionally necessary to demonstrate the presence of apoptosis by additional means, such as caspase-3 immunohistochemistry staining in the TUNEL-positive specimen *(8)*. Tissue preparation and fixation are critical steps to successful immunohistochemistry for both DNA and protein detection. Tissues are very complex, nonuniform structures; therefore, their response to preservatives and other reagents during immunohistochemistry staining varies. Applying various fixatives, optimizing reagents, and addressing other aspects of immunohistochemistry can reduce background staining, eliminate false signals, and lead to optimal detection of apoptotic cells *(9)*. Animal models are currently widely used in medical research to understand the molecular bases of diseases *(10, 11)*. Our laboratories have investigated a wide range of animal tissues both for the occurrence of apoptosis and for other cellular events leading to disease. It was necessary to optimize and invent tissue treatment procedures that allowed both for apoptosis-related and general immunostaining in the same tissue. The observations we believe are most informative are presented in this chapter.

2. Materials

2.1. Tissue Preparation, Fixation, and Sectioning

1. 2-Methylbutane (also known as isopentane; Fisher Scientific).
2. TFM Tissue Freeze Medium, TBS (American Master*Tech Scientific).
3. Base Mold – Disposable (American Master*Tech Scientific).
4. Superfrost® Plus slides (American Master*Tech Scientific).
5. Calcium acetate, zinc chloride, zinc acetate, if making zinc fixative from scratch (Sigma-Aldrich).
6. Unisette-tissue-cassettes (American Master*Tech Scientific).
7. Ethanol (Pharmaco).
8. 2-Propanol (Fisher Scientific).
9. Xylene (American Master*Tech Scientific).
10. IHC Zinc fixative (formalin free), if not making from scratch (BD Pharmingen™).
11. 10% buffered formalin (American Master*Tech Scientific).

12. Paraffin (American Master*Tech Scientific).

13. Shandon Citadel™ tissue processor (Thermo Scientific).

14. Humidity chamber (Sigma-Aldrich).

15. TEC™ tissue embedding center (TBS®).

2.2. TUNEL Assay

1. Paraformaldehyde (Electron Microscopy Sciences).

2. Proteinase K lyophilizate (Roche).

3. ApopTag® Peroxidase In Situ Apoptosis Detection Kit (Chemicon International, now Millipore).

4. Hydrogen peroxide (Sigma-Aldrich).

5. GelBond® Film agarose gels support medium (Lonza).

6. DAB Substrate Kit, 3,3′-diaminobenzidine (Vector Laboratories Inc.).

7. Gill's hematoxylin II (American Master*Tech Scientific).

8. Cytoseal™ 280 (Richard-Allan Scientific®).

9. VasoTACS™ In Situ Apoptosis Detection Kit (Trevigen, Inc.).

2.3. Immunohisto-chemistry

1. Acetone (Fluka® Analytical).

2. Normal horse serum (Vector Laboratories, Inc.).

3. Cleaved caspase-3 (Asp-175) antibody (Cell Signaling Technology®).

4. Avidin/Biotin Blocking kit (Vector Laboratories, Inc.).

5. Tween® 20 (Sigma-Aldrich).

6. Biotinylated goat anti-rabbit secondary antibody (Vector Laboratories, Inc.).

7. VECTASTAIN® ABC kit. (Vector Laboratories, Inc.).

8. Triton® X-100 (Sigma-Aldrich).

9. Sucrose (Sigma-Aldrich).

10. Alexa Fluor® 594 donkey anti-rabbit IgG (Molecular Probes®).

11. Hoechst 33258, 10 mg/mL solution in water (Invitrogen™).

12. Polyvinyl alcohol mounting medium (Fluka® Analytical).

---

3. Methods

3.1. Tissue Preparation, Fixation, and Sectioning

3.1.1. Frozen Tissue Preparation

1. Fill a 1-L beaker with 500 mL of 2-methylbutane. Place the beaker in a styrofoam box large enough to accommodate the beaker. Fill the styrofoam box with liquid nitrogen up to the level of 2-methylbutane in the baker. Wait until the 2-methylbutane cools down so that a frost layer forms on the inside surface of the beaker (approximately $-165°C$; takes approximately 5–10 min).

2. Harvest and cut the tissues into pieces not thicker than 5 mm. Do not rinse tissues, but place them on a gauze pad for a few seconds. Put one drop of tissue freeze medium in the base mold, put a tissue piece on top and cover with the same tissue freeze medium; avoid air bubble formation.

3. Drop the mold into the cooled 2-methylbutane and wait 5 min.

4. Place the mold on dry ice until dry and store at –80°C until sectioning.

5. Section tissues at 5 μm and mount 5-μm sections on Superfrost Plus glass slides and allow to airdry for 2–3 h before performing the TUNEL assay.

3.1.2. Preparation of Zinc-Fixed, Paraffin-Embedded Tissues

1. Prepare the zinc fixative from scratch by dissolving 0.5 g calcium acetate, 5.0 g zinc chloride, and 5.0 g zinc acetate in 1 L of 0.1 M Tris-HCl buffer, pH 7.6 (*see* **Note 1**). The final pH will be approximately 6.6. Do not readjust the pH, as this will cause the zinc to precipitate out from the fixative solution. Store the zinc fixative at room temperature (*see* **Note 2**).

2. Harvest and cut the tissues into approximately 4 mm thick segments; tissue thickness should not exceed 4 mm. Drop the segments into the zinc fixative for 36 h (±4 h) at room temperature.

3. After fixation, put the tissues into pencil-marked tissue cassettes, and dehydrate them in a tissue processor at room temperature as follows:
   (a) 2 × 45 min in 50% ethanol.
   (b) 2 × 30 min in 70% ethanol.
   (c) 1 × 45 min in 95% ethanol.
   (d) 3 × 40 min in 100% 2-propanol.
   (e) Clear tissues in xylene for 1 h (2 changes, 30 min each).
   (f) Infiltrate tissues with paraffin at 58–60°C for 1 h (2 changes, 30 min each).

4. Remove tissues promptly from processor and embed them in paraffin for sectioning, following the routine histologic procedure.

5. Section tissues at 5 μm and mount 5-μm sections on Superfrost Plus glass slides and allow to airdry overnight before performing the TUNEL assay.

6. Put slides in a standard laboratory oven at 56°C for 20 min. Do not allow the temperature to rise above 65°C because this can cause DNA damage, resulting in high background and nonspecific staining (*see* **Note 3**).

3.1.3. Preparation of Formalin-Fixed, Paraffin-Embedded Tissues

1. Harvest tissues and cut into approximately 5-mm thick segments, fix harvested tissues in 10% phosphate-buffered formalin for 16–24 h.

2. Dehydrate, clear, and paraffin-infiltrate tissues in a tissue processor using the routine 8 h protocol as follows:

    (a) 1 × 40 min in 70% ethanol.
    (b) 1 × 40 min in 80% ethanol.
    (c) 2 × 30 min in 95% ethanol.
    (d) 3 × 45 min in 100% ethanol.
    (e) 3 × 35 min xylene.
    (f) 2 × 50 min paraffin at 58–60°C.

3. Embed tissues in paraffin for sectioning, following the routine histologic procedure.

4. Section tissues at 5 μm and mount 5-μm sections on Superfrost Plus glass slides and allow to airdry at least overnight to ensure adherence.

5. Put slides in a standard laboratory oven at 56°C for 20 min. Do not allow the temperature to rise above 65°C because this can cause DNA damage, resulting in high background and nonspecific staining.

3.2. TUNEL Assay

3.2.1. ApopTag® Peroxidase In Situ Apoptosis Detection in Zinc-Fixed and Formalin-Fixed Paraffin-Embedded Tissues

Processing and Pretreatment of Zinc-Fixed, Paraffin-Embedded Tissues

1. Deparaffinize zinc-fixed, paraffin-embedded tissue sections by placing the slides in two changes of xylene for 4 min each (*see* **Note 4**).

2. Wash the slides three times in 100% 2-propanol for 4 min each.

3. Wash the slides twice in 95% ethanol for 2 min each.

4. Rinse the slides twice in distilled water for 1 min each.

5. Postfix sections on slides in 2% freshly prepared paraformaldehyde in 1× PBS at room temperature for 20 min.

6. Wash slides three times with 1× PBS for 3 min each.

7. Pretreat tissues with freshly diluted proteinase K (20 μg/mL) at room temperature for 10 min.

8. Rinse the slides three times in distilled water for 3 min each.

9. Put the slides in 1× PBS for 5 min.

Processing and Pretreatment of Formalin-Fixed, Paraffin-Embedded Tissue Sections

1. Deparaffinize formalin-fixed, paraffin embedded tissue sections by placing the slides in two changes of xylene for 5 min each.

2. Wash the slides twice in 100% ethanol for 5 min each.

3. Wash the slides twice in 95% ethanol for 2 min each.

4. Rinse the slides twice in distilled water for 1 min each.

5. Put the slides in 1× PBS for 5 min.

6. Pretreat tissues with freshly diluted proteinase K (20 μg/mL) at room temperature for 30 min.

7. Rinse the slides three times in distilled water for 3 min each.

8. Put the slides in 1× PBS for 5 min.

**DNA End Labeling and
Signal Detection**

The following steps are identical for formalin and zinc-fixed paraffin-embedded tissues.

1. Inactivate endogenous peroxidase by incubating tissue sections with 3% hydrogen peroxide in 1× PBS at room temperature for 5 min.

2. Rinse the sections twice with 1× PBS for 5 min each.

3. Immediately apply 15 µL/cm² equilibration buffer to the sections and incubate at room temperature at least for 30 s.

4. Apply 11 µL/cm² working strength TdT enzyme to the sections. Cover with Gel Bond film. Incubate in a humidity chamber at 37°C for 1 h.

5. Put the slides in a coplin jar containing Stop/Wash Buffer, agitate for 15 s, and incubate for 10 min at room temperature.

6. Wash slides in three changes of 1× PBS for 2 min each.

7. Apply 30 µL/cm² HRP conjugated anti-digoxigenin antibody to the sections and incubate in the humidified chamber for 30 min at room temperature.

8. Wash the slides in four changes of 1× PBS for 1 min each.

9. Apply DAB-peroxidase substrate to the sections in humidified chamber and incubate for 3–6 min at room temperature to develop color. Determine the optimal length of staining by carefully monitoring color development under the microscope. When sections are optimally stained, as determined by careful monitoring of color development under the microscope, proceed to **step 10**.

10. Wash the slides in three changes of water for 1 min each.

11. Lightly counterstain slides in hematoxylin at room temperature for approximately 10 s.

12. Rinse slides in water until the water is clear (2–3 changes of water).

13. Dip slides 10 times quickly into acid-alcohol composed of 70% ethanol and 1% HCl.

14. Rinse slides in water once.

15. Immediately place slides into a bluing solution composed of 2% ammonium hydroxide in water for 20–30 s.

16. Wash slides with two changes of distilled water for 1 min each.

17. Wash slides with 95% ethanol twice for 2 min each.

18. Wash slides with 100% ethanol twice for 3 min each.

19. Clear slides in xylene for 3 min and cover each with Cytoseal™ 280 mounting medium and a cover slip.

3.2.2. VasoTACS™ In Situ Apoptosis Detection in Formalin-Fixed, Paraffin-Embedded, and Frozen Tissues

Processing and Pretreatment of Formalin-Fixed and Paraffin-Embedded Tissues

1. Deparaffinize formalin-fixed, paraffin embedded tissue sections by placing the slides in two changes of xylene for 5 min each.
2. Wash the slides twice in 100% ethanol for 5 min each.
3. Wash the slides twice in 95% ethanol for 2 min each.
4. Rinse the slides twice in distilled water for 1 min each.
5. Put the slides in 1× PBS for 5 min.
6. Pretreat tissues with freshly diluted proteinase K (20 μg/mL) at room temperature for 30 min.
7. Rinse the slides three times in distilled water for 3 min each.
8. Put the slides in 1× PBS for 5 min.

Processing and Pretreatment of Frozen Tissues

1. Dry slides for 2 h at 37°C. Put slides into 100% for 1 min, then 95% ethanol for 1 min.
2. Wash the slides once with distilled water for 1 min.
3. Put slides into 1× PBS for 3 min.
4. Post-fix tissues on slides in freshly prepared 2% paraformaldehyde for 15 min.
5. Wash slides twice in 1× PBS for 5 min each.
6. Pretreat the sections with approximately 50 μL of proteinase K solution (20 μg/mL) at room temperature for 20 min.
7. Wash slides four times with distilled water for 2 min each.
8. Inactivate endogenous peroxidase by incubating tissue sections with 3% hydrogen peroxide in PBS or methanol at room temperature for 5 min.

DNA End Labeling and Signal Detection

The following steps are identical for formalin-fixed paraffin-embedded and frozen tissues.

1. Wash slides twice in distilled water for 5 min each.
2. Immerse slides in 1× TdT labeling buffer at room temperature for 5 min.
3. Cover sections with 50 μL of labeling Reaction Mix, followed by hydrophobic cover slips, and incubate at 37°C for 1 h in a humidified chamber after covering.
4. Immerse sections in 1× TdT stop buffer for 5 min at room temperature.
5. Wash samples twice with distilled water for 5 min each.
6. Cover the sections with 50 μL of Strep-HRP solution and incubate for 10 min at room temperature.
7. Wash samples three times with distilled water for 5 min each.
8. Cover the sections with Blue Label solution for 2–6 min.
9. Wash slides three times in distilled water for 5 min each.

10. Apply red counterstaining for 30–40 s.

11. Wash slides by dipping them 10 times in two changes of distilled water.

12. Wash slides with 95% ethanol twice for 2 min each.

13. Wash slides with 100% ethanol twice for 3 min each.

14. Clear slides by dipping them 10 times in two changes of xylene.

15. Cover sections with Cytoseal™ 280 mounting medium and cover slip. Keep slides in the dark because the substrate is light-sensitive (*see* **Note 5**).

3.3. Immunohistochemistry on Frozen Tissues

3.3.1. DAB-Based Immunohistochemistry

1. Section tissues at 5 μm and mount 5-μm frozen sections on Superfrost Plus glass slides and let them dry at room temperature at least for 30 min.

2. Prepare the following acetone fixative: mix 100 mL of pre-cooled (4°C) acetone with 5 mL of 10% buffered formalin.

3. After 10 min, place the slides into the precooled (4°C) acetone fixative for 3 min. Do not move slides.

4. Place slides into precooled (4°C) 1× PBS for 5 min. Do not move slides.

5. Wash slides with 1× PBS at room temperature twice for 5 min each.

6. Block tissues with 7% horse serum in 1× PBS for 30 min at room temperature.

7. Place caspase-3 primary antibody diluted 1:300 with 1× PBS on the tissues, incubate at 4°C overnight.

8. Wash slides with 1× PBS at room temperature once for 3 min.

9. Block slides, following the instructions in the Avidin–Biotin blocking kit.

10. Block tissue peroxidase activity with a 1:100 dilution of hydrogen peroxide in 1× PBS for 10 min.

11. Wash slides with 0.05% solution of Tween® 20 in 1× PBS for 5 min.

12. Place biotinylated goat anti-rabbit secondary antibody diluted 1:800 with 1× PBS on the tissues at room temperature for 1 h.

13. Prepare the Avidin–Biotin–HRP complex (AB-complex) in 1× PBS as suggested by the manufacturer.

14. Wash slides with with 0.05% solution of Tween® 20 in 1× PBS for 5 min.

15. Place the AB-complex on the tissues for 30 min at room temperature.

16. Wash slides twice with 0.05% solution of Tween® 20 in 1× PBS for 5 min.

17. Develop color with DAB-substrate kit for 3–5 min, monitoring color development under microscope.

18. Wash in water twice for 1 min each.

19. Lightly counterstain slides in hematoxylin at room temperature for approximately 10 s.

20. Wash slides with two changes of distilled water.

21. Wash slides with 95% ethanol twice for 2 min each.

22. Wash slides with 100% ethanol twice for 3 min each.

23. Wash slides with two changes of xylene.

24. Cover sections with Cytoseal™ 280 mounting medium and coverslip (*see* **Note 6**).

3.3.2. AlexaFluor® 594 Immunofluorescence Staining of Frozen Tissues

1. Section tissues at 5 μm and mount 5-μm frozen sections on Superfrost Plus glass slides, dry at room temperature for at least for 10 min, and fix them in fresh 2% paraformaldehyde solution for 15 min.

2. Wash with 1× PBS twice for 5 min each.

3. Treat tissues with 0.5% Triton® X-100 plus 0.05% Tween® 20 in 1× PBS for 12 min.

4. Block with 5% horse serum, 2% BSA, and 0.05% Tween® 20 in 1× PBS at room temperature for 30 min.

5. Without washing slides, place the caspase-3 primary antibody diluted 1:300 with 1× PBS on the tissues. Incubate at 4°C overnight.

6. Wash slides twice in 1× PBS for 5 min each.

7. Put 0.5% sucrose in 1× PBS on the slides and wait 5 min.

8. Tap down slides, and immediately place AlexaFluor® 594 secondary antibody diluted 1:300 with 1× PBS on the tissues at room temperature for 1 h. The antibody is labeled with AlexaFluor 594.

9. Wash slides in 1× PBS for 2 min.

10. Dilute Hoechst 33258 stock 1:10,000 in 1× PBS and place on tissues for 3 min at room temperature.

11. Wash in 1× PBS twice for 5 min each.

12. Wash in water once for 1 min.

13. Dry slides for 5 min at room temperature.

14. Put one drop of polyvinyl alcohol mounting medium on the tissues (~30 μL) and coverslip.

3.3.3. Imaging

Capture fluorescence images with the ApoTome Imaging System using an Axiovert 200 inverted microscope from Carl Zeiss (*see* **Note** 7). Examine DAB-stained cells on an Olympus BX-51 microscope and capture digital images with an Olympus DP71 camera (*see* **Notes 8** and **9**).

4. Notes

1. Ready to use zinc fixative is also commercially available from BD Pharmingen, as indicated in the **Subheading 2**.

2. The panel of methods detailed here is focused on preparation of tissues for TUNEL assays for demonstrating the presence of apoptotic cells in different tissues. Snap frozen - paraformaldehyde-fixed, formalin-fixed paraffin-embedded and zinc-fixed paraffin-embedded tissues are widely used in medical research *(12–14)*. Although not uniform in this aspect, all provide good tissue morphology and DNA preservation. In our hands, each of these tissue fixation methods proved to be excellent for demonstrating TUNEL-positive cells.

3. As to protein immunoreactivity either directly or after antigen retrieval *(15)*, in our experience, zinc fixation is an excellent tissue fixation protocol, especially for lung and fat tissues, which are conventionally known to be refractory to satisfactory immunohistochemistry staining.

4. When TUNEL assays are being set up for the first time, it is worth using positive control tissues known to contain apoptotic cells or cleaved DNA. Mouse or rat intestinal tissue sections (*see* **Fig. 1a, b**) or nuclease-treated target tissues (*see* **Fig. 1c, d**) can provide excellent signals if the TUNEL assay is successfully established.

5. We observed that the signal intensity of the TUNEL assay kits we routinely use in our apoptosis studies varies with the tissue fixation protocol: one kit may stain tissues prepared a given way better than another, so it is worth trying both kits in case one gives weak signals or ambiguous results. In addition, it is important – especially in animal models – to be "at the right place at the right time" for catching apoptotic cells. Apoptotic cells are not always there where they are theoretically expected to be, and it is also easy to miss the time point of their optimal detection, so if the apoptotic cells are not detectable, it is not always due to a TUNEL assay-related technical issue.

6. In our experience, snap-frozen tissues work best for both for TUNEL and immunohistochemistry staining, followed by zinc-fixed and formalin fixed tissues. However, we wish to emphasize the importance of zinc fixation because we have determined that the resulting tissue sections can be used very efficiently not only for TUNEL assays but for analyzing the expression of proteins by immunohistochemistry with many antibodies *(16–18)*.

7. Because of ongoing technical innovations, well-established detection kits are occasionally replaced with new ones.

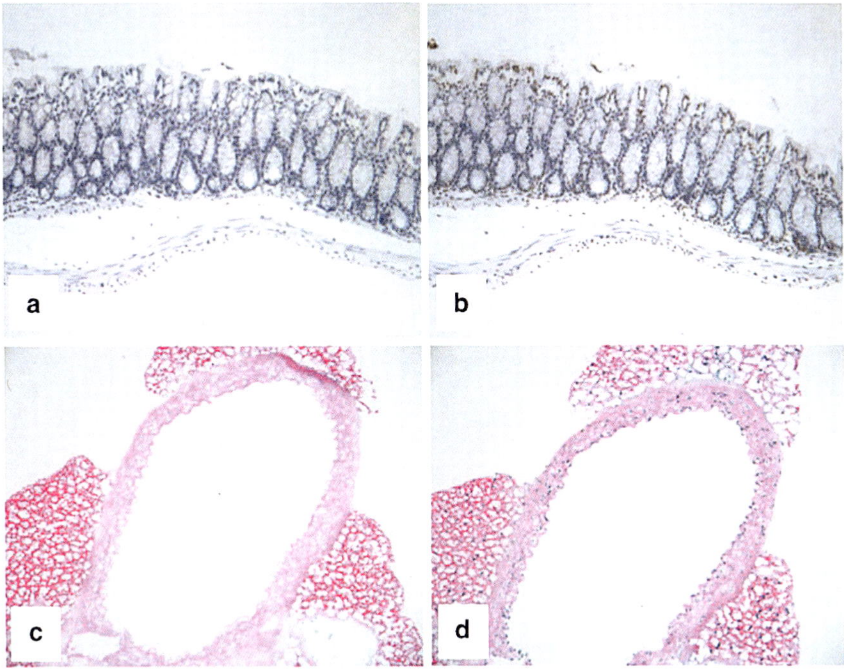

Fig. 1. Demonstration of apoptosis in positive control tissues. TUNEL-based detection of fragmented DNA in formalin-fixed paraffin embedded tissues. Staining of apoptotic intestinal cells by using the ApopTag Peroxidase In Situ Apoptosis Detection Kit. (**a**) Staining without the presence of the TdT enzyme. (**b**) Staining in the presence of TdT enzyme of the kit. (**c**) Frozen rat neck carotid artery without nuclease digestion. (**d**) Paraformaldehyde-fixed frozen rat neck carotid artery digested with TACS-Nuclease. In (**c**) and (**d**) the VasoTACS In Situ Apoptosis Detection Kit was used to demonstrate the presence of fragmented DNA in the tissues. (**a**) and (**b**) were counterstained with hematoxylin, (**c**) and (**d**) with eosin.

Unfortunately, their utility for tissue staining can not be verified unless they are tested on the tissues of interest, so it is not possible to give reliable technical advice in this situation. One of the latest products that may be important to future searches for improved immunohistochemistry staining is Vector's VECTASTAIN® ABC kit. This is a new variant of the DAB-based immunohistochemistry signal detection system. We are currently evaluating it in our laboratories, with promising preliminary results.

8. Although the importance of zinc fixation is deservedly emphasized for multiple reasons, we do not yet have extensive experience with using this type of fixation for immunofluorescence staining. Our current favorite immunofluorescence staining protocol is performed most successfully on snap-frozen tissue sections that are paraformaldehyde fixed before fluorescence staining. The result was identical to that of DAB-based immunostaining of snap-frozen acetone-fixed tissue sections targeted for the presence of cleaved caspase-3 in liver tissues

collected from our mouse sepsis model, where TUNEL positive cells were also detectable (*see* **Figs. 2** and **3**) *(19)*.

9. Once the assay is established, it has broad potential applications, as the accompanying figures to this chapter illustrate: TUNEL-positive tubular epithelial cells in the rat kidney after cold ischemia (*see* **Fig. 4**) *(20)*; apoptosis of myocardiac

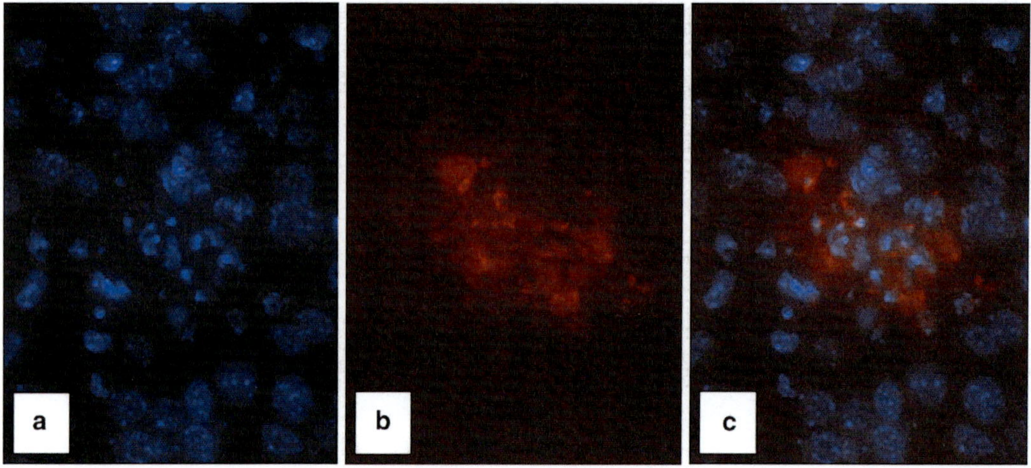

Fig. 2. Caspase-3-specific immunofluorescence staining of apoptotic cells. The presence of apoptotic cells is shown by caspase-3-specific AlexaFluor594 fluorescence-based immunohistochemistry in snap frozen, paraformaldehyde-fixed liver tissues collected from a mouse sepsis model. (**a**) Hoechst 33258 nuclear staining of the apoptosis-positive liver region. (**b**) Immunohistochemistry for cleaved caspase-3 protein in the same apoptosis-positive liver region. (**c**) Merged images from (**a**) and (**b**) (each at ×63 magnification).

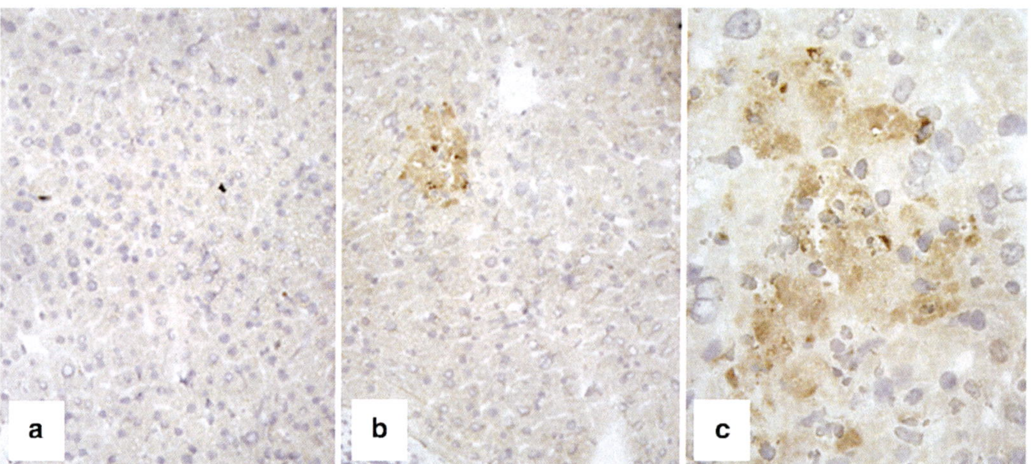

Fig. 3. Caspase-3-specific DAB-immunohistochemistry staining of apoptotic cells. The presence of apoptotic cells is shown by caspase-3-specific, DAB-based immunohistochemistry in snap frozen liver tissues, fixed in acetone, and collected from the same mouse sepsis model as in Fig. 2. (**a**) Control liver. (**b**) Immunohistochemistry for cleaved caspase-3 protein in an apoptotic liver region using a rabbit anti-caspase-3 polyclonal antibody (×40 magnification). (**c**) The same region as in (**b**), but at ×100 magnification. Tissues were counterstained with hematoxylin.

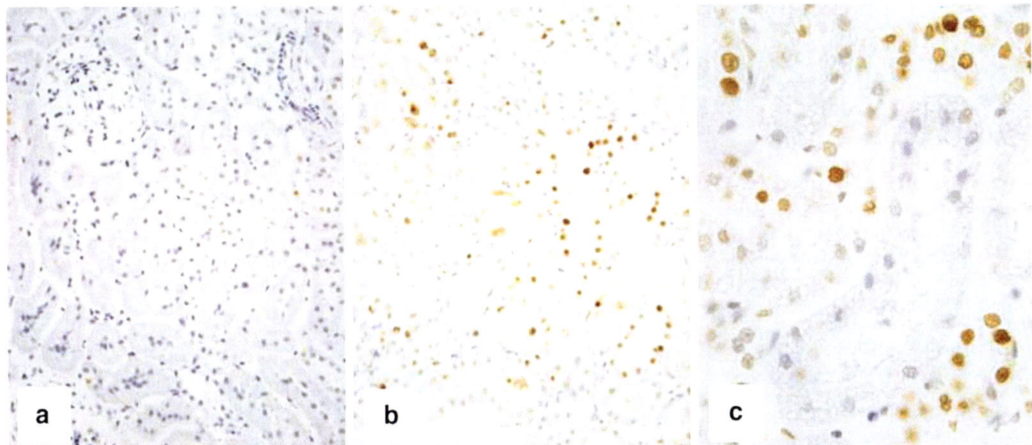

Fig. 4. Demonstration of the presence of apoptotic cells after cold ischemia in the rat kidney. (**a**) Normal rat kidney (×40 magnification). (**b**) Rat kidney exposed to cold ischemia (×40 magnification). (**c**) Rat kidney exposed to cold ischemia (×100 magnification). For apoptosis detection, the ApopTag Peroxidase In Situ Apoptosis detection Kit was used on zinc-fixed paraffin-embedded tissue sections. Tissues were counterstained with hematoxylin.

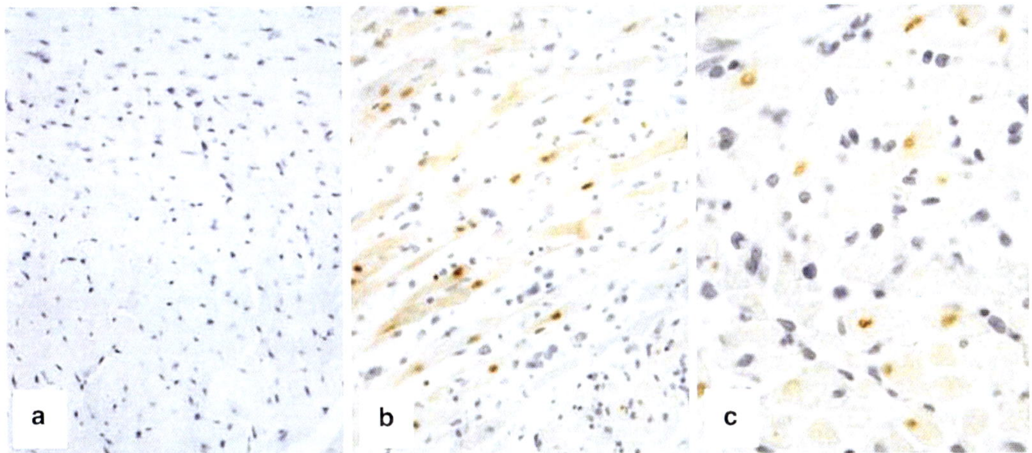

Fig. 5. Apoptosis of myocardiac cells in transplanted hearts. In this cardiac xenograft model, a mouse heart was transplanted to a rat recipient. (**a**) Normal heart. (**b**) Transplanted mouse heart 10 days posttransplantation (×40 magnification). (**c**) The same transplanted heart as in (**c**), but with ×100 magnification. For apoptosis detection, the ApopTag Peroxidase In Situ Apoptosis Detection Kit was used on formalin-fixed, paraffin-embedded tissue sections. Tissues were counterstained with hematoxylin.

cells in a mouse-to-rat xenotransplantation model (*see* **Fig. 5**) *(21)*; TUNEL-positive endothelial cells in the atrium of the HO-1 knockout mouse heart (*see* **Fig. 6**) *(22)*. These results show that immunohistochemistry staining and TUNEL assays work reliably in a broad range of well-prepared tissue specimens. Since our understanding of the molecular details of the mechanism of apoptosis is not yet complete, we still need to investigate the molecular events in the

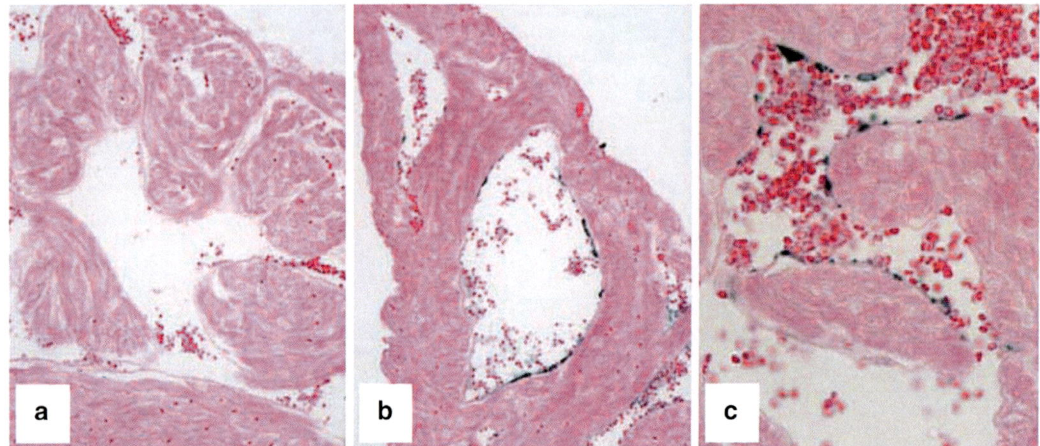

Fig. 6. Endothelial cell apoptosis in the atrium of HO-1 knockout mice. (**a**) A tissue section from the atrium of a normal mouse heart. (**b**) The same tissue from an HO-1 knockout mouse (×40 magnification). (**c**) Tissue section from the atrium of an HO-1 knockout mouse (×100 magnification). For apoptosis detection, the VasoTACS In Situ Apoptosis Detection Kit was used on formalin-fixed, paraffin-embedded tissue sections. Tissues were counterstained with eosin.

apoptotic cells and in their surroundings. In these investigations, the role of immunohistochemistry staining will be central, especially in in-vivo studies designed to identify or validate new biomarkers and regulatory molecules that contribute to the apoptotic process and its regulation. We hope the protocols described here will help future researchers in these efforts.

Acknowledgments

The authors thank Christiane Ferran, Leo Otterbein, and Beek Chin at Beth Israel Deaconess Hospital of Harvard Medical School for their continuous support of the work presented in this publication. The work was financially supported by RO1 grants HL08013, DK063275. We thank also Alexis Khalil for critical reading of the manuscript.

References

1. Taylor, R. C., Cullen, S. P., and Martin, S. J. (2008). Apoptosis: controlled demolition at the cellular level, *Nature reviews* **9**, 231–241.

2. Willingham, M. C. (1999). Cytochemical methods for the detection of apoptosis, *J Histochem Cytochem* **47**, 1101–1110.

3. Labat-Moleur, F., Guillermet, C., Lorimier, P., Robert, C., Lantuejoul, S., Brambilla, E., and Negoescu, A. (1998). TUNEL apoptotic cell detection in tissue sections: critical evaluation and improvement, *J Histochem Cytochem* **46**, 327–334.

4. Fondevila, C., Shen, X. D., Tsuchiyashi, S., Yamashita, K., Csizmadia, E., Lassman, C., Busuttil, R. W., Kupiec-Weglinski, J. W., and Bach, F. H. (2004). Biliverdin therapy protects rat livers from ischemia and reperfusion injury, *Hepatology (Baltimore, Md)* **40**, 1333–1341.

5. Ito,Y.,Shibata,M.A.,Kusakabe,K.,andOtsuki,Y. (2006). Method of specific detection of apoptosis using formamide-induced DNA denaturation assay, *J Histochem Cytochem* **54**, 683–692.

6. Allen, R. T., Hunter, W. J., III, and Agrawal, D. K. (1997). Morphological and biochemical characterization and analysis of apoptosis, *J Pharmacol Toxicol Meth* **37**, 215–228.

7. Barrett, K. L., Willingham, J. M., Garvin, A. J., and Willingham, M. C. (2001). Advances in cytochemical methods for detection of apoptosis, *J Histochem Cytochem* **49**, 821–832.

8. Gown, A. M., and Willingham, M. C. (2002). Improved detection of apoptotic cells in archival paraffin sections: immunohistochemistry using antibodies to cleaved caspase 3, *J Histochem Cytochem* **50**, 449–454.

9. Miething, F., Hering, S., Hanschke, B., and Dressler, J. (2006). Effect of fixation to the degradation of nuclear and mitochondrial DNA in different tissues, *J Histochem Cytochem* **54**, 371–374.

10. Hakem, R., and Mak, T. W. (2001). Animal models of tumor-suppressor genes, *Annu Rev Genet* **35**, 209–241.

11. Griffin, J. L. (2006). Understanding mouse models of disease through metabolomics, *Curr Opin Chem Biol* **10**, 309–315.

12. Srinivasan, M., Sedmak, D., and Jewell, S. (2002). Effect of fixatives and tissue processing on the content and integrity of nucleic acids, *Am J Pathol* **161**, 1961–1971.

13. Wester, K., Asplund, A., Backvall, H., Micke, P., Derveniece, A., Hartmane, I., Malmstrom, P. U., and Ponten, F. (2003). Zinc-based fixative improves preservation of genomic DNA and proteins in histoprocessing of human tissues. Laboratory investigation. *J Tech Meth Pathol* **83**, 889–899.

14. Beckstead, J. H. (1994). A simple technique for preservation of fixation-sensitive antigens in paraffin-embedded tissues, *J Histochem Cytochem* **42**, 1127–1134.

15. Taylor, C. R. (2006). Standardization in immunohistochemistry: the role of antigen retrieval in molecular morphology, *Biotech Histochem* **81**, 3–12.

16. Banz, Y., Hess, O. M., Robson, S. C., Csizmadia, E., Mettler, D., Meier, P., Haeberli, A., Shaw, S., Smith, R. A., and Rieben, R. (2007). Attenuation of myocardial reperfusion injury in pigs by Mirococept, a membrane-targeted complement inhibitor derived from human CR1, *Cardiovascular Res* **76**, 482–493.

17. Fondevila, C., Shen, X. D., Tsuchiyashi, S., Yamashita, K., Csizmadia, E., Lassman, C., Busuttil, R. W., Kupiec-Weglinski, J. W., and Bach, F. H. (2004). Biliverdin therapy protects rat livers from ischemia and reperfusion injury, *Hepatology (Baltimore, Md)* **40**, 1333–1341.

18. Patel, V. I., Daniel, S., Longo, C. R., Shrikhande, G. V., Scali, S. T., Czismadia, E., Groft, C. M., Shukri, T., Motley-Dore, C., Ramsey, H. E., Fisher, M. D., Grey, S. T., Arvelo, M. B., and Ferran, C. (2006). A20, a modulator of smooth muscle cell proliferation and apoptosis, prevents and induces regression of neointimal hyperplasia, *Faseb J* **20**, 1418–1430.

19. Onishi, S., Miyata, H., Inamoto, T., Qi, W. M., Yamamoto, K., Yokoyama, T., Warita, K., Hoshi, N., and Kitagawa, H. (2007). Immunohistochemical study on the delayed progression of epithelial apoptosis in follicle-associated epithelium of rat Peyer's patch, *J Vet Med Sci* **69**, 1123–1129.

20. Bartels-Stringer, M., Kramers, C., Wetzels, J. F., Russel, F. G., Groot, H., and Rauen, U. (2003). Hypothermia causes a marked injury to rat proximal tubular cells that is aggravated by all currently used preservation solutions, *Cryobiology* **47**, 82–91.

21. Soares, M. P., Lin, Y., Anrather, J., Csizmadia, E., Takigami, K., Sato, K., Grey, S. T., Colvin, R. B., Choi, A. M., Poss, K. D., and Bach, F. H. (1998). Expression of heme oxygenase-1 can determine cardiac xenograft survival, *Nat Med* **4**, 1073–1077.

22. Evans, P. C., Taylor, E. R., and Kilshaw, P. J. (2001). Signaling through CD31 protects endothelial cells from apoptosis, *Transplantation* **71**, 457–460.

Chapter 5

Detection of Apoptosis in Cell-Free Systems

Dhyan Chandra and Dean G. Tang

Summary

Apoptosis is a fundamental process required for proper embryonic development. Various methods have been described to detect apoptosis both in vitro as well as in vivo. Activation of caspases represents the key event in the apoptotic process. To dissect the molecular events leading to caspase activation, we have been using cell-free systems that recapitulate the mitochondrial death pathway. In the cell-free apoptosis assays, we either detect caspase activation in stimulated cells by utilizing subcellular fractions or reconstitute various components in cytosol (or mitochondria) to study molecular mechanisms of caspase activation. In either case, we utilize Western blot and/or substrate cleavage to monitor caspase activation. Using in-vitro reconstitution approach of caspase activation, we have discovered various factors that regulate caspase activity. Therefore, cell-free system not only is an invaluable tool to study apoptosis signaling but also provides molecular insight on caspase activation patterns and inhibitor specificities.

Key words: Apoptosis, Apoptosome, Cytochrome *c*, Cell-free reconstitution, Substrate cleavage, Caspase activation, Mitochondria, Cytoplasm, Apaf-1

1. Introduction

Apoptosis plays an essential role in animal development and in maintaining the homeostasis of adult tissues *(1)*. Deficiency in apoptosis is a hallmark of cancer and autoimmune diseases whereas excessive apoptosis is implicated in neurodegenerative diseases, strokes, and cardiac diseases. The family of caspases (cysteine aspartic acid-specific protease) is the key effectors in the execution of apoptotic cell death *(2)*. Caspases are synthesized as inactive proenzymes, which become proteolytically cleaved during apoptosis to generate active enzymes. Activated caspases then cleave cellular proteins such as poly(ADP-ribose) polymerase (PARP) to dismantle the dying cells *(3)*. In response to stress,

Peter Erhardt and Ambrus Toth (eds.), *Apoptosis*, Methods in Molecular Biology, vol. 559
DOI 10.1007/978-1-60327-017-5_5
© Humana Press, a part of Springer Science + Business Media, LLC 2004, 2009

cells release cytochrome *c* from the intermembrane space of the mitochondria to the cytosol. The released cytochrome *c* binds to and activates the adaptor protein Apaf-1, which in turn activates the initiator procaspase-9 in the presence of ATP, leading to the formation of apoptosome and subsequent activation of "executioner" caspases such as caspase-3, 6, or 7 *(4)*.

We have been using cell-free systems to detect apoptotic activity/caspase activation in cytosolic or mitochondrial extracts *(5–10)*. We generally use two approaches to detect apoptosis. First, we isolate cytosolic or mitochondrial extracts from cells that have been treated in culture with an apoptosis-inducing agent. Second, purified cytosolic extracts from untreated cells is used in reconstitution experiments with addition of bovine cytochrome *c* or recombinant active caspases. Apoptotic activity in these extracts can be examined by the measurement of enzymatic caspase activity, and/or by Western blots of proteins processed during apoptosis (i.e., caspases and their substrates). It was in 1993 when the first paper described that a cell-free system could mimic characteristic features of apoptosis in intact cells *(11)*. Later, many other investigators have used cell-free systems successfully for dissection of biochemical mechanisms during the apoptotic process, such as the identification and characterization of the "apoptosome," AIF (apoptosis-inducing factor), and the DNA fragmentation factor ICAD *(12–14)*. Here we describe our protocols for the detection of caspase activation in cell-free systems *(5–10)*.

2. Materials

2.1. Cell Culture and Subcellular Fractionation

1. For cell culture we used Dulbecco's Modified Eagle's Medium (DMEM) (Gibco Grand Island, NY) supplemented with 10% fetal bovine serum (FBS, HyClone) and 1% Penicillin and Streptomycin (*see* **Note 1**).

2. Staurosporine (Sigma), dissolved in tissue-culture grade dimethyl sulfoxide (DMSO) at 1 mM, stored in aliquots at –20°C, and then added to cell-culture dishes as required.

3. Solution of trypsin (0.25%) and ethylenediamine tetraacetic acid (EDTA) (1 mM) from Gibco/BRL used for harvesting cells from the dishes.

4. Phosphate buffered saline (1×) (PBS): 137 mM sodium chloride (NaCl), 2.7 mM potassium chloride (KCl), 4.3 mM disodium hydrogen phosphate (Na_2HPO_4), 1.4 mM potassium dihydrogen phosphate (KH_2PO_4).

5. Teflon cell scrapers (Fisher Scientific).

6. Homogenizing (hypotonic) buffer: 20 mM 4-(2-hydroxyethyl)-1-piperazineethanesulfonic acid (HEPES), pH 7.5, 10 mM KCl, 1.5 mM MgCl$_2$, 1 mM sodium EDTA, 1 mM sodium EGTA, 1 mM DTT, 250 mM sucrose and mixture of protease inhibitors (Sigma).

7. Dounce homogenizer using high clearance pestle from Fischer Scientific.

8. TNC buffer: 10 mM Tris-acetate, pH 8.0, 0.5% NP-40, 5 mM CaCl$_2$.

9. Small-volume ultracentrifugation tubes (i.e., less than 5 mL; Beckman Coulter, Inc.)

2.2. SDS-Polyacrylamide Gel Electrophoresis (SDS-PAGE)

1. Micro-BCA Protein Assay Kit (Pierce Biotechnology, Inc.)

2. For resolving gel: 1.5 M Tris-HCl, pH 8.8, 10% sodium dodecyl sulfate (SDS). Store at room temperature (*see* **Note 2**).

3. For stacking gel: 1.0 M Tris-HCl, pH 6.8, 10% SDS. Store at room temperature.

4. Thirty percent acrylamide/bis solution (in 29:1 ratio in deionized distilled water) and *N,N,N,N'*-tetramethyl-ethylenediamine, TEMED (Bio-Rad) (*see* **Note 3**).

5. Ammonium persulfate: prepare 10% solution in distilled water and immediately freeze in single use (200 µL) aliquots at –20°C.

6. Running buffer: 25 mM Tris-HCl, 250 mM glycine, 0.1% (w/v) SDS. Prepare 5× or 10× and store at room temperature or alternatively could be purchased from BioRad.

7. Prestained molecular weight markers: Low-range markers (Bio-Rad).

8. SDS gel-loading buffer (6×): 350 mM Tris-HCl, pH 6.8, 10% (w/v) SDS, 30% (w/v) glycerol, 9.25% dithiothreitol (DTT), 0.02% (w/v) bromophenol blue. Make 0.5 mL aliquots and store at –80°C (*see* **Note 4**).

2.3. Western Blotting

1. Transfer Buffer: 24 mM Tris (do not adjust pH), 192 mM glycine, 20% (v/v) methanol.

2. Supported nitrocellulose membrane from BioRad, 3 MM chromatography paper from Fisher Scientific.

3. Tris-buffered saline with Tween 20 (TBS-T): Prepare 10× stock with 1.37 M NaCl, 200 mM Tris-HCl, pH 7.5; store at room temperature. Before using, make 1× solution in distilled water with addition of 0.1% Tween-20.

4. Blocking buffer: 5% (w/v) nonfat dry milk in TBS-T.

5. Primary and secondary antibody dilution buffer: TBS-T supplemented with 3% (w/v) nonfat dry milk.

6. Secondary antibody: Anti-rabbit or mouse IgG (depending on the primary antibody) conjugated to horseradish peroxidase (Amersham Biosciences).

7. Enhanced chemiluminescence (ECL) reagents from Amersham Biosciences.

8. Autoradiography X-ray film from Fisher Scientific.

2.4. Stripping and Reprobing Blots for Caspase-3 and Actin

1. Stripping buffer: 62.5 mM Tris-HCl, pH 6.8, 2% (w/v) SDS. Store at room temperature. Warm to working temperature of 55°C and add 100 mM β-mercaptoethanol (*see* **Note 5**).

2. Primary antibody: Anti-caspase-9 (Chemicon), anti-caspase-3 (Biomol), and anti-actin (ICN).

2.5. Substrate Cleavage Assay for Caspases

1. Caspase reaction buffer: 50 mM HEPES, pH 7.4, 100 mM NaCl, 0.1% CHAPS, 10 mM DTT, 1 mM EDTA, 10% glycerol. Always prepare fresh reaction buffer.

2. Ac-DEVD-AFC and Ac-LEHD-AFC (Biomol) dissolved in DMSO to the stock concentration of 10 mM and make aliquots and store at –80°C (*see* **Note 6**).

3. 7-Amino-4-trifluoromethyl-coumarin (AFC) from Sigma.

3. Methods

During apoptosis, procaspase-9 (~46 kDa) is processed to generate the p37/p35 kDa fragments. In our experiments, we have utilized an antibody that recognizes both the proform and the p37/p35 bands. As illustrated in **Fig. 1**, in GM701 fibroblasts treated with staurosporine (STS), the p37/p35 fragments were detected in cytosolic as well as in the mitochondrial fractions. Similarly, procaspase-3 (~32 kDa) is processed to the p20/p17 bands, the latter representing catalytically active caspase-3 (*5–10*; **Fig. 1**). In such Western blotting assays, apoptosis should preferably be quantified side by side using DAPI staining to identify apoptotic nuclei (**Fig. 1**, bottom). This is important because cleavage of procaspase-9 does not indicate that the enzyme is active, although procaspase-3 cleavage *does* suggest its proteolytic activation.

In vitro reconstitution experiment is a relatively novel approach to mimic and study caspase activation in vivo. Using freshly purified cytosol, we could readily reconstitute caspase activation with the addition of cytochrome *c* alone (**Fig. 2**). On the one hand, many other investigators have used dATP or ATP (around 1 mM) together with cytochrome *c* to initiate caspase

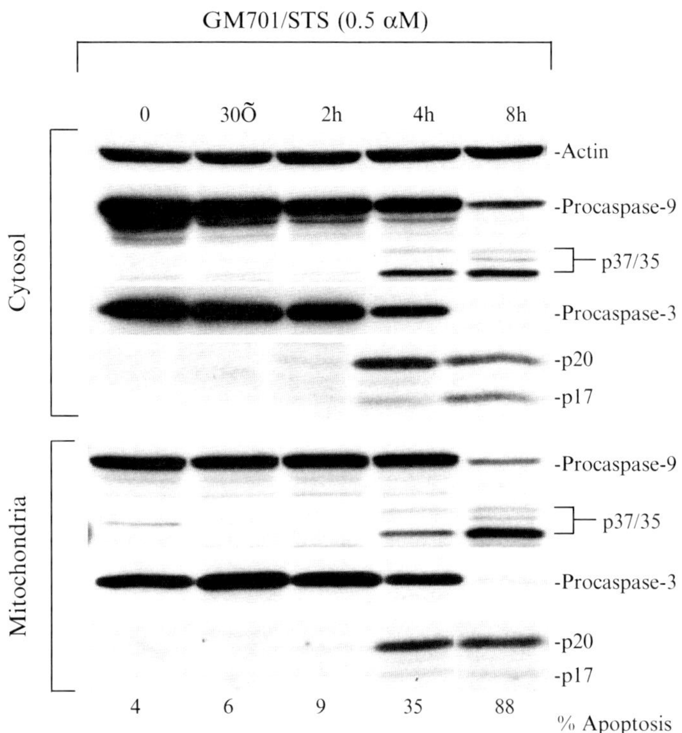

Fig.1. Proteolytic processing of procaspase-9 and -3 in GM701 cells treated with STS. 30 (cytosol) or 60 (mitochondria) μg of proteins was used in Western blotting for caspase-9, caspase-3, or actin (modified from ref.*6*).

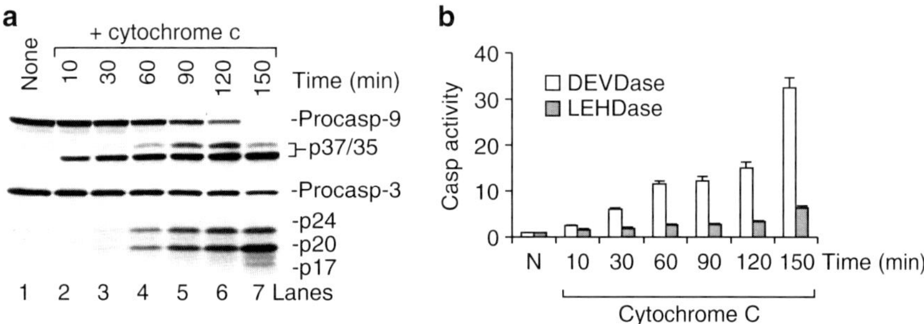

Fig. 2. Cytochrome *c* initiates caspase activation without addition of dATP or ATP. Fresh GM701 cytosol (3 μg/μL) was incubated with cytochrome *c* (15 μg/mL) for the time periods indicated. At the end, Western blotting was performed to detect procaspase-9 and -3 processing (**a**). 50 μg of reaction mixture was also used to determine LEHDase and DEVDase activities (**b**). Modified from **ref.** *9*.

processing in such reconstitution systems. We, on the other hand, have found that *freshly* purified cytosols contain sufficient amount of dATP or ATP (generally in m*M* range) to support cytochrome *c*-initiated caspase activation. Below we describe our general protocol for cell-free caspase activation analyzed by Western blotting and/or LEHDase/DEVDase activity assays (**Fig. 2**).

3.1. Subcellular Fractionation

1. Treat cultured cells (e.g., GM701; ~10 million) with an apoptotic stimulus (e.g., staurosporine) or vehicle control. Harvest (using a cell scraper or trypsin/EDTA) and wash both treated and mock-treated cells twice with ice-cold 1× PBS.

2. Suspend washed cells in 600 µL of homogenizing (hypotonic) buffer and incubate on ice for 30 min.

3. Homogenize the cell suspension with a Dounce homogenizer using high clearance pestle (140 strokes) (*see* **Note 7**).

4. Centrifuge at 1,000 × g for 5 min to remove nuclei and unbroken cells (*see* **Note 8**).

5. Centrifuge the resulting supernatant again at 10,000 × g for 20 min at 4°C to obtain the pellet, which is enriched in mitochondria.

6. The resulting supernatant is further subjected to ultracentrifugation at 100,000 × g for 1 h at 4°C to obtain cytosol (or S100).

7. Mitochondrial fractions are washed thrice in homogenizing buffer and then solubilized in 60 µL of TNC buffer containing protease inhibitors (*see* **Note 9**).

8. Measure protein concentrations of the prepared mitochondrial and cytosolic fractions using Micro BCA Protein Assay Kit.

3.2. Cell-Free Reconstitution Experiments

1. Cell-free reactions are performed in homogenizing buffer in a total volume of 100 µL.

2. Purified cytosols (3 mg/mL) are activated by adding bovine cytochrome c (15 µg/mL; Sigma) without (d)ATP and incubated at 37°C for 150 min (*see* **Note 10**).

3. After incubation, samples are used for either substrate cleavage assays for caspase-9 (LEHDase) and caspase-3 (DEVDase) or procaspase cleavage by Western blotting.

3.3. Preparation of SDS-PAGE Gels

1. Clean the glass plates thoroughly with a rinsable detergent, rinse extensively with distilled water, and assemble according to the manufacturer's instructions.

2. Depending upon the size of apparatus, prepare 10 mL reaction mix for 15% resolving gel by mixing in a 50 mL disposable plastic tube or conical flask in following order: 2.3 mL distilled water, 5.0 mL of 30% acrylamide solution, 2.5 mL of 1.5 M Tris-HCl, pH 8.8, 0.1 mL SDS, and 0.1 mL ammonium persulfate; mix and then add 4 mL of TEMED. Mix immediately and proceed to the next step. Polymerization begins as soon as TEMED is added.

3. Using Pasteur pipette, pour the above acrylamide solution into the gap between the glass plates. Leave 1 cm space below the length of the comb for stacking gel. Gel should be in vertical position and overlay a thin layer of distilled water. Leave the gel at room temperature for 30 min to polymerize.

4. Pour off the water and wash several times with water to remove unpolymerized acrylamide and drain all the liquid using paper towels.

5. Depending on the size of gel, prepare stacking gel by mixing 2.7 mL distilled water, 0.67 mL acrylamide, 0.5 mL 1.0 M Tris, pH 6.8, 40 μL of 10% SDS, 40 μL of ammonium persulfate, and then add 4 μL of TEMED. Mix immediately and, without delay, pour the stacking gel solution directly on the polymerized resolving gel. Immediately insert the comb while avoiding air bubbles, add more stacking gel to fill the spaces of the comb completely, and leave it at room temperature for 30 min to polymerize.

3.4. Preparation of Samples and Running Gels

1. While stacking gel is polymerizing, take 20–50 μg protein (from mitochondrial or cytosolic fraction) per lane in a total volume of 30 μL (for 18-well gel) or 40 μL (for 12-well gel). Make up the volume with 1× PBS. For Western blotting of the reconstitution experiments, 30–40 μL of reaction samples after incubation is used.

2. Add 6 μL (for 18-well gel) or 8 μL (for 12-well gel) of 6× SDS-loading buffer. Boil for 5–10 min in a heating block and centrifuge for 1 min to collect samples to the bottom of the tubes.

3. Once the stacking gel has set, carefully remove the comb and use a 3-mL syringe fitted with a 22-gauge needle to wash the wells with running buffer. Mount the gel in electrophoresis apparatus and add the Tris-Glycine running buffer to the upper and lower chambers of the gel unit and remove any trapped air bubbles at the bottom of the gel or in the wells.

4. Load 30 or 40 μL sample (depending on the capacity of the wells) onto 15% SDS-PAGE gels. Use one well for low-range prestained protein markers. Also load 1× SDS sample-loading buffer in any empty wells.

5. Attach the electrophoresis apparatus to power supply and first run at 80 V for 20–30 min. When the bromophenol blue dye has moved to resolving gel increase the voltage to 120 V and run the gel until the dye reaches the bottom. This process generally takes 2–3 h.

3.5. Transfer of Proteins and Western Blotting

1. While the SDS-PAGE gel is still running, prepare transfer buffer, and keep in cold room.

2. Soak chromatography paper and fiber pads in transfer buffer 10–20 min before start of transfer and also soak nitrocellulose

membrane in transfer buffer. If PVDF membrane is to be used, soak in 100% methanol for 5–10 min.

3. Once the bromophenol blue dye has reached the bottom of the gel, disconnect power supply, remove the gel from the gel holding apparatus and cut out the stacking gel. Detach the dye-containing gel at the bottom and wash for 5 min in transfer buffer on a rotating shaker.

4. Arrange the transfer cassettes in following order: Black side of cassette on bottom, fiber pad, single sheet of same-sized chromatography paper, gel (marker side of the gel on right), nitrocellulose membrane, one sheet of chromatography paper, and then fiber pad. Close the cassette and avoid and remove air bubbles in every step.

5. Submerge the resulting cassette sandwich in a transfer tank that contains transfer buffer. The cassette is placed into the transfer tank such that the nitrocellulose membrane is between the gel and the anode. This orientation is very critical otherwise the proteins will be lost from the gel into the buffer rather than transferred to the nitrocellulose membrane.

6. Insert a small magnet and ice pack in the transfer tank and run at 100 V with slow stir of the magnetic stirrer for 1–2 h depending on the molecular weights of the proteins to be transferred. For proteins up to 50 kDa, a 75-min transfer should be sufficient.

7. After completion of the transfer, cut the lower right-hand side of membrane before taking it out and this will become lower left-hand side to mark the transfer side (i.e., protein transfer side is up) and wash two times with 1× TBS-T for 5 min each.

8. Block the membrane with 5% non-fat dry milk in 1× TBS-T for 1 h at room temperature on a rocking platform. At the end of incubation, wash the membrane one time with 1× TBS-T for 5 min.

9. Probe with rabbit polyclonal antibody for caspase-9 from Chemicon diluted (1,000×) in 1× TBS-T containing 3% non-fat dry milk for 2 h at room temperature. At the end of incubation, wash the membrane four times with 1× TBS-T for 10 min each.

10. Probe with secondary antibody, rabbit IgG conjugated to horseradish peroxidase, diluted (5,000×) in 1× TBS-T containing 3% non-fat dry milk, for 1 h at room temperature. After incubation, wash the membrane four times with 1× TBS-T for 10 min each.

11. During washing, 2 mL aliquots of ECL reagents (i.e., 2 mL of ECL A and B) are warmed separately at room

temperature. Just before use mix ECL reagents in equal ratio, pour directly on the membrane, and incubate for 1 min and then immediately expose to X-ray film to detect signals. ECL incubation and detection should be performed at room temperature in a dark room having safe red light (*see* **Note 11**).

3.6. Reprobing the Membrane Blots for Caspase-3 and Marker Proteins

1. After the completion of caspase-9 Western blotting and once a satisfactory exposure for the result of the processed caspase-9 has been obtained, the membrane is stripped and then reprobed one by one with antibodies that recognize the processed caspase-3 and actin, respectively, for a loading control that confirms equal recovery of the samples through the procedure. If the molecular weights of the target molecules are very different, the two (or more) antibodies can be added simultaneously for reprobing.

2. Stripping buffer (50 mL per blot – *see* **Note 5**) is warmed to 55°C and then β-mercaptoethanol is added. The blot is incubated for 30 min with continuous slow agitation.

3. Once the blot is stripped, it is extensively washed in TBS-T buffer (three times with 50 mL for each wash for 10 min), and then blocked again in blocking buffer.

4. The membrane is then ready to be reprobed with anti-caspase-3 (1:3,000 in TBS-T) with washes, secondary antibody, and ECL detection as described above. This process is repeated for actin (1:5,000) or any other molecule(s). When properly done, the stripping-reprobing process can be repeated for up to 5–8 times. Some examples are shown in **Figs. 1** and **2**.

3.7. LEHDase (for Caspase-9) and DEVDase (for Caspase-3) Activity Measurement

1. For caspase activity measurement, 30–50 μg of mitochondrial or cytosolic proteins is added to a reaction mixture containing 30 μM fluorogenic peptide substrates, Ac-DEVD-AFC or Ac-LEHD-AFC in a total volume of 100 μL.

2. Similarly, at the end of reconstitution experiments, 30–50 μg of reconstituted sample is added to the reaction mixture described earlier.

3. Production of 7-amino-4-trifluoromethyl-coumarin (AFC) is monitored in a spectrofluorimeter (Hitachi F-2000 fluorescence spectrophotometer) using excitation wavelength 400 nm and emission wavelength 505 nm (*see* **Note 12**).

4. The fluorescent units are converted into nanomoles of AFC released per hour per mg of protein using a standard curve. The results are generally presented as fold activation over the control (**Fig. 2**).

4. Notes

1. Fetal bovine serum should be heat-inactivated in waterbath prior to use at 56°C for 30 min and make aliquots in a 50 mL disposable plastic tube.

2. Wear gloves and mask while handling SDS powder to prevent inhalation of the fine powder. Alternatively, premade polyacrylamide gels can be purchased from commercial sources.

3. Acrylamide is highly hazardous (neurotoxic) and should not be purchased in a powder form unless absolutely necessary. It is now available in premixed form from various suppliers. Always take precaution while handling unpolymerized acrylamide. TEMED should be stored at room temperature in a desiccator.

4. The 6× SDS-loading buffer, when stored as aliquots at –80°C, is stable for up to one year. Repeated freezing and thawing is not recommended.

5. 2-β-mercaptoethanol is toxic and gives a very unpleasant smell in the laboratory. Use tight container and proper care while handling it.

6. Ac-DEVD-AFC, Ac-LEHD-AFC, and AFC are light sensitive.

7. When homogenizing cells, take care not to overhomogenize because this will damage mitochondria and cytochrome c will leak out in control cells also. To prevent overhomogenization, monitor cells under a microscope every 50 strokes to achieve an optimal 60–80% of cell breakage. Do not try to achieve 100% cell breakage.

8. Take 2 µL of supernatant and observe under a microscope. If some nuclei or unbroken cells are observed in the supernatant, recentrifuge for 5 min at $1,000 \times g$.

9. Decrease or increase the amount of TNC buffer to obtain desired concentration of mitochondrial lysates.

10. Various investigators use 1 mM dATP or ATP to reconstitute caspase activation in cell-free system. We find that dATP or ATP is not required for cytochrome c-initiated caspase activation when fresh cytosol is used in such assays.

11. If signal is very weak with ECL, ECL plus could be used as alternative detection reagent.

12. It is very important to use proper filter for excitation (400 nm) and emission (505 nm) for caspase activity measurement.

Acknowledgements

This work was supported in part by an NIH K01 award to DC (7K01CA123142) and by grants from NIH (R01-AG023374, R01-ES015888, and R21-ES015893–01A1), Department of Defense (W81XWH-07–1–0616 and PC073751), and Elsa Pardee Foundation to DGT.

References

1. Horvitz, H. R. (1999). Genetic control of programmed cell death in the nematode *Caenorhabditis elegans*. *Cancer Res.* **59**, 1701S–1706S.

2. Salvesen, G. S. and Dixit, V. M. (1997). Caspases: intracellular signaling by proteolysis. *Cell* 91, 443–446.

3. Thornberry, N. A. and Lazebnik, Y. (1998). Caspases: enemies within. *Science* **281**, 1312–1316.

4. Wang, X. (2001). The expanding role of mitochondria in apoptosis. *Genes Dev.* 15, 2922–2933.

5. Chandra, D., Liu, J. W., and Tang, D. G. (2002). Early mitochondrial activation and cytochrome c up-regulation during apoptosis. *J. Biol. Chem.* **277**, 50842–50854.

6. Chandra, D., and Tang, D. G. (2003). Mitochondrially localized active caspase-9 and caspase-3 result mostly from translocation from the cytosol and partly from caspase-mediated activation in the organelle. Lack of evidence for Apaf-1-mediated procaspase-9 activation in the mitochondria. *J. Biol. Chem.* 278, 17408–17420.

7. Chandra, D., Choy, G., Deng, X., Bhatia, B., Daniel, P., and Tang, D. G. (2004). Association of active caspase 8 with the mitochondrial membrane during apoptosis:potential roles in cleaving BAP31 and caspase 3 and mediating mitochondrion-endoplasmic reticulum cross talk in etoposide-induced cell death. *Mol. Cell. Biol.* 24, 6592–6607.

8. Chandra, D., Choy, G., Daniel, P. T., and Tang, D. G. (2005). Bax-dependent regulation of Bak by voltage-dependent anion channel 2. *J. Biol. Chem.* 280, 19051–19061.

9. Chandra, D., Bratton, S. B., Person, M. D., Tian, Y., Martin, A. G., Ayres, M., et al. (2006). Intracellular nucleotides act as critical prosurvival factors by binding to cytochrome C and inhibiting apoptosome. *Cell* **125**, 1333–1346.

10. Chandra, D., Choy, G., and Tang, D. G. (2007). Cytosolic accumulation of HSP60 during apoptosis with or without apparent mitochondrial release: evidence that its pro-apoptotic or pro-survival functions involve differential interactions with caspase-3. *J. Biol. Chem.* **282**, 31289–31301.

11. Lazebnik, Y. A., Cole, S., Cooke, C. A., Nelson, W. G., and Earnshaw, W. C. (1993). Nuclear events of apoptosis in vitro in cell-free mitotic extracts: a model system for analysis of the active phase of apoptosis. *J. Cell. Biol.* **123**, 7–22.

12. Zou, H., Henzel, W. J., Liu, X., Lutschg, A., and Wang. X. (1997). Apaf-1, a human protein homologous to C. elegans CED-4, participates in cytochrome c-dependent activation of caspase-3. *Cell* **90**, 405–413.

13. Susin, S. A., Lorenzo, H. K., Zamzami, N., Marzo, I., Snow, B. E., Brothers, G. M., et al. (1999). Molecular characterization of mitochondrial apoptosis-inducing factor. *Nature* 397, 441–446.

14. Enari, M., Sakahira, H., Yokoyama, H., Okawa, K., Iwamatsu, A., and Nagata, S. (1998). A caspase-activated DNase that degrades DNA during apoptosis, and its inhibitor ICAD. *Nature* **391**, 43–50.

Part II

Detection of Non-Apoptotic Cell Death

Chapter 6

Methods to Analyze Cellular Necroptosis

Benchun Miao and Alexei Degterev

Summary

Necroptosis is a mechanism of necrotic cell death induced by external stimuli in the form of death domain receptor (DR) engagement by their respective ligands, TNF-alpha, Fas ligand (FasL) and TRAIL, under conditions when apoptotic cell death execution is prevented, e.g. by caspase inhibitors. Although it occurs under regulated conditions, necroptotic cell death is characterized by the same morphological features as unregulated necrotic death. RIP1 kinase activity is a key step in the necroptosis pathway. We have previously identified specific and potent small-molecule inhibitors of necroptosis, necrostatins, which efficiently prevent execution of this form of cell death. Herein, we describe the methods to analyze cellular necroptosis, and the methods to analyze the inhibitory effects of anti-necroptosis compounds (necrostatin-1).

Key words: Necroptosis, Death domain receptor, RIP1 kinase, TNF-alpha, Electron microscopy, Immunoprecipitation, Western blotting

1. Introduction

The mechanism of apoptosis has been extensively characterized over the past decade, but little is known about alternative forms of regulated cell death. Although stimulation of the Fas/TNF receptor family triggers a canonical "extrinsic" apoptosis pathway, multiple studies *(1–6)* have demonstrated that in the absence of intracellular apoptotic signaling DR engagement is capable of activating a common nonapoptotic death pathway, which we termed necroptosis *(2)*. We have previously identified small molecules, termed necrostatins, which potently and selectively inhibit this form of cell death *(2, 7–10)*.

Peter Erhardt and Ambrus Toth (eds.), *Apoptosis*, Methods in Molecular Biology, vol. 559
DOI 10.1007/978-1-60327-017-5_6
© Humana Press, a part of Springer Science + Business Media, LLC 2004, 2009

Necroptosis is triggered by the same stimuli that normally activate apoptosis, underscoring the notion that it is a regulated process of cell death. However, necroptosis is clearly distinct from apoptosis as it does not involve key apoptosis regulators, such as caspases, Bcl-2 family members or cytochrome *c* release from mitochondria. Furthermore, the cell morphology of necroptotic demise, including early loss of plasma membrane integrity, lack of nuclear fragmentation, mitochondrial dysfunction, and oxidative stress, is very similar to that of necrosis. Indeed, necrostatin-1 was demonstrated to be of protective effect in inhibiting necrotic injury during ischemia/reperfusion injury in the heart *(11)* and brain *(2)*.

The serine/threonine kinase activity of DR associated molecule RIP1 was identified as an upstream and key step in Fas ligand or TNF-induced necroptosis. Cells deficient in RIP kinase fail to undergo necroptosis, while restoring this protein, but not its kinase dead mutant, was shown to result in acquisition of the sensitivity to the necrotic cell death *(3)*. Here, we describe cell survival and RIP1 kinase assays to evaluate necroptosis activation and changes in RIP1 kinase activity.

2. Materials

2.1. Reagents

1. Human TNF-alpha (Cell Sciences).

2. Mouse monoclonal agonistic Fas antibody (clone EOS9.1) (Biolegend).

3. TRAIL (KillerTRAIL) and Pan-caspase inhibitor zVAD.fmk can be purchased from Axxora.

4. Necrostatin-1 (Calbiochem).

5. All chemicals are obtained from Sigma.

2.2. Cell Culture and Lysis

1. Dulbecco's Modified Eagle's Medium (DMEM) or RPMI 1640 are supplemented with 10% (v/v) heat-inactivated fetal bovine serum (FBS) and antimycotic–antibiotic mix (Invitrogen). All cell lines are available from ATCC.

2. Solution of trypsin (0.25%) and ethylenediamine tetraacetic acid (EDTA) (1 mM) are from Invitrogen.

3. 96-well plates (white, black, or clear plates) are from Corning Incorporated.

4. 1 × cell lysis buffer: 20 mM Tris-HCl, pH 7.5, 150 mM NaCl, 1 mM EDTA, 1 mM EGTA, 1% (v/v) Triton X-100, 2.5 mM sodium pyrophosphate, 1 mM b-glycerophosphate,

1 mM Na_3VO_4, 1 μg/mL leupeptin, 1 mM PMSF (*see* **Notes 1** and **2**).

5. Teflon cell scrapers (Fisher Scientific).

2.3. Immunoprecipitation and In Vitro Kinase Assay

1. Protein A agarose beads: add 5 mL of 1 × PBS to 1.5 g of protein A agarose beads (Pierce). Agitate for 2 h at 4°C; pellet by centrifugation at 14,000 × *g* for 1 min. Wash pellet twice with PBS. Resuspend beads in 1 volume of PBS (can be stored for 2 weeks at 4°C).

2. 1 × kinase lysis buffer (TL buffer): 20 mM HEPES, pH 7.3, 150 mM NaCl, 1% (v/v) Triton X-100, 5 mM EDTA, 5 mM NaF, 0.2 mM Na_3VO_4 (ortho) and complete protease inhibitor cocktail (Roche).

3. 1 × kinase reaction buffer: 20 mM HEPES, pH 7.3, 5 mM $MgCl_2$ and 5 mM $MnCl_2$.

2.4. SDS-Polyacrylamide Gel Electrophoresis (SDS-PAGE)

1. 1 × separating buffer: 1.5 M Tris-HCl, pH 8.8, 0.4% (w/v) SDS. Store at room temperature.

2. 1 × stacking buffer: 1 M Tris-HCl, pH 6.8, 0.4% (w/v) SDS. Store at room temperature.

3. Forty percent acrylamide/bis solution (29:1) (this is a neurotoxin when unpolymerized and so care should be taken not to receive exposure) and N, N, N, N'-tetramethyl-ethylenediamine (TEMED, Bio-Rad) (*see* **Notes 3** and **4**).

4. Ammonium persulfate: prepare 10% (w/v) solution in water and immediately freeze in single use (200 μL) aliquots at −20°C.

5. 5 × running buffer: 125 mM Tris (do not adjust pH), 960 mM glycine, 0.5% (w/v) SDS. Store at room temperature.

6. Prestained protein molecular weight markers (New England BioLabs).

2.5. Western Blotting

1. 1 × transfer buffer: 25 mM Tris (do not adjust pH), 190 mM glycine, 20% (v/v) methanol, 0.05% (w/v) SDS. Store at room temperature (with cooling during use).

2. Nitrocellulose membrane is from Millipore and 3 MM Chr chromatography paper from Whatman.

3. Tris-buffered saline with Tween (TBS-T): Prepare 10 × stock buffer with 250 mM Tris-HCl, pH 7.4, 1.37 M NaCl, 27 mM KCl, 1% (v/v) Tween-20. Dilute 100 mL with 900 mL water for use.

4. 1 × blocking buffer: 5% (w/v) fraction V bovine serum albumin (BSA) in TBS-T.

5. 1 × primary antibody dilution buffer: TBS-T supplemented with 2% (w/v) fraction V BSA.

6. Enhanced chemiluminescent (ECL) reagents are from Kirkegaard and Perry and Bio-Max ML film from Kodak (*see* **Note 5**).

3. Methods

3.1. Cell Survival Assays

3.1.1. Preparation of Samples

1. Seed cells (mouse fibrosarcoma L929, mouse fibroblast Bal-bc3T3, human Jurkat T cells, or FADD-deficient variant of Jurkat cells) in 96-well plates (white plates for luminescent assays; black plates for fluorescent assays; clear plates for MTT assay) at the density of $5–10 \times 10^3$ cells per well for adherent cells, or $2–5 \times 10^4$ cells per well for suspension cells in 100 µL of the appropriate media.

2. Add human TNF-alpha at final concentration of 10 ng/mL, agonistic Fas antibody (100 ng/mL) or TRAIL (5 ng/mL) and zVAD.fmk at 100 µM, Cycloheximide (CHX) at 1 µg/mL. RIP-deficient Jurkat cells, available from ATCC, can be used as a negative control as these cells are insensitive to necroptosis. Necroptosis in L929 and FADD-deficient Jurkat cells can be induced by addition of 10 ng/mL human TNF-alpha alone.

3. Add DMSO (control) or increasing concentrations of test compound (e.g., necrostatin-1 at 0.029, 0.058, 0.12, 0.23, 0.46, 0.93, 1.9, 3.7, 11.1, 33.3, and 100 µM) dissolved in DMSO (the final concentration of DMSO is 0.5%) to each well for EC_{50} determination. Three/four parallel samples at each concentration should be tested.

4. Incubate cells for 24 h at 37°C in a humidified incubator with 5% CO_2, followed by cell viability assessment using one or more of the methods described below.

3.1.2. ATP Assay

1. For the ATP assay, use a luminescence-based commercial kit (CellTiter-Glo, Promega). Add 30 µL of the cell lysis/ATP detection reagent to each well.

2. Incubate the plates on a rocking platform for 10 min at room temperature.

3. Measure the luminescence using a Wallac Victor 3 plate reader (Perkin-Elmer) or similar.

4. Cell viability is expressed as a ratio of the signal in the wells treated with both TNF-alpha and compound to the signal in the wells treated with compound alone. This is done to account for nonspecific toxicity, which in most cases is <10%.

3.1.3. Sytox Assay

1. For the Sytox assay, incubate cells with 1 μM Sytox Green reagent (Molecular Probes) for 30 min at 37°C.

2. Perform fluorescent measurement using a fluorescence plate reader (Victor 3 or similar) with excitation at 485 nm and emission at 538 nm.

3. Add 5 μL of 20% (v/v) Triton X-100 solution into each well to produce maximal lysis and incubate for 1 h at 37°C, and then perform the second reading.

4. Calculate the ratio of values (percentage of dead cells in each well) before and after Triton X-100 treatment and normalize it to the relevant controls not subjected to cytotoxic stimuli.

3.1.4. MTT Assay

1. For the methylthiazolyldiphenyl-tetrazolium bromide (MTT) assay, use the CellTiter 96 AQueous Non-Radioactive Cell Proliferation Assay Kit (Promega). Add 10 μL of MTT solution to each well.

2. Mix by tapping gently on the side of the tray or shake briefly on an orbital shaker, and incubate at 37°C until brown color develops (usually < 4 h).

3. Measure the absorbance on an ELISA plate reader at 490 nm.

3.1.5. Cellular EC$_{50}$ Determination

Calculate EC$_{50}$ values for the inhibitors using nonlinear regression analysis of sigmoid dose response (variable slope) curves from plots of log (inhibitor concentration) vs. viability values in GraphPad Prizm software package using values from replica samples.

3.2. Flow Cytometry Analysis

3.2.1. Preparation of Samples

1. Seed cells in a 6-well plate at the density of 3–5×10^5 cells/well (adherent cells) or 5–10×10^5 cells/well (suspension cells) in 4 mL of corresponding phenol red free media containing 10% FBS.

2. Induce necroptosis as described above (*see* **Subheading 3.1.1**).

3. Add appropriate concentrations of test compound (e.g., necrostatin-1) or equivalent volume of DMSO as a control (the final concentration of DMSO is 0.5%) to each well. Perform analysis in triplicate wells.

4. After an incubation for 24 h at 37°C in a humidified incubator with 5% CO_2, for suspension cells, wash cells once with PBS (pH 7.4), and resuspend cells in 500 μL of phenol free media supplemented with 10% FBS. For adherent cells, trypsinize cells, gently wash the cells with serum-containing media once and PBS for another time. Then resuspend cells in 500 μL of media.

3.2.2. Cell Death Measurement Using Propidium Iodide (PI)

1. Add PI at final concentration of 2 μg/mL (Roche) and incubate at room temperature for 30 min in the dark.

2. Analyze samples with FACS (FACSCalibur, BD Biosciences) using FL3 channel to determine percentage of PI-positive (dead) cells.

3.2.3. Annexin V/PI Staining Assay

1. Use an ApoAlert Annexin V Apoptosis Kit (Clontech) in this assay. Resuspend the cells in 200 μL of 1× binding buffer, and add 5 μL of 20 μg/mL Annexin V and 10 μL of 50 μg/mL PI (*see* **Note 6**).

2. Incubate at room temperature for 5–15 min in the dark.

3. Wash cells twice with PBS. Analyze cells by FACSCalibur using a single laser emitting excitation light at 488 nm, and analyze the percentages of cells with low PI and high annexin V, high PI and high annexin V, high PI and low annexin V using ModFit software (Verity Software House) (*see* **Note 7**). An example of the results produced is shown in **Fig. 1**.

3.2.4. Mitochondrial Membrane Potential ($\Delta\Psi_m$) Analysis

1. Incubate cells with 40 nM $DiOC_6$ for 30 min at 37°C in the dark.

2. Wash cells twice with PBS. Analyze cells by FACSCalibur using FL1 channel and calculate the percentages of cells with low $DiOC_6$ fluorescence (low mitochondrial membrane potential).

3.2.5. Reactive Oxygen Species Analysis

1. Incubate cells with 5 μM DCFDA (Molecular Probes) for 30 min at 37°C in the dark.

2. Wash cells twice with PBS. Analyze cells by FACSCalibur using FL1 channel.

In addition, samples prepared above (*see* **Subheading 3.2.1**) are also subjected to Western Blotting analysis (*see* **Subheading 3.5**) after cell lysates preparation (*see* **Subheading 3.4.3**) and SDS-PAGE separation (*see* **Subheading 3.4.6**) as described below to analyze changes of autophagy-related proteins in cell necroptosis.

3.3. Electron Microscopy Analysis (Cell Morphologic Analysis)

1. Prepare cell samples as mentioned above (*see* **Subheading 3.2.1**), using 12 or 24-well plates instead of 6-well plates. Seed adherent cells directly on Thermanox Coverslips (Nalge Nunc International, supplied through Fisher Scientific).

2. Induce necroptosis as described above (*see* **Subheading 3.1.1**).

3. Fix cells in 0.1 M phosphate buffer (pH 7.4) containing 2.5% glutaraldehyde and 2% paraformaldehyde for 1 h at room temperature.

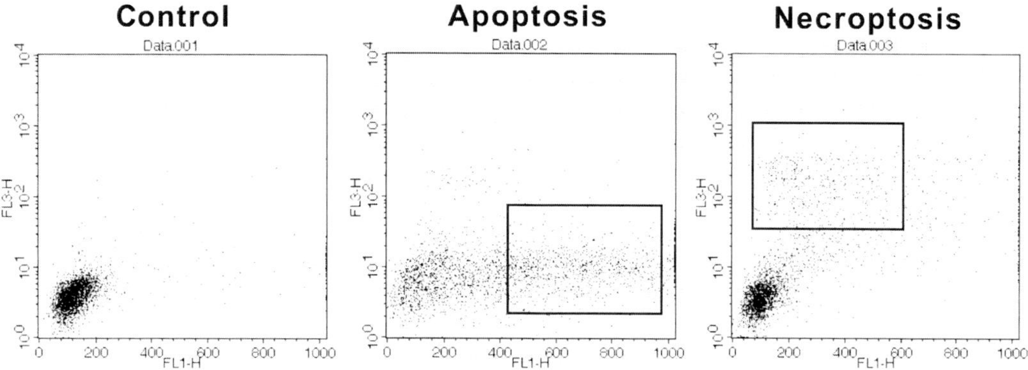

Fig. 1. Annexin V/PI staining of apoptotic and necroptotic cells. Wild type Jurkat cells were treated with anti-Fas antibody in combination with cycloheximide (to induce apoptosis) or cycloheximide/zVAD (to induce necroptosis) for 10 h. Cells were stained with Annexin V-EGFP and PI, and analyzed by FACS using FL1 (Annexin) and FL3 (PI) channels. Appearance of characteristic Annexin+/PI− apoptotic and Annexin-/PI+ necroptotic cells is highlighted with the *boxes*.

4. After fixation, discard fixation solution, rinse cells three times (each 5 min duration) with 0.1 M PBS to remove the fixation solution leaving on the surface of specimens.

5. After washing, post-fix cells with 1% osmium tetroxide in 0.1 M phosphate buffer (pH 7.4) for 1 h at room temperature in the dark.

6. Rinse cells three times (each 5 min duration) with 0.1 M PBS.

7. Dehydrate samples through a series of graded ethanol solutions from 70 to 100%. The schedule is as follows: 70% ethanol for 10 min, 90% ethanol for 10 min, and three changes of 100% ethanol for 5 min each. Make sure the coverslips are not allowed to dry out.

8. After dehydration, the infiltration process requires steps through an intermediate solvent. Place samples in 100% propylene oxide and incubate for 15 min, and change propylene oxide once and incubate for another 15 min.

9. Put samples to propylene oxide/epoxy resin mixture (1:1) and incubate for 1 h.

10. Place samples in the embedding resin (Embed 812, Electron Microscopy Sciences) for 12–18 h.

11. Transfer samples to a flat embedding mold with freshly prepared 100% resin and incubate for a minimum of 1 h.

12. Place mold containing samples in 60°C oven and polymerize for 48 h.

13. After polymerization, cut the resin blocks first into thick sections at 1–2 μm with glass knives using an Ultracut UCT

(Leica) or a MT 5000 (RMC) and stain with toluidine blue. These sections are used as a reference to trim blocks for thin sectioning.

14. Then cut the appropriate blocks containing cells into thin sections at 70–90 nm using a diamond knife (Diatome, Electron Microscopy Sciences) and place the sections on either copper or nickel mesh grids.

15. After drying on filter paper for a minimum of 1 h, stain the sections with uranyl acetate (saturated aqueous solution) and lead citrate for contrast.

16. Acquire images of the cells using an electron microscope (Philips Tecnai 12 electron microscope). An example result is shown in **Fig. 2**.

3.4. In Vitro Kinase Analysis

3.4.1. Transient Electroporation of Jurkat Cells (see Notes 8)

This method is useful for overexpressing genes of interest to rapidly screen for modulators of necroptosis, including RIP1 kinase mutants.

1. Resuspend 5×10^6 FADD-deficient Jurkat cells in 100 µL of reagent V (Amaxa).

2. Mix cells with 2.7 µg of the expression vector and 0.3 µg of pEGFP-N1 vector (Clontech) and transfer the mixture to cuvette. Perform electroporation using Nucleofector II (program G-10, Amaxa).

3. Following electroporation, immediately add 500 µL of 37°C full media to cuvette and pipette the cells into 24-well plate.

4. Twenty-four hours after electroporation, dilute the cells to 5 mL with full media and incubate in 6-well plate for additional 24 h.

5. Forty-eight hours after electroporation, separate live cells by centrifugation ($350 \times g$, 40 min) over a layer of Ficoll-PAQUE (Amersham), and wash cells twice with full media following separation.

6. Resuspend cells in media containing agents to induce necroptosis, e.g., Fas antibody/cycloheximide/zVAD.fmk, and analyze cell death 24 h after treatment.

7. Add 2 µg/mL of PI to each sample and determine percentages of GFP positive cells (FL1 channel) and PI negative cells amongst the GFP-positive cells (FL3 channel, cells gated in FL1 channel) using FACSCalibur. Cell death results in decrease in the percentage of GFP-positive cells and increase in PI-positive cells in the GFP-positive cell population.

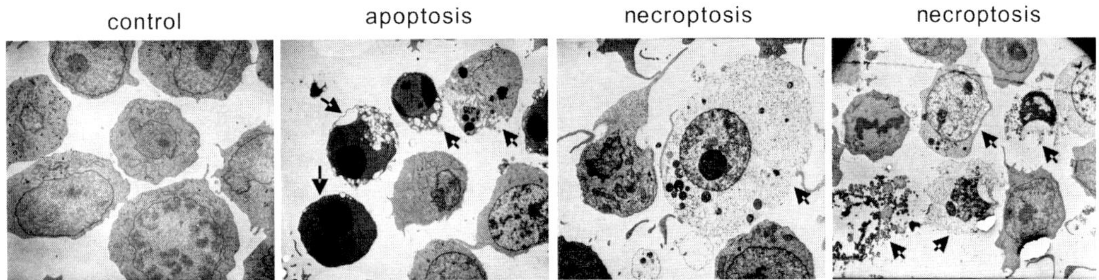

Fig. 2. Electron micrographs of the apoptotic and necroptotic Jurkat cells. For the induction of apoptosis, wild type Jurkat cells were treated with FasL and cycloheximide for 16 h. For the induction of necroptosis, FADD-deficient Jurkat cells were treated with TNF-alpha for 16 h. *Arrows* indicate dying cells.

3.4.2. Transient Transfection of Balbc 3T3 or 293T Cells (35 mm Wells)

1. Approximately 24 h prior to transfection, plate Balbc 3T3 or 293T cells at a cell density of $1–3 \times 10^5$ cells in 2 mL of complete growth medium per 35 mm well to obtain 50–70% confluence the following day.

2. Culture the cells overnight.

3. In a sterile plastic tube, add the TransIT-LT1 Transfection Reagent (6 µL per 2 µg DNA, MirrusBio), drop-wise into serum-free medium (100 µL per 2 µg DNA). Mix by gentle pipetting.

4. Incubate at room temperature for 5 min.

5. Add diluted TransIT-LT1 Transfection Reagent to plasmid DNA and mix by gentle pipetting.

6. Incubate at room temperature for 15 min.

7. Add the TransIT-LT1 Transfection Reagent/DNA complex mixture prepared above, dropwise to the cells. Gently rock the dish back and forth and from side to side to distribute the complexes evenly (*see* **Note 9**).

8. Incubate cells for 48 h, followed by downstream analysis. Transfection of 293T cells with pcDNA-FLAG-RIP1 vector provides a convenient source of RIP1 kinase for kinase assay (see below).

3.4.3. Preparation of Cell Lysates

1. For suspension cells (Jurkat cells), remove the medium and rinse the cells twice with ice-cold PBS by centrifugation at 240 $\times g$ for 5 min each time.

2. Remove PBS and add 0.5 mL of 1 $\times$ ice-cold TL buffer to each tube and keep on ice, and go to **step 6** below.

3. For adherent cells (Balbc 3T3 or 293T cells), remove the medium and rinse the cells twice with ice-cold PBS.

4. Remove PBS and add 0.5 mL of 1 × ice-cold TL buffer to each well and incubate the plate on ice for 5 min.

5. Scrape cells off the plate and transfer to microcentrifuge tubes. Keep on ice for 20 min.

6. Microcentrifuge for 10 min at 4°C at 14,000 × g and transfer the supernatant to a new tube. The supernatant is the cell lysate. If necessary, lysate can be stored at –80°C.

3.4.4. RIP1
Immunoprecipitation

1. For FLAG-RIP1 expressed in 293T cells: add 2–10 µL of anti-Flag M2 agarose primary antibody beads (Sigma) to 500 µL cell lysate, and incubate with gentle rocking overnight at 4°C.

2. For endogenous RIP1 immunoprecipitation from 5 × 10⁶ Jurkat cells: add 5 µL (1.25 µg) of monoclonal RIP1 antibody (clone G322–2, BD Biosciences) to lysate and incubate overnight at 4°C.

3. In case of endogenous RIP1 immunoprecipitation, add protein A agarose beads (20 µL of 50% bead slurry, Pierce) and incubate with gentle rocking for 2 h at 4°C.

4. Microcentrifuge for 30 s at 4°C at 3,000 × g. Wash pellet three times with 500 µL of 1 × TL buffer and twice with 500 µL of 1 × 20 mM HEPES, pH 7.3. Keep on ice during washes.

3.4.5. In Vitro RIP1 Auto-
phopshorylation Kinase
Assay

1. Suspend bead pellet in 15 µL of the kinase reaction buffer (20 mM HEPES, pH 7.3, 5 mM $MgCl_2$, 5 mM $MnCl_2$).

2. Initiate kinase reaction by addition of 10 µM cold ATP and 1 µCi of γ-³²P-ATP, and carry out the reaction for 30 min at 30°C.

3. Terminate reaction with the same volume of 2 × SDS sample buffer. Vortex, then microcentrifuge for 30 s.

4. Heat the samples to 95–100°C for 2–5 min; cool on ice.

5. Microcentrifuge for 5 min.

6. Load the sample (15–20 µL) on 8% SDS-PAGE gel (see SDS-PAGE method below).

3.4.6. SDS-PAGE

1. Prepare a 1.5-mm thick, 8% gel by mixing 2.5 mL of 1.5 M Tris-HCl buffer (pH 8.8) with 2 mL of 40% acrylamide/bis solution, 5.3 mL water, 50 µL of 20% SDS, 100 µL of 10% ammonium persulfate solution and 10 µL TEMED (*see* **Note 4**). Swirl gently to mix. Pour the gel immediately, leaving space for a stacking gel, and gently overlay with water. The gel should polymerize in about 30 min (*see* **Note 3**).

2. Pour off water and rinse once with water.

3. Prepare the stacking gel by mixing 500 μL of 1.0 M Tris-HCl buffer (pH 6.8) with 500 μL acrylamide/bis solution, 2.9 mL water, 20 μL of 20% SDS, 40 μL of 10% ammonium persulfate solution, and 4 μL TEMED. Swirl gently to mix. Pour stacking solution on top of separating gel, and insert comb into stacking gel (taking care to avoid forming bubbles on the end of the teeth). Allow gel to polymerize 30 min (*see* **Note 10**).

4. Once the stacking gel has set, carefully remove the comb (pulling straight out to leave precise wells) and use a syringe to wash the wells with running buffer.

5. Add running buffer to the upper and lower chambers of the gel unit and load 20 μL of each sample in each well. Include one well for prestained protein molecular weight marker.

6. Complete the assembly of the gel unit and connect to a power supply. The gel is run at 20 mA through the stacking gel and 30 mA through the separating gel until dye front reaches the bottom of the gel. Watch the dye bands so that the samples do not run off the bottom of the gel. The dye front (blue and pink) can be run off the gel if desired.

3.4.7. Autoradiography

1. When samples mentioned above are done running, remove gel from the apparatus, and soak in fixation buffer containing 30% methanol and 10% acetic acid for 15 min. Pat dry.

2. Cut a piece of Whatman 3 MM filter paper to the size of gel, and place the gel on the filter paper. The thickness of paper is important to avoid cracking of the gel.

3. Cover gel/paper with plastic wrap and place in the protein gel dryer.

4. Open the right c.w. spigot and press the rubber mat (atop the gel) at the corners to aid in the creation of a vacuum.

5. Set timer of dryer for two hours, and the dryer will turn off automatically.

6. After the dryer turns off, break seal on dryer cover by pulling rubber hose off gel dryer and open up before turning off the water generating the vacuum.

7. Approximately 75 kDa RIP1 band is visualized by analysis in Storm 8200 or similar Phosphorimager (GE Healthcare) or by exposing the dried gel to X-ray film overnight at –80°C. An example result is shown in **Fig. 3**.

3.5. Western Blotting

Separate a portion of the beads prior to radioactive kinase assay to confirm equal RIP1 amounts in the samples and analyze by Western blotting.

Fig. 3. RIP kinase assay. For the kinase assay, 293T cells were transfected with the wild type or kinase dead (K45R) mutant of RIP kinase in pcDNA-FLAG-RIP vector (generous gift of Dr. Jurg Tschopp, University of Lausanne), followed by immunoprecipitation with anti-FLAG beads. Alternatively, endogenous RIP immunoprecipitated from Jurkat cells using anti-RIP antibody was used.

1. Transfer the samples separated by SDS-PAGE (*see* **Subheading 3.4.6**) to nitrocellulose membranes electrophoretically. Soak four sheets of 3 MM filter paper in 1× transfer buffer. Wet a sheet of nitrocellulose membrane just larger than the size of the separating gel with methanol, rinse with water, and then soak in transfer buffer.

2. Place two sheets of 3 MM filter paper onto semi-dry transfer apparatus electrode plate (Bio-Rad). Submerge the nitrocellulose membrane in the transfer buffer on top of the 3 MM paper.

3. Discard the stacking gel and cut one corner from the separating gel to allow its orientation to be tracked. Soak the separating gel in transfer buffer, and then carefully lay on top of the nitrocellulose membrane.

4. Carefully lay another two sheets of 3 MM paper on top of the gel, ensuring that no bubbles are trapped in the resulting sandwich. Then close the lid of the transfer apparatus (*see* **Note 11**).

5. Thus the nitrocellulose membrane is between the gel and the anode. It is vitally important to ensure this orientation or proteins will be lost from the gel to the filter paper rather than transferred to the nitrocellulose.

6. Carry out the transfer for **90 min** at room temperature at 200–250 mA (BioRad) with prechilled buffer. Larger proteins take longer to transfer.

7. After transfer, remove membrane from transfer apparatus. Wash nitrocellulose membrane with TBS-T for 5 min at room temperature with gentle rocking.

8. Block membrane in TBS-T containing 5% BSA (blocking buffer) for 1 h at room temperature on a rocking platform with gentle rocking.

9. Wash membrane three times for 5 min with TBS-T.

10. Incubate membrane with primary antibody (e.g. RIP1 antibody, clone G322–2, BD Biosciences, 1:1,000 dilution) overnight at 4°C with gentle rocking (*see* **Note 12**). Remove the primary antibody, and wash the membrane three times for 5 min with TBS-T.

11. Incubate membrane with HRP-conjugated secondary antibody (SouthernBiotech, 1:2,000 dilution) in blocking buffer with gentle agitation for 1 h at room temperature.

12. Remove the secondary antibody and wash the membrane three times for 5 min with TBS-T.

13. During the final wash, warm 2 mL aliquots of each of the two components of the ECL reagent (ECLGold, Signagen Labs) separately to room temperature, and do the remaining steps in a dark room under safe light conditions. Remove the final wash from the blot, mix the ECL reagents together (1:1) and immediately add to the blot (2 mL of ECL detection mix per a membrane), which is then rotated by hand for 1 min to ensure even coverage.

14. Remove the membrane from the ECL reagents, and wrap it in plastic wrap that has been cut to the size of an X-ray film cassette.

15. Place the membrane in an X-ray film cassette with a piece of X-ray film for a suitable exposure time, typically a few minutes (*see* **Note 13**).

16. Quantify results of the Western blotting using Scion Image software (Scion Corporation).

4. Notes

1. Unless stated otherwise, all solutions should be prepared in Milli-Q or equivalently purified water that has a resistivity of 18.2 MΩ-cm and total organic content of less than five parts per billion. This standard is referred to as "water" in this text.

2. Add 1 mM PMSF immediately prior to use.

3. Unpolymerized acrylamide is a neurotoxin. Wear gloves and take care when handling, once polymerized, gels can be disposed off in the regular trash, keep excess solution to monitor polymerization process of gel.

4. TEMED is best stored at room temperature in a desiccator. Buy small bottles as it may decline in quality (gels will take longer to polymerize) after opening.

5. Quantification of data may be desired and this can be done by scanning densitometry of the films, providing that care is taken to ensure that the signal has not saturated. Alternatively, the chemiluminescent signal can be captured digitally with an instrument such as a FujiFilm LAS-1000 plus.

6. If you plan to fix your cells, incubate them with annexin V before fixation, because cell membrane disruption can allow annexin V to bind to PS on the inner surface of the cell membrane. Rinse unbound annexin V with binding buffer before fixation. Do not expose reagent to strong light during storage and incubation. PI is toxic. Handle with extreme caution.

7. The signal generated by Annexin V can be detected in the FITC signal detector, and the signal generated by PI can be monitored by the detector reserved for phycoerythrin emission. In particular, in annexin V/PI staining assay, necroptotic cells appear PI positive/Annexin negative, which is totally different from apoptosis (PI negative/ Annexin positive).

8. Transient electroporation can also be performed using Gene Pulser II electroporator (Bio-Rad) and hypoosmolar lectroporation buffer (Eppendorf), although this method will typically result in lower electroporation efficiency and higher nonspecific cell death.

9. To increase protein expression level in Balbc 3T3 cells, 4–5 h after transfection add equal volume of the media with 2 × BoosterExpress 2 or 3 reagent (Genlantis).

10. Glycerol in equilibration buffer helps prevent cracking during drying of high percentage gels. Insert comb into stacking gel at an angle to help avoid trapping air bubbles at the end of comb. "Smiling" and "frowning" of gels largely due to unequal heat distribution/salt concentration across gel. Fill empty wells with extract buffer.

11. All steps including placing nitrocellulose membrane, separating gel, and filter paper should be done gently from one side to another side with the help of buffer to make sure that no bubbles appear. A 50 or 15-mL tube can be used to expel possible bubbles from one side to another.

12. The primary antibodies can be stored at 4°C by addition of 0.02% final concentration sodium azide (exercise caution since azide is highly toxic), and they can be used for up to 20 blots over several months, with the only adjustment required to increase length of exposure time to film at the ECL step.

13. ECL reagent can be further diluted if signal response is too fast. Drain membrane of excess developing solution (do

not let dry), wrap in plastic wrap, and expose to X-ray film. An initial 10-s exposure should indicate the proper exposure time. Because of the kinetics of the detection reaction, signal is most intense immediately following incubation and declines over the following 2 h.

References

1. Chan, F. K., Shisler, J., Bixby, J. G., Felices, M., Zheng, L., Appel, M., Orenstein, J., Moss, B. and Lenardo, M. J. (2003). A role for tumor necrosis factor receptor-2 and receptor-interacting protein in programmed necrosis and antiviral responses. *J. Biol. Chem.* **278**, 51613–51621.

2. Degterev, A., Huang, Z., Boyce, M., Li, Y., Jagtap, P., Mizushima, N., Cuny, G. D., Mitchison, T. J., Moskowitz, M. A. and Yuan, J. (2005). Chemical inhibitor of nonapoptotic cell death with therapeutic potential for ischemic brain injury. *Nat. Chem. Biol.* **1**, 112–119.

3. Holler, N., Zaru, R., Micheau, O., Thome, M., Attinger, A., Valitutti, S., Bodmer, J. L., Schneider, P., Seed, B. and Tschopp, J. (2000). Fas triggers an alternative, caspase-8-independent cell death pathway using the kinase RIP as effector molecule. *Nat. Immunol.* **1**, 489–495.

4. Khwaja, A. and Tatton, L. (1999). Resistance to the cytotoxic effects of tumor necrosis factor alpha can be overcome by inhibition of a FADD/caspase-dependent signaling pathway. *J. Biol. Chem.* **274**, 36817–36823.

5. Matsumura, H., Shimizu, Y., Ohsawa, Y., Kawahara, A., Uchiyama, Y. and Nagata, S. (2000). Necrotic death pathway in Fas receptor signaling. *J. Cell Biol.* **151**, 1247–1256.

6. Vercammen, D., Brouckaert, G., Denecker, G., Van de Craen, M., Declercq, W., Fiers, W. and Vandenabeele, P. (1998). Dual signaling of the Fas receptor: initiation of both apoptotic and necrotic cell death pathways. *J. Exp. Med.* **188**, 919–930.

7. Jagtap, P. G., Degterev, A., Choi, S., Keys, H., Yuan, J. and Cuny, G. D. (2007). Structure-activity relationship study of tricyclic necroptosis inhibitors. *J. Med. Chem.* **50**, 1886–1895.

8. Teng, X., Degterev, A., Jagtap, P., Xing, X., Choi, S., Denu, R., Yuan, J. and Cuny, G. D. (2005). Structure-activity relationship study of novel necroptosis inhibitors. *Bioorg. Med. Chem. Lett.* **15**, 5039–5044.

9. Teng, X., Keys, H., Jeevanandam, A., Porco, J. A., Jr., Degterev, A., Yuan, J. and Cuny, G. D. (2007). Structure-activity relationship study of [1,2,3]thiadiazole necroptosis inhibitors. *Bioorg. Med. Chem. Lett.* **17**, 6836–6840.

10. Wang, K., Li, J., Degterev, A., Hsu, E., Yuan, J. and Yuan, C. (2007). Structure-activity relationship analysis of a novel necroptosis inhibitor, Necrostatin-5. *Bioorg. Med. Chem. Lett.* **17**, 1455–1465.

11. Smith, C. C., Davidson, S. M., Lim, S. Y., Simpkin, J. C., Hothersall, J. S. and Yellon, D. M. (2007). Necrostatin: a potentially novel cardioprotective agent? *Cardiovasc. Drugs Ther.* **21**, 227–233.

Chapter 7

Detection of Cell Death by Autophagy

Narasimman Gurusamy and Dipak K. Das

Summary

Autophagy (Greek: Self digestion) is an intracellular process involved in removal of damaged or mis-folded proteins or organelles. Damaged or misfolded proteins or organelles are first engulfed in a membraneous structure called autophagosome, and then the autophagosome fuse with lysosome to form autophagolysosome, where the contents are digested. Autophagy is a catabolic process induced during nutritional depletion via phosphatidylinositol 3 kinase pathway. Autophagy is induced in several diseases such as various cancers, heart failure, etc. When autophagy is induced, several autophagic genes are upregulated that help the formation of autophagosome. Several autophagosome specific marker proteins have been identified, among them MAP1LC3-II protein, which is cleaved from MAP1LC3-I, is specifically incorporated into the autophagosomal membrane. The formation of MAP1LC3-II can be analyzed by Western immunoblotting or immunofluorescence. Detailed methods of detection of MAP1LC3-II by Western immunoblotting and immunofluorescence are described.

Key words: Autophagy, LC3, Autophagosome

1. Introduction

Under normal conditions, autophagy occurs for the recycling of long lived macromolecules and organelles to maintain the homeostasis, differentiation, and tissue remodeling *(1, 2)*. Autophagy has been implicated in several diseases like cancers, neurodegenerative diseases, and in cardiac injury *(3, 4)*. However, under pathological conditions, autophagy may contribute for either cell survival or cell death *(3, 5)*. There are three different forms of autophagy, namely, micro-autophagy, macroautophagy, and chaperone-mediated autophagy *(1)*. Microautophagy refers to proteins that are locally taken up by the lysosomes. During macroautophagy, first an isolation membrane is formed, which then surrounds the target protein or organelle *(6, 7)*.

Peter Erhardt and Ambrus Toth (eds.), *Apoptosis*, Methods in Molecular Biology, vol. 559
DOI 10.1007/978-1-60327-017-5_7
© Humana Press, a part of Springer Science + Business Media, LLC 2004, 2009

Thus formed structure known as autophagosome then fuses with lysosome (autophagolysosome), where the engulfed proteins or organelles are degraded *(6, 7)*. In chaperone-mediated autophagy, heat shock protein 73 mediates the delivery of proteins to lysosome *(8)*. Autophagy is shown to be regulated via phosphoinositol-3 kinase pathways *(9)*.

Macroautophagy is the most active form of autophagy. Several research findings show that during the induction of autophagy several autophagic genes are regulated *(1, 10)*. Autophagic genes are denoted as *Atg*, and further they are numbered. During the initiation of autophagy, binding between Atg12 and Atg5 is initiated *(10)*. Dimers of Atg16 bind with Atg12-Atg5 *(11)*. This complex then associates with curve-shaped phagophore. Development of phagophore into autophagosome requires the involvement of microtubule associated protein light chain 3 (LC3) *(12)*. LC3, the mammalian homologue of Atg8, is synthesized as a precursor protein and present in the cytosol. Atg4 cleaves the c-terminal amino acids from LC3 to produce cytosolic LC3-I *(13)*. LC3-I is activated by Atg7 and Atg3, and finally conjugated to phosphatidylethanolamine (LC3-II) *(11)*. LC3-II is a tightly membrane bound protein specifically associated with autophagosome *(12, 14)* (**Fig. 1**).

Formation of double layered autophagosomes can be monitored by electron microscopy *(15)*, but lack of morphological expertise can lead to misidentify other cellular structures such as

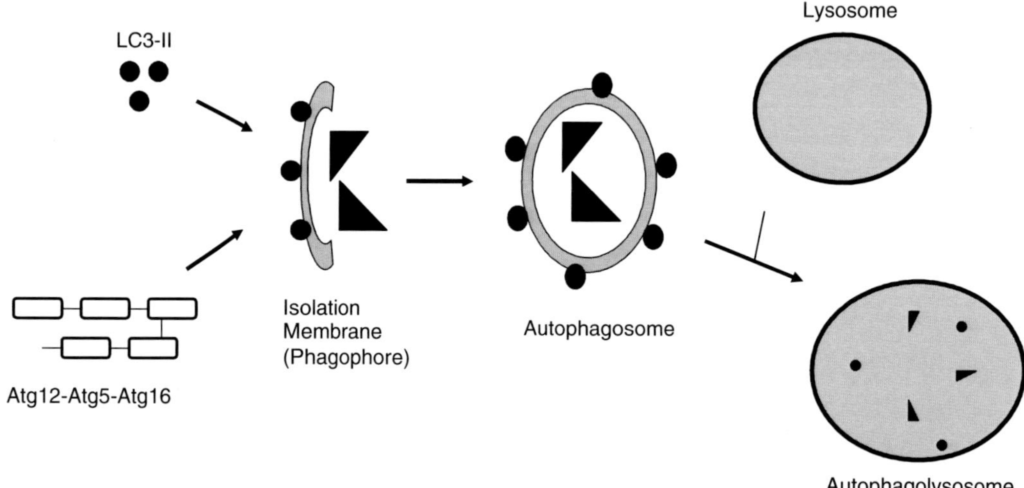

Fig. 1. Autophagosome formation. When autophagy is initiated, autophagic genes (*Atg*) are upregulated. Complex formation between Atg12 and Atg5 is initiated, followed by the binding of dimer of Atg16. In another pathway, precursor LC3 is cleaved to form LC3-I, which is further conjugated with phosphatidylethanolamine to form LC3-II. Binding between the complex Atg12-Atg5-Atg16 and LC3-II initiate phagophore formation, which surrounds and engulfs the target protein or organelle in the cell. Thus formed vesicle (autophagosome) finally fuses with lysosome and forms autophagolysosome, where the target protein or organelle is digested.

swollen endoplasmic reticulum (in dying cells) as autophagosomes. In some cases, monodansylcadaverine has been used as a fluorescent marker for autophagic vacuoles *(16)*; however, this method is limited to in vivo conditions, and not specific for autophagosome *(2)*. Lysosome associated membrane protein-2A (LAMP-2A) has been used as a marker to study chaperone-mediated autophagy *(17)*. However, at present, only LC3 is the credible marker to study autophagosome formation. LC3 is found to be present in both isolation membrane and autophagosome *(14)* and can be studied by immunofluorescence and by Western blotting, in which the conversion of LC3 to LC3-I and LC3-II can be clearly seen.

2. Materials

2.1. Detection of MAP1-LC3 by Western Immunoblotting

Formation of LC3-I and LC3-II can be well studied by Western immunoblotting by allowing proteins to separate in a 15% gel.

1. Tissue homogenization buffer: 25 mM Tris-HCl, pH 8.0, 25 mM NaCl, 1 mM sodium orthovanadate, 10 mM sodium fluoride, 10 mM sodium pyrophosphate, 10 nM okadaic acid, 0.5 mM EDTA, 1 mM PMSF, and protease inhibitors (mixture of protease inhibitor cocktail is available from Sigma).
2. RIPA buffer (Cell lysis buffer): 50 mM Tris-HCl, pH 7.4, 150 mM NaCl, 1% NP-40, 0.5% sodium deoxycholate, 0.1% SDS. Addition of protease inhibitors to RIPA buffer is recommended.
3. Protein estimation reagents (BCA, Pierce).
4. Standard 15% polyacrylamide gel for SDS-PAGE.
5. SDS-PAGE running buffer: 0.025 M Tris base, 0.192 M glycine, 0.1% SDS.
6. Transfer buffer: 0.025 M Tris-HCl, 0.192 M glycine, pH 8.5, 20% methanol.
7. Tris buffered saline/Tween 20 (TBS/T): 25 mM Tris-HCl, 2.7 mM KCl, 0.137 M NaCl, pH 7.4, 0.1% Tween-20.
8. Blocking buffer: 5% non fat dry milk in TBS/T.
9. MAP1LC3 antibody (Santa Cruz).
10. Horse radish peroxidase or alkaline phosphatase conjugated secondary antibody.
11. Luminol detection reagent (from Amersham or Santa Cruz).
12. Autoradiography film.

2.2. Detection of MAP1LC3 by Immunofluorescence

Detection of cytoplasmic staining of LC3 can be combined with the results of Western immunoblotting to study the autophagic activity in a cell or tissue.

1. Tissue slides.

2. 4% Paraformaldehyde.

Preparation of 4% paraformaldehyde fixative from powder: Mix 4.0 g of paraformaldehyde powder with 100 mL of PBS. Heat the solution at 60°C with stirring in a hood (*see* **Note 1**). Add 1 M NaOH drop by drop while stirring to depolymerize the paraformaldehyde (or clear the solution). Aliquot and store at –20°C. Discard the solution after thawing.

3. 10× Phosphate buffered saline (PBS): Dissolve 80 g NaCl, 2 g KCl, 14.4 g Na_2HPO_4, and 2.4 g KH_2PO_4 in 800 mL distilled water, and make up the volume up to 1 L. When diluted to 1× PBS with distilled water, the pH should be 7.4.

4. Staining dishes with lids.

5. PAP pen.

6. 0.1% Triton X-100 in PBS.

7. MAP1LC3 antibody (Santa Cruz).

8. Fluorescent conjugated secondary antibody like Alexa Fluor® 488 (Green emission) or Alexa Fluor® 594 (Red emission) (Molecular Probes).

9. 5% bovine serum albumin (BSA) in PBS for blocking.

10. 50% glycerol in PBS as mounting medium.

11. Cover slips.

3. Methods

3.1. Detection of MAP1-LC3 by Western Immunoblotting

1. For tissue: About 100 mg of tissue is cut into small pieces with a scissors, and transferred to 1 mL of homogenization buffer taken in a polystyrene tube. Homogenize the tissue in a Polytron homogenizer at a high speed for 10–15 s (*see* **Note 2**). Centrifuge the tubes at $1,100 \times g$ for 10 min; the resultant pellet can be separately treated as nuclear pellet and the supernatant from the above centrifugation should be further centrifuged at $28,000 \times g$ for 20 min. The resultant supernatant is known as cytosolic extract, where the autophagosomal membrane proteins should be present.

2. For cells: Wash cultured cells twice with 8–10 mL of PBS (for 100 mm dish) and add 500 µL (for 100 mm dish) of RIPA buffer containing protease inhibitors. Volume of RIPA buffer should be adjusted according to the size of cell culture plate. For six, 12, or 96-well plates, add 200, 100, or 50 µL volumes of RIPA buffer. Collect the cells using a cell scraper and transfer the content to an eppendorff tube.

3. Total proteins in tissue homogenate or cell lysate can be estimated by using BCA protein assay kit (Pierce).

4. Preparation of 15% separating gel (*see* **Note 3**): Mix the following reagents in the order mentioned: 4.7 mL distilled water, 5.0 mL of 1.5 M Tris-HCl, pH 8.8, 10.0 mL of 30% acrylamide:bisacrylamide (19:1 ratio), 200 µL 10% SDS, 100 µL of 10% ammonium per sulfate, 12 µL tetramethyl-ethylenediamine. The above volumes are enough for casting two gels of 1.5 mm thickness. Mix gently, and pour into the gel casting assembly (Biorad) slowly to avoid formation of any air bubbles. Overlay with water to make the surface uniform. Allow it to polymerize for about 45–60 min till two separate layers are visible. Discard the overlaid water, and rinse the layer with distilled water. Remove adhering water using Kim-Wipes®.

5. Preparation of stacking gel: 6.1 mL distilled water, 2.5 mL of 0.5 M Tris-HCl, pH 6.8, 1.33 mL of 30% acrylamide/bisacrylamide (19:1 ratio), 100 µL of 10% SDS, 50 µL of 10% ammonium per sulfate, 10 µL tetramethyl-ethylenediamine. Add the above reagents in the order mentioned, mix gently, and pour on top of the separating gel. Place a comb slowly to avoid trapping air bubbles, and allow it to polymerize for about 20–30 min. After polymerization is complete, remove the comb, and rinse the wells thoroughly but gently with distilled water.

6. Mix protein samples with 1/3 volume of 4× protein loading buffer (250 mM Tris-HCl pH 6.8, 8% SDS, 40% glycerol, 8% β-mercaptoethanol, 0.02% bromophenol blue), and heat at 95°C for 5 min. Cool down, spin, and mix before loading into gel (*see* **Note 4**).

7. Gel electrophoresis: Assemble gels in to the electrophoresis apparatus (Biorad), and fill with SDS-PAGE running buffer (about 400 mL for Biorad mini apparatus). Load equal amount of total protein (for example, 50 µg) from each sample into wells. Load protein marker in one of the lanes. Run the electrophoresis at a constant volt of 200 till the dye front reaches to the end (*see* **Note 5**).

8. Transfer of protein: All materials such as filter pad, 3MM Whatman filter paper, and nitrocellulose membrane (0.2 µm pore size) should be soaked in transfer buffer. Filter papers and the membrane should be cut to the size of the gel. Assemble in the following order specified into the gel holding cassette from cathode to anode (black side to white in the Biorad apparatus): fiber pad, three Whatman filter papers, gel, nitrocellulose membrane, three Whatman filter papers, and fiber pad. Roll over a glass rod between each layer to remove any air bubbles formed. Place the cassette in the transfer unit, fill it with transfer buffer, and set the transfer at 100 V for 60 min (*see* **Note 6**).

9. Once transfer is finished, mark the membrane and rinse with TBS/T for 5 min with gentle shaking.

10. Block the membrane with blocking buffer for 60 min at room temperature.

11. Rinse the membrane with TBS/T, and incubate with MAP1LC3–1 antibody (1:500 dilution in TBS/T) at 4°C for overnight.

12. Wash the membrane three times for 5 min each in TBS/T with gentle agitation.

13. Incubate the membrane in horse radish peroxidase or alkaline phosphatase conjugated secondary antibody (1:2,000 or appropriate dilution in blocking buffer) for 60 min at room temperature. Selection of secondary antibody should be based on the host source of primary antibody. For example, if the primary antibody source is rabbit, the secondary antibody should be anti-rabbit.

14. Wash the membrane three times for 10 min each in TBS/T with gentle agitation (*see* **Note** 7).

15. Prepare ECL solution according to the manufacture's instruction and add the solution to cover the membrane and incubate at room temperature for 3–5 min.

16. Drain the ECL solution; wrap the membrane in a plastic wrap (Saran Wrap). Expose the membrane with an autoradiography film for 10 s–5 min.

17. Develop the film in an X-ray processor. MAP1LC3 bands: In most of the cases, two bands such as LC3-I (16 KDa) and LC3-II (14 kDa) will appear as shown in **Fig. 2**.

3.2. Detection of MAP1LC3 by Immunofluorescence

1. For tissue: Tissue section frozen with OCT compound works better than paraffin fixed samples. Cut 4 µm thick sections and place on the treated surface of glass slide.

For cells cultured in chamber-glass slide: Aspirate the medium, wash cells twice with PBS to remove the adhering medium

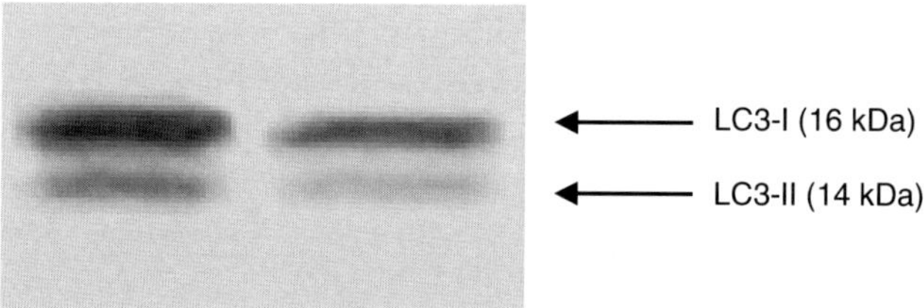

Fig. 2. Detection of MAP1LC3-I and MAP1LC3-II by Western immunoblotting. Cytosolic extract from cardiac tissue homogenates were separated by 15% SDS-PAGE and transferred to nitrocellulose membrane. LC3-I and LC3-II bands were detected by specific antibody obtained from Santa Cruz.

(Caution: Be careful while aspiration so that the adherent cells are not disturbed).

2. Place the slides containing tissue section or adherent cells into a jar containing 4% paraformaldehyde in PBS at room temperature for 15 min (*see* **Note 8**).

3. Wash the slides with three changes of PBS for 5 min each.

4. Dry the area around tissue sections with a clean Kim-Wipes® and draw a circle around those with a PAP pen to prevent the loss of solution by spreading.

5. Permeabilize the cells by treating them with 0.1% Triton X-100 in PBS for 1 min.

6. Wash the slides with two changes of PBS for 5 min each.

7. Block the samples with 5% BSA in PBS to avoid nonspecific binding of antibody, and incubate the slides for 2 h (for tissue) or 1 h (for cells) at room temperature.

8. Wash the slides with three changes of PBS for 5 min each.

9. Dilute the primary antibody to the recommended concentration (1:25 in PBS with 1% BSA). Incubate with primary antibody solution for 1 h at room temperature.

10. Wash the slides with two changes of PBS for 5 min each.

11. Dilute the fluorescent conjugated secondary antibody to 1:400 in PBS containing 1% BSA. Incubate the slides with secondary antibody solution for 1 h at room temperature in a dark place.

12. Wash the slides with three changes of PBS for 5 min each.

13. Counter staining of nucleus can be done with Hoechst 33342 (Molecular Probes) at the concentration of 2 μg/mL for 10 min.

14. Wash the slides with three changes of PBS for 10 min each.

15. Apply 50% glycerol in PBS or any mounting medium to the middle of the section. Mount cover slip slowly to avoid entrapping air. Seal the edges of cover slip with nail polish.

16. Examine the slides using either a fluorescence microscope with appropriate filters or a confocal microscope with appropriate excitation wavelength (**Fig. 3**).

17. Store slides in a dark location at 4°C.

4. Notes

1. Paraformaldehyde fumes are toxic. To avoid breathing it, stirring should be done in a hood.

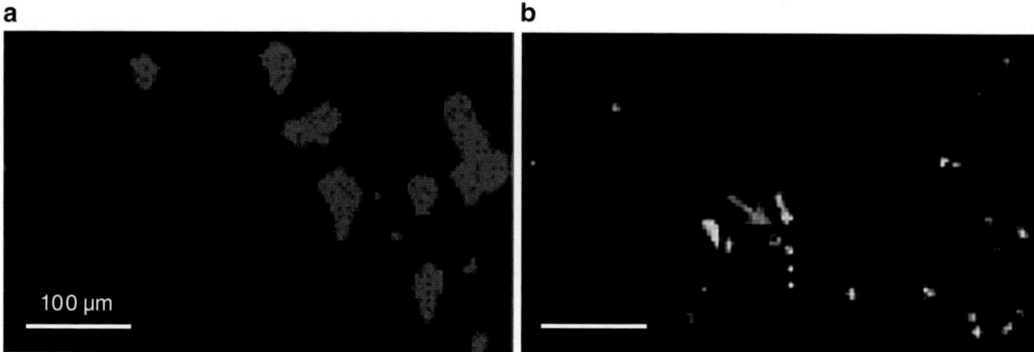

Fig. 3. Detection of LC3 by immunofluorescence. Myocardial tissue sections showing the staining of nucleus with Hoechst 33342 (**a**) and MAP1LC3 followed by Alexa Fluor® 488 (**b**). The slides were examined under a fluorescent microscope with appropriate filter.

2. Cutting the tissue and homogenization should be done as quickly as possible to avoid the induction of endogenous protease activity, which may degrade proteins. Polytron homogenization can be done in two or three short bouts and allowing cooling of tubes on ice in between. (Caution: High speed of homogenization would cause production of heat, which can activate endogenous protease present in cells).

3. MAP1LC3-II is well separated from LC3-I in a 15% SDS-polyacrylamide gel electrophoresis.

4. Do not keep protein samples out at room temperature without heating because endogenous proteases are still active in sample buffer.

5. We have experienced that LC3-I is well separated from LC3-II by running the gel at 200 V for 60 min.

6. Remove air between layers by rolling a glass rod on top of each layer. Entrapment of air would affect the transfer of protein.

7. Washing after secondary antibody incubation is critical for reducing the background signal. So the membrane should be washed for at least 15 min with TBS/T.

8. We have experienced that fixation using 4% paraformaldehyde works well for membrane associated components like LC3.

References

1. Cuervo A. M. (2004). Autophagy: many paths to the same end. *Mol. Cell. Biochem.* **263**, 55–72.

2. Klionsky D. J. and Emr S. D. (2000). Autophagy as a regulated pathway of cellular degradation. *Science* **290**, 1717–21.

3. Huang J., Klionsky D. J. (2007). Autophagy and human disease. *Cell Cycle.* **6**, 1837–49

4. Takagi H., Matsui Y., Sadoshima J. (2007). The role of autophagy in mediating cell survival and death during ischemia and reperfusion in the heart. *Antioxid. Redox. Signal.* **9**, 1373–81.

5. Shintani T., Klionsky D. J. (2004). Autophagy in health and disease: a double-edged sword. *Science* **306**, 990–5.

6. Ravikumar B., Vacher C., Berger Z., Davies J. E., Luo S., Oroz L. G., Scaravilli F., Easton D. F., Duden R., O'Kane C. J., Rubinsztein D. C. (2004). Inhibition of mTOR induces autophagy and reduces toxicity of polyglutamine expansions in fly and mouse models of Huntington disease. *Nat. Genet.* **36**, 585–95.

7. Xue L., Fletcher G. C., Tolkovsky A. M. (2001). Mitochondria are selectively eliminated from eukaryotic cells after blockade of caspases during apoptosis. *Curr. Biol.* **11**, 361–5.

8. Cuervo A. M., Dice J. F. (1996). A receptor for the selective uptake and degradation of proteins by lysosomes. *Science* **273**, 501–3.

9. Petiot A., Ogier-Denis E., Blommaart E. F., Meijer A. J., Codogno P. (2000). Distinct classes of phosphatidylinositol 3"-kinases are involved in signaling pathways that control macroautophagy in HT-29 cells. *J. Biol. Chem.* **275**, 992–8

10. Klionsky D. J., Cregg J. M., Dunn W. A., Jr., Emr S. D., Sakai Y., Sandoval I. V., Sibirny A., Subramani S., Thumm M., Veenhuis M., Ohsumi Y. (2003). A unified nomenclature for yeast autophagy-related genes. *Dev. Cell* **5**, 539–45.

11. Kuma A., Mizushima N., Ishihara N., Ohsumi Y. (2002). Formation of the approximately 350-kDa Apg12-Apg5.Apg16 multimeric complex, mediated by Apg16 oligomerization, is essential for autophagy in yeast. *J. Biol. Chem.* **277**, 18619–25.

12. Tanida I., Ueno T., Kominami E. (2004). LC3 conjugation system in mammalian autophagy. *Int. J. Biochem. Cell Biol.* **36**, 2503–18.

13. Kirisako T., Ichimura Y., Okada H., Kabeya Y., Mizushima N., Yoshimori T., Ohsumi M., Takao T., Noda T., Ohsumi Y. (2000). The reversible modification regulates the membrane-binding state of Apg8/Aut7 essential for autophagy and the cytoplasm to vacuole targeting pathway. *J. Cell Biol.* **151**, 263–76.

14. Mizushima N., Yamamoto A., Hatano M., Kobayashi Y., Kabeya Y., Suzuki K., Tokuhisa T., Ohsumi Y., Yoshimori T. (2001). Dissection of autophagosome formation using Apg5-deficient mouse embryonic stem cells. *J. Cell Biol.* **152**, 657–68.

15. Mizushima N. (2004). Methods for monitoring autophagy. *Int. J. Biochem. Cell Biol.* **36**, 2491–502.

16. Biederbick A., Kern H. F., Elsasser H. P. (1995). Monodansylcadaverine (MDC) is a specific in vivo marker for autophagic vacuoles. *Eur. J. Cell Biol.* **66**, 3–14.

17. Dice J. F. (2007). Chaperone-mediated autophagy. *Autophagy* 3, 295–9.

Part III

Modifications of Apoptotic Proteins during Apoptosis

Chapter 8

Methods to Analyze Transglutamination of Proteins Involved in Apoptosis

Zoltán Nemes and László Fésüs

Summary

Enhanced expression of transglutaminases is a frequent, though not obligatory phenomenon in apoptosis, which is associated with cells dying in steady interaction with their tissue environment. Modification of cellular proteins by transglutamination is a tightly controlled procedure. If transglutamination is dysregulated, it has profound and potentially detrimental consequences on cellular functioning. Under conditions normally occurring in living cells transglutaminase activity is usually undetectably low (latent) and can only be tested by careful preselection of proteins of interest. In late stages of apoptosis, transglutaminases can become rampant in dying cells and a minuscule fraction of dead cells may overshadow many more living ones, which may cause inherent and severe methodological and interpretation bias. Therefore, in this chapter, we describe the analysis of dead cell remnants for protein-bound transglutaminase-mediated cross-link content. In the techniques described below, we rely on the increasing availability and user-friendliness of mass spectrometric equipments.

Key words: Apoptosis, Transglutaminase, Transglutamination, GGEL isodipeptide cross-linking, Mass spectrometry

---

1. Introduction

Transglutaminases are enzymes capable of utilizing the high-energy bond of protein-bound γ-carboxamido groups from glutamine for forming isopeptide bonds with amines, ester bonds with alcohols, or glutamine hydrolysis with water (*see* **Fig. 1**). Unexpectedly, a corresponding catalytic activity for peptide-bound β-aspartyl carboxamido residues has not been found, though transglutaminases are present in various species, including prokaryotes, fungi, plants, and animals *(1)*.

Peter Erhardt and Ambrus Toth (eds.), *Apoptosis*, Methods in Molecular Biology, vol. 559
DOI 10.1007/978-1-60327-017-5_8
© Humana Press, a part of Springer Science + Business Media, LLC 2004, 2009

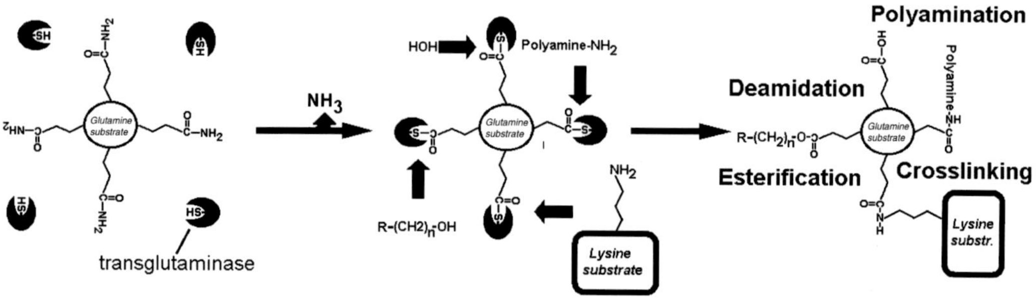

Fig. 1. Protein transglutamination and release of γ-glutamyl-ε-lysine (GGEL) cross-links from proteins by enzymatic hydrolysis. Using ammonia as a leaving group, the high energy protein-γ-glutamyl-transglutaminase intermediate can transfer the activated carboxy residue on various small nucleophiles. It is assumed that in living cells a prealigned lysine (acceptor) substrate may suppress small substrate molecules.

Mammalian transglutaminases (a family of nine homologous genes expressing eight enzymes in multiple splice variants) all require calcium ions for attaining the catalytically active conformation (2). In vitro the optimal calcium ion concentration of these enzymes is by several orders of magnitude higher than those occurring freely in the cytoplasm of normal cells (3). Therefore, it is assumed that transglutaminases exert their nominal activity in the extracellular space (e.g., factor XIII transglutaminase in the blood plasma, or prostate transglutaminase in the seminal fluid) or during cell death, when cytoplasmic Ca^{2+} levels rise irreversibly (e.g., keratinocyte, tissue, and epidermal-type transglutaminases). Though the most ubiquitous tissue-type transglutaminase ("TG2") has well-known signaling and scaffolding activities independent of transglutamination and therefore might modulate cell death at earlier steps of the cell death scenario (4). Calcium surges in the reversible (signaling) range may also elicit cross-linking activity in living cells through regulatory proteins or substrate-dependent conformational changes (5).

Given that protein-bound γ-glutaminyl carboxamido groups, as well as the γ-glutamyl esters tend to show spontaneous hydrolysis under mild conditions (6), hitherto a reliable method for the quantitative analysis of these reaction products could only be developed in debased in vitro systems (7). The isopeptide bonds, however, are remarkably stable and resistant to both spontaneous as well as enzymatic hydrolysis. According to our current knowledge, the γ-glutamyl isopeptide bonds can be eliminated only by two catalytic routes, one is the reverse activity of the creator enzyme, the other requires degradation of the protein and release of the γ-glutamyl-lysine isodipeptide (8).

Polyamines were suggested to be the most abundant amine acceptor substrates available for transglutaminases, though spermine and spermidine adducts are typical in plant cells where their amount varies with the cell cycle (9). In mammalian cells the most remaining products of transglutamination are intracellular protein polymers joined together by isopeptide cross-bridges *(10)*.

Protein cross-linking by transglutaminases creates γ-glutamyl-ε-lysine (GGEL) isopeptide bonds, i.e., the primary amino group of protein bound lysines transamidates the γ-glutaminyl carboxamido group.

GGEL isopeptide cross-links though are not unique tools for covalently linking two proteins together. Not even mentioning the disulphide bridges; attachment of ubiquitin and Small Ubiquitin-like MOdifying (SUMO) proteins also creates isopeptide cross-links. Extracellular matrix fibers, such as basement membranes, collagen, and elastin fibers are also replete with covalent cross-links formed from oxidized lysine side chain derivatives. Oxidative damage to membrane lipids and sugars may also give rise to bivalent aldehydes (such as malondialdehyde), which can cross-link proteins. The formation of dityrosine cross-links via oxidation of *para*-phenolic OH groups is also known. This variability of protein–protein cross-linking makes the specific identification and analysis of GGEL cross-links unavoidable in complex (biological) samples.

The quantitative determination of GGEL cross-link density utilizes the unique resistance of the isopeptide bond to generic broad-specificity proteinases as well as the distinctive chemical properties and mass of the GGEL isodipeptide *(11)*.

---

2. Materials

2.1. Cell and Tissue Lysis, Protein Extraction

1. PBS-EDTA: 10 mM phosphate-Na, 150 mM NaCl, 1 mM ethylenediamine tertaacetic acid-Na (from 0.5 M pH 8.0 stock), pH 7.6 at 25°C (*see* **Note 1**).

2. TriPure reagent (Roche). Toxic. Consult MSDS provided with this product. Store in the dark.

3. Chloroform, absolute ethanol, 95% ethanol, isopropanol, guanidine-HCl (analytical grade).

4. Polytron tissue homogenizer (Cole-Palmer).

2.2. Protease Digestion in [¹⁸O]-Water

1. 95% [^{18}O]-water (Aldrich, *see* **Note 2**).

2. NH$_4$HCO$_3$ (analytical grade) from a freshly opened container is weighed into a sterile wide-mouth screw-cap bottle and poured over with water to give a 1 M stock. The bottle is tightly closed and the salt is then dissolved by agitation (*see* **Note 3**). This stock solution is kept at +4°C and can be used within a month without sterilization.

3. γ-Glutamyl-ε-Lysine (Sigma) is dissolved in water at 0.1 mM and stored in 50-μL aliquots at –20°C.

4. Proteinase K (from *Tritirachium album*): 20,000 U/mL glycerol stock solution (Biochemika, Fluka). Store at –20°C (*see* **Note 4**).

5. Leucine aminopeptidase (porcine, sequencing grade, Sigma), carboxypeptidase A (PMSF-treated, bovine, Fluka), and carboxypeptidase Y (yeast, BioChemika, Fluka) are dissolved at 1 mg/mL in water and stored in single use aliquots at –80°C.

6. *N*-9-Fluorenylmethoxycabonyl (Fmoc) derivatization: 9-Fluorenylmethoxycarbonyl chloride (Fmoc-Cl, Fluka, derivatization grade) is stored at +4°C in unopened factory package until use (*see* **Note 5**). Fmoc derivatization solution:ethanol: water:triethylamine (reagent purity):Fmoc-Cl 7:1:1:1 (by volume). Prepare freshly. Conduct all procedures with Fmoc-Cl with regard to its extreme caustic properties: work in fume hood; wear gloves, protective apron, and glasses.

2.3. MALDI MS and Data Assessment

1. Formic acid, acetonitrile are toxic and should be handled accordingly. Aqueous solutions are stable at RT without limitations.

2. Matrix solution: 7 mg/mL α-cyano-4-hydroxycinnamic acid (Sigma), dissolved in 50% v/v acetonitrile and 0.1% TFA in water.

3. C18 ZipTips (Millipore).

4. 4700 MALDI TOF/TOF Proteomics Analyzer (Applied Biosystems).

3. Methods

The method given here is based on equal ionization and flight opportunities of "light" GGEL molecules given to the samples as internal standards and "heavy" GGEL molecules liberated from cross-linked proteins by enzymatic hydrolysis with [^{18}O]-water.

Since the determination of GGEL cross-links in term gives absolute amounts (usually pM), the most acceptable measurement for GGEL amount is to relate it to the total amount of protein in the sample. Nevertheless, commonly used dye photometric protein assays do not provide precise data on the absolute amount of protein found in the samples, though can be used to provide reliable assessment of relations between samples. Both dye-binding ("Bradford-type") and inorganic chromogen-based (Folin-Ciocalteau-type) assays can be used, though the latter method gives more reliable measurements in the presence of detergents. Protein assays allow us to give the frequency of GGEL as percent of the standard, which usually means mock treated cells. However, if an absolute quantitation of GGEL cross-link density is needed, or in tissues, where some fractions of the total proteins are always insoluble, amino acid analysis is the method of choice, which allows defining the frequency of GGEL per total amino acids.

GGEL isodipeptide is uniquely resistant to broad-spectrum protease attack, not even proteinase K elicits significant decay in its amount over several days. Proton attack, however, can split the isopeptide bonds and increases the rate of ^{18}O exchange to ^{16}O *(12)*, therefore it is prudent to avoid reaction and chromatographic media below pH 2 during digestion and chromatography and process samples as quickly as possible.

3.1. Cell and Tissue Lysis, Protein Extraction

1. Cells and tissues are washed from medium and cross-linking is terminated by washing twice in ice-cold PBS-EDTA in a 15-mL Falcon centrifuge tube (*see* **Note 6**).

2. The washing fluid is discarded and ice cold TriPure reagent is added in a ratio of 3 mL per 10^7 cells or 8 mL/g of wet tissue.

3. Cells are homogenized into the TriPure reagent by passing through a pipette tip several times, whereas tissues require thorough mincing by Polytron type homogenizer. During lysis the material is allowed to warm up to room temperature (RT).

4. Transfer the homogenate into 1.5-mL tubes – by putting 1 mL into each – and spin for 10 min at 12,000 × *g* at +4°C. Remove oily top layer (this step is especially important when processing brain or fatty tissues) and carefully isolate the supernatant from the insoluble pellet. Resuspend and wash pellet once in the original volume of TriPure reagent and next in 100% ethanol.

5. Incubate each homogenized sample for 5 min at RT to ensure the complete dissociation of nucleoprotein complexes.

6. Add chloroform to each sample. Use 0.2-mL chloroform for each 1-mL TriPure Isolation Reagent required in the initial homogenization.

7. Cap tube securely and shake it vigorously for 15 s (*see* **Note 7**).

8. Incubate tube at RT for 15 min.

9. To separate the solution into three phases, centrifuge tube at 12,000 × *g* for 15 min at +4°C.

10. Remove the colorless upper aqueous phase from the interphase and lower red organic phase and discard (as hazardous chemical waste) (*see* **Note 8**).

11. Add 100% ethanol to the interphase and organic phase. Use 0.3 mL 100% ethanol for each 1-mL TriPure Isolation Reagent required in the initial homogenization. Cap the sample, then invert it several times to mix it thoroughly.

12. Incubate sample for 2–3 min at RT to allow the DNA precipitate to form and centrifuge the sample at 2,000 × g for 5 min at +4°C.

13. Pipette the supernatant (containing phenol, ethanol, and protein) from each sample into a fresh tube. These samples are stable at +4°C.

14. Add isopropanol to the phenol–ethanol supernatant. Use 1.5-mL isopropanol for each 1-mL TriPure Isolation Reagent required in the initial homogenization. Cap the tube and invert it several times to mix it thoroughly.

15. Incubate the sample for a minimum of 10 min at RT to allow the protein precipitate to form. Centrifuge the sample at 12,000 × g for 10 min at + 4°C. Wash away phenol by resuspending the pellet first with 0.3 M guanidine–HCl in 95% ethanol and next in 100% ethanol after pelleting again at 12,000 × g (see **Note 9**). Use 2-mL guanidine hydrochloride/ethanol for each 1-mL TriPure Isolation Reagent required in the initial homogenization procedure. Samples can be stored under ethanol for months even at RT.

3.2. Proteinase Digestion in [^{18}O]-Water

1. Pellet proteins from **steps 4** and **5** of **Subheading 3.1** by centrifugation at 12,000 × g at +4°C; discard the ethanol supernatant and air dry.

2. Add 0.3 mL of 0.1 M NH_4HCO_3 (pH 8.3) dissolved in water containing 60–80% [^{18}O] (see **Note 2**) and 0.5 µM GGEL isodipeptide (from 0.1 mM stock) for each mL of TriPure reagent used in **Subheading 3.1, step 2**.

3. Add 10 µL proteinase K solution for each mL of TriPure.

4. Digest overnight at 60°C.

5. Check whether digestion is complete by centrifuging 15 min at 15,000 × g. If there is a visible pellet, repeat **steps 3–5**.

6. Autoclave samples at 121°C for 20 min.

7. Add 1.5 µg leucine-aminopeptidase, 0.4 µg carboxypeptidase A and 0.3 µg carboxypeptidase Y per mL TriPure used in **step 2** of **Subheading 3.1** and incubate at 37°C for 36 h.

8. Take out a fraction of the samples (3 × 10%) for amino acid analysis (see **Note 10**).

9. Lyophilize samples.

10. Flush sample tubes with N_2 and dissolve in 100 µL Fmoc derivatization solution to convert amino acids into their N-FMOC derivatives (see **Note 11**).

11. Close lid and incubate for 1 h at room temperature.

12. Lyophilize.

3.3. MALDI MS and Data Assessment

1. Dissolve samples in 100 μL 1% formic acid.

2. Suck 5 μL of sample up into a ZipTip, wash three times with 1% formic acid and elute with 5 μL of 60% acetonitrile containing 1% formic acid (*see* **Note 12**).

3. Pipette 1 μL matrix solution onto the MALDI target plate and immediately mix with 1 μL of desalted sample from the previous step. Let the matrix and the Fmoc-aminoacids (Nα,N α-diFmoc-GGEL) co-crystallize by air-drying at RT.

4. Collect MALDI-TOF ion spectra including masses 720–724 Da (*see* **Note 13**).

5. Compare ion counts of 730, 732, and 734 Da (x, y, z) and calculate sample GGEL content by the following equation, where r is the final $[^{18}O]/[^{16}O + {}^{18}O]$ ratio in the sample GGEL carboxyl OH-s:

$$r = \frac{2[z - 0.12(y - 0.12x)]}{y - 0.12x + 2[z - 0.12(y - 0.12x)]}.$$

G and S are multiples of sample GGEL and unlabeled internal GGEL standard, respectively.

$$G = \frac{z - 0.12(y - 0.12x)}{r^2},$$

$$S = x - G(1 - r)^2.$$

Sample GGEL content is thus calculated from the ratio of G/S and related to total protein mass or amino acid content (*see* **Note 14**). An example for spectrum evaluation is shown in **Fig. 2**.

4. Notes

1. If not otherwise stated, solutions are made in water, which means sterile water with a resistivity >18.2 MΩ cm and organic content <5 parts per billion ("~MilliQ water").

2. The ^{18}O is incorporated into the OH of the C-terminal carboxylic groups liberated by the proteinase digestion. For economic reasons, the steps ensuing after the digestion are carried out in media containing plain ($[^{16}O]$) water; a part of the ^{18}O incorporated during hydrolysis of the peptide bond will be replaced with ^{16}O. In order to keep the random error low, the $[^{18}O]$ water should not be diluted below 50%, and

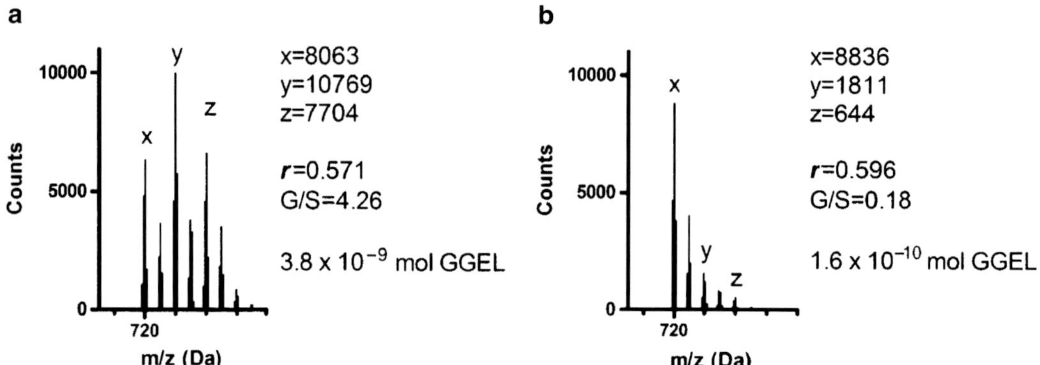

Fig. 2. MALDI mass spectra detecting the relative intensities (amounts) of masses corresponding to all-[^{16}O], mono[^{18}O] and di[^{18}O] labeled αN, α N -diFmoc-GGEL + H$^+$. 2 × 10^6 HL60 cells were treated with (**a**) 1 μM all-trans retinoic acid + 1% DMSO, or (**b**) 1% DMSO control for 72 h to induce transglutaminase 2 induction during abortive myeloid maturation leading to massive apoptosis. Cell proteins were extracted and enzymatically hydrolyzed with [^{18}O]water in the presence of 0.9 pM light (all[^{16}O] GGEL as internal standard), then derivatized with Fmoc-Cl. *x*, *y*, and *z* values are calculated, subpeak areas belonging to monoisotopic masses 720, 722, and 724 (±100 ppm), respectively. *G/S* values and sample GGEL content (0.9 pM × *G/S*) were calculated as given in **Subheading 3.3** (*see also* **Note 14**).

thus r (*see* **Subheading 3.3.5**) should be above 0.4 to keep the signal/noise ratio at acceptable levels.

3. Ammonium bicarbonate solutions tend to lose more CO_2 than NH_3 upon standing in open air or not tightly capped bottles and thus become increasingly alkaline. The NH_4HCO_3 stock solution can be sterilized by pressing (not sucking!) the ice-cold liquid through a syringe filter.

4. Proteinase K is a mixture of aggressive fungal proteases. Self-digestion and decay in proteinase activity is kept low by maintaining a few μM higher Ca^{2+} ion concentration in the reaction buffer than that of Ca-chelators including the target proteins and some types of plastic. The Fluka's stock solution contains enough Ca^{2+} for the reaction, thus supplementation is unnecessary.

5. We regularly reopen and reuse Fmoc-Cl from the factory glass vessel by opening, handling, and closing it under a stream of N_2 gas streamed to the outlet through a Pasteur pipette in a fume hood. Beware that manipulating Fmoc-Cl like this needs two people. A P_2O_5 exsiccator in the cold is recommended for storage of opened Fmoc-Cl vessels.

6. In dying cells the activity of so far latent transglutaminases might easily become rampant during cell harvesting, centrifugation, and lysis. To obtain reliable results, therefore, it is important to chelate calcium ions with EDTA in the cell medium before harvesting either by centrifugal removal of the culture medium or by scraping off cells. If adherent cells

are harvested by trypsin, it is enough to add the EDTA to the trypsin solution. Since the sample collection procedures are carried out on ice, inhibition of endogenous proteases is not necessary.

7. Solvents tend to leak out through the regular closures of Eppendorf tubes. Screw cap tubes with silicone sealing rings are safe in this respect.

8. Carryover of a small fraction of the watery phase causes less problems later on than removing from the interphase, which contains most of the proteins.

9. Carryover of phenol should be kept to a minimum, since it interferes with the liophylization and desalting of samples, and increases background noise during the final mass spectrum collection step.

10. Amino acid analysis is the only way to determine the amount of insoluble proteins. To check the completeness of digestion, an aliquot can be directly analyzed without acid hydrolysis. A blank with enzymes shall be subtracted.

11. An incomplete dissolution of material usually does not alter final results.

12. Do not store sample materials in acidic solvents for longer than a few minutes. If you cannot process them right away, lyophilize samples in a rotary evaporator ("Speedvac").

13. The use of the named device is clearly a luxury for such pristine determination. A "simple" LC/ESI/Q-Trap MS would also be suitable, although finding the optimal separation and ionization parameters requires somewhat more exercise and time. A MALDI run is definitely easier and the analysis of multiple samples is faster than working with LC/ESIMS devices. Though MALDI spectra are generally thought to be "noisy" below $1,000\,m/z$ owing to ions from the matrix, such interference is not detected with our peaks of interest. Exclusion of the peaks between 600 and $1,000\,m/z$ are primarily exercised to protect statistical data processing algorithms (such as Mascot and Sequest) from confounding by "junk peaks."

14. The following basic connections are necessary to understand the given formulas: (a) At a compound with an atomic formula $C_{41}H_{41}N_3O_9 + H^+$ the height of the afternext peak is ~12% of the monoisotopic peaks. This ratio (to the second decimal) is determined by the frequency of naturally occurring ^{13}C. (b) With r ratio of ^{18}O substitution at COOH the distribution of monoisotopic, singly and doubly substituted GGEL will be $(1-r)^2 : 2(r-r^2) : r^2$.

Acknowledgments

The authors would like to thank to Marteen Aerts and Bart Devreese for technical expertise. This work was supported by the Hungarian Scientific Research Fund (OTKA TS 044798), the Hungarian Ministry of Health (ETT115/03), the Hungarian Ministry of Education (OM00427/04) the Flemish-Hungarian intergovernmental research grant (B33/04, B/06093), and the Fund for Scientific Research/Flanders (G0422.98).

References

1. Lóránd, L. and Graham, R. M. (2003). Transglutaminases: crosslinking enzymes with pleiotropic functions. *Nat Rev Mol Cell Biol* **4**, 140–156.

2. Mehta, K. (2005). Mammalian transglutaminases: a family portrait. *Prog Exp Tumor Res* **38**, 1–18.

3. Mariani, P., Carsughi, F., Spinozzi, F., Romanzetti, S., Meier, G., Casadio, R., et al. (2000). Ligand-induced conformational changes in tissue transglutaminase: Monte Carlo analysis of small-angle scattering data. *Biophys J* **78**, 3240–3251.

4. Fésüs, L. and Piacentini, M. (2002). Transglutaminase 2: an enigmatic enzyme with diverse functions. *Trends Biochem Sci* **27**, 534–539.

5. Nemes, Z., Petrovski, G., Cs'sz, É. and Fésüs, L. (2005). Structure–function relationships of transglutaminases – a contemporary view. *Prog Exp Tumor Res* **38**, 19–36.

6. Robinson, N. E. and Robinson, A. B. (2001). Deamidation of human proteins. *Proc Natl Acad Sci USA* **98**, 12409–12413.

7. Nemes, Z., Petrovski, G. and Fésüs, L. (2005). Tools for the detection and quantitation of protein transglutamination. *Anal Biochem* **342**, 1–10.

8. Loewy, A. G., Blodgett, J. K., Blase, F. R. and May, M. H. (1997). Synthesis and use of a substrate for the detection of isopeptidase activity. *Anal Biochem* **246**, 111–117.

9. Del Duca, S. and Serafini-Fracassini, D. (2005). Transglutaminases of higher, lower plants and fungi. *Prog Exp Tumor Res* **38**, 223–247.

10. Fésüs, L. and Szondy, Z. (2005). Transglutaminase 2 in the balance of cell death and survival. *FEBS Lett* **579**, 3297–3302.

11. Nemes, Z., Mádi, A., Marekov, L. N. Piacentini, M., Steinert, P. M. and Fésüs, L. (2001). Analysis of protein transglutamylation in apoptosis. *Methods Cell Biol* **66**, 111–133.

12. Stewart, II, Thomson, T. and Figeys, D. (2001). 18O labeling: a tool for proteomics. *Rapid Commun Mass Spectrom* **15**, 2456–2465.

Chapter 9

Methods to Analyze *S*-nitrosylation of Proteins Involved in Apoptosis

Neelam Azad, Anand Krishnan V. Iyer, and Yon Rojanasakul

Summary

Nitric oxide (NO) is an important signaling molecule that plays a key role in various physiological and pathological processes. One of the well-established mechanisms by which NO regulates the function of various target proteins is through *S*-nitrosylation. NO readily reacts with thiol (SH) groups in the cysteine residues of target proteins to form nitrosothiol (S-NO) groups. This posttranslational modification of proteins can positively or negatively regulate various signaling pathways including apoptosis. Likewise, *S*-nitrosylation of various apoptosis-regulatory proteins has been demonstrated to modify the apoptotic response to various stimuli. We have shown that NO nitrosylates important antiapoptotic proteins, such as Bcl-2 and FLIP, and prevents their downregulation via the ubiquitin-proteasomal degradation pathway. To detect protein *S*-nitrosylation, we isolated the protein by immunoprecipitation and analyzed cysteine nitrosylation by Western blotting or spectrofluorometry.

Key words: *S*-nitrosylation, Apoptosis, Nitric oxide, Bcl-2, Immunoprecipitation, Western blot

1. Introduction

In 1987, nitric oxide (NO) was discovered as an ubiquitous cell-signaling molecule that exerted pleitropic effects including vasodilation, synaptic transmission, immune regulation, cellular proliferation, and apoptosis *(1–3)*. In 1992, Stamler and coworkers discovered that the redox state and chemistry of NO facilitates its interaction with various proteins, thus regulating various intracellular and intercellular signaling events *(4)*. NO reacts with cysteine thiols in target proteins in the presence of an electron acceptor to form nitrosothiols (S-NO). Over the past 15 years, several substrates of

Peter Erhardt and Ambrus Toth (eds.), *Apoptosis*, Methods in Molecular Biology, vol. 559
DOI 10.1007/978-1-60327-017-5_9
© Humana Press, a part of Springer Science + Business Media, LLC 2004, 2009

S-nitrosylation in cellular activities, metabolic processes, and signaling systems including apoptosis have been reported, establishing the biological significance of this mechanism *(5)*. *S*-nitrosylation of various pro- and antiapoptotic proteins has been demonstrated to either induce or inhibit apoptosis. For instance, nitrosylation of caspases including caspase-8, caspase-9, and caspase-3 at their active site has been shown to inhibit their enzyme activity and nitrosylation of antiapoptotic proteins including FLIP and Bcl-2 has been reported to stabilize and activate these proteins *(6–10)*.

Due to the reversible regulation of protein function, *S*-nitrosylation has now become the prototypic posttranslational modification analogous to phosphorylation, but it is difficult to quantify *(11)*. Various chemical assays for the detection of nitrosylated proteins including triiodide chemiluminescence, photolysis chemiluminescence, and Griess-based colorimetric and fluorometric assays are very complex for widespread application and unreliable in intricate biological systems *(12–15)*. Biotin switch assay is considered as the most reliable method that is more easily adapted to the study of *S*-nitrosylation. However, it is a tedious procedure and its sensitivity may be limiting as it can detect protein glutathionylation *(16, 17)*. We have recently shown that NO nitrosylates antiapoptotic proteins Bcl-2 and FLIP, preventing their downregulation via ubiquitin-proteasomal degradation *(6, 7)*. Nitrosylation stabilizes these proteins and increases their expression levels. Therefore, NO may inhibit apoptosis through its ability to nitrosylate important antiapoptotic proteins. We used a direct immunoprecipitation method to pull down the protein of interest and analyzed its *S*-nitrosylation either by Western blotting or by spectrofluorometry.

2. Materials

2.1. Cell Culture and Lysis

1. RPMI-1640 medium (Sigma) supplemented with 10% fetal bovine serum (FBS, Atlanta Biologicals), 2 mM L-glutamine (Sigma), 100 units/mL penicillin and 100 µg/mL streptomycin (Gibco/BRL).

2. Solution of 0.25% trypsin-ethylenediamine tetraacetic acid (EDTA) from Gibco/BRL.

3. Proteasome inhibitor lactacystin (LAC, Alexis Biochemicals).

4. Sterile phosphate buffered saline (PBS, Sigma).

5. Sodium dichromate ($Na_2Cr_2O_7 \cdot 2H_2O$) [Cr(VI)] (Sigma) was dissolved in sterile PBS at 1 mM, stored in aliquots at 4°C, and then added to tissue culture dishes as required to induce apoptosis.

6. NO donor sodium nitroprusside (SNP, Sigma) was dissolved in sterile PBS at 1 mg/mL, stored in aliquots at –20°C, and then added to tissue culture dishes as required.

7. NO donor dipropylenetriamine (DPTA) NONOate (Alexis Biochemicals) was dissolved in sterile PBS at 10 mM, stored in aliquots at –20°C, and then added to tissue culture dishes as required.

8. NO inhibitor aminoguanidine (AG, Sigma) was dissolved in sterile PBS at 3 mg/mL, stored in aliquots at –20°C, and then added to tissue culture dishes as required.

9. NO inhibitor 2-(4-carboxyphenyl)-4,4,5,5-tetramethyl-imidazoline-1-oxy-3-oxide (PTIO, Sigma) was dissolved in sterile PBS at 50 μg/mL, stored in aliquots at –20°C, and then added to tissue culture dishes as required.

10. Dithiothreitol (DTT, Sigma), an inhibitor of protein *S*-nitrosylation, was dissolved in sterile PBS at 1 M, stored in aliquots at –20°C, and then added to tissue culture dishes as required.

11. Immunoprecipation (IP) cell lysis buffer: 20 mM Tris–HCl, pH 7.4, 150 mM NaCl, 10% glycerol, 0.2% NP-40, 100 mM phenylmethylsulfonyl fluoride (PMSF), and protease inhibitor. Store at 4°C (*see* **Note 1**).

12. Cell scrapers (Fisher Scientific).

2.2. Bicinchoninic Acid (Bca) Protein Concentration Assay

1. BCA protein assay kit (Pierce Biotechnology).

2.3. Immunoprecipitation

1. Protein A-agarose beads (Santa Cruz Biotechnology).
2. Anti-myc agarose beads (Santa Cruz Biotechnology).
3. IP lysis buffer (*see* **Subheading 2.1, item 11**).

2.4. Sodium Dodecyl Sulfate-Polyacrylamide Gel Electrophoresis

1. Separating buffer (4×): 1.5 M Tris–HCl, pH 8.8, and 0.4% SDS. Store at room temperature.
2. Stacking buffer (4×): 0.5 M Tris–HCl, pH 6.8, and 0.4% SDS. Store at room temperature.
3. Forty percent acrylamide/bis solution (37.5:1 with 2.6% C) (Bio-Rad).
4. *N,N,N,N′*-Tetramethyl-ethylenediamine (TEMED, Bio-Rad).
5. Ammonium persulfate (APS): Prepare a 10% solution by dissolving 0.01 g of APS in 1 mL of water. Store in aliquots at –20°C (*see* **Note 2**).
6. Prepare 0.1% aqueous solution of SDS. Store at room temperature.

7. Running buffer (10×): 30.3 g Tris base, 144.2 g glycine, 10 g SDS, and water up to 1,000 mL. Adjust pH to 8.8. Store at room temperature (*see* **Note 3**).

8. Laemmli sample buffer (6×): 7 mL 4× Tris–HCl/SDS, pH 6.8, 3-mL glycerol (30% final concentration), 1 g SDS (10% final concentration), 0.93 g dithiothreitol (DTT) (0.6 M in final solution), 1.2 mg bromophenol blue (0.012% in final solution), and water up to 10 mL (if needed). Store aliquots at –80°C.

9. BenchMark® Pre-stained Protein Ladder (Invitrogen).

2.5. Western Blotting for Nitrosylated Protein

1. Transfer buffer (10×): 30.3 g Tris base, 144.2 g glycine, 3.75 g SDS, and water up to 1,000 mL. Store at room temperature (*see* **Note 3**).

2. Nitrocellulose membrane (0.45 μm) from Pierce Biotechnology, Rockford, IL.

3. Chromatography paper (3MM) from Whatman.

4. Tris-buffered saline with Tween (TBST): Prepare 1× solution by mixing 20 mL of 1 M Tris, 27.5 mL of 5 M NaCl, and water up to 1,000 mL. Add 1 mL of Tween-20 (Research Organics) to this solution.

5. Blocking buffer: 5% (w/v) nonfat dry milk in TBST.

6. Primary antibody: Anti-*S*-nitroso-cysteine antibody (Sigma).

7. Secondary antibody: Anti-rabbit IgG conjugated to horseradish peroxidase (Sigma).

8. Primary and secondary antibody dilution buffer: 5% nonfat dry milk in TBST.

9. Supersignal® West Pico chemiluminescence kit from Pierce Biotechnology.

2.6. Stripping and Reprobing Blots for Total Bcl-2

1. Stripping buffer: 62.5 mM Tris–HCl, pH 6.8, 2% (w/v) SDS. Store at room temperature. Warm to working temperature of 70°C and add 100 mM β-mercaptoethanol.

2. Primary antibody: Anti-Bcl-2 (Santa Cruz Biotechnology).

3. Secondary antibody: Anti-mouse IgG conjugated to horseradish peroxidase (Sigma).

4. Primary and secondary antibody dilution buffer: 5% nonfat dry milk in TBST.

2.7. Fluorometric Measurement of S-Nitrosylation

1. Mercury (II) chloride ($HgCl_2$, Sigma): Prepare a 10 mM stock solution in water. Pipette 20 μL in 1-mL medium for a final concentration of 200 μM. Store at –20°C.

2. 2,3-Diaminonaphthalene (DAN, Sigma): Prepare a 10 mM stock solution in dimethylsulfoxide (DMSO). Pipette 20 μL in 1-mL medium for a final concentration of 200 μM. DAN is light sensitive so store in amber colored tubes at –20°C.

3. Sodium hydroxide (NaOH, Sigma): Prepare a 10 M stock solution in water. Pipette 100 µL in 1-mL medium for a final concentration of 1 M. Store at room temperature.

3. Methods

3.1. Preparation of Samples for Immuno-precipitation

1. Human lung epithelial H460 cells (ATCC) stably transfected with myc-tagged Bcl-2 plasmid were cultured in 100-mm cell culture dishes and passaged when approaching confluence with trypsin-EDTA to provide experimental cultures on 60-mm dishes. A 1:4 split will provide experimental cultures that were confluent after 24 h with each dish providing a single data point.

2. Cells were rinsed once with serum free RPMI-1640 and treated as required in serum-free RPMI-1640 containing 10 µM lactacystin (LAC) (*see* **Note 4**).

3. After 3 h incubation at 37°C, cell supernatant was aspirated out and cells were washed with ice-cold PBS. Cells were then incubated with 200 µL of IP lysis buffer on ice for 5–10 min (*see* **Note 5**). Cells scraped from the culture plates were collected in appropriately labeled tubes.

4. The tubes were closed and centrifuged at $14{,}000 \times g$ for 15 min at 4°C. The precipitate (cell debris) was discarded and the supernatants were collected in a new set of labeled microcentrifuge tubes.

5. Protein content was determined using BCA method.

3.2. BCA Protein Content Assay

1. Preparation of Protein Standard Solution: Dilute the content of one protein standard ampule containing bovine serum albumin (BSA), which is provided in the kit, into a clean vial using the same diluent as the sample (PBS can also be used) so that the final concentration is 1 mg/mL.

2. Preparation of the BCA working reagent (WR)
   (a) Total volume of WR required: (number of standards + number of unknowns) × (number of replicates) × (volume of WR per sample) = total volume WR required (*see* **Note 6**).
   (b) Prepare WR by mixing 50 parts of BCA Reagent A with 1 part of BCA Reagent B, provided in the kit (50:1, Reagent A:B) (*see* **Note 7**).

3. Procedure
   (a) Pipette 200 µL of the WR into each well of a 96-well plate.
   (b) Add BSA standard solution (0, 1, 2, 3, 4, 5 µL) or unknown sample (2 µL) into the wells containing the WR. Adjust each well to the final volume of 300 µL with PBS.

(c) Cover plate and incubate at 37°C for 30 min.

(d) Measure the absorbance at 562 nm on a microplate reader. Calculate the protein concentration to be used accordingly.

3.3. Immunoprecipitation

1. Pipette 12 µL of protein A-agarose beads into labeled microcentrifuge tubes kept on ice. Add 100 µL of ice cold PBS and mix gently by inverting 5–6 times (*see* **Note 8**).

2. Centrifuge at $1,000 \times g$ for 1 min at 4°C (*see* **Note 9**).

3. Aspirate out PBS gently using a small syringe needle connected to the vacuum (*see* **Note 10**).

4. Add 8 µL of anti-myc-agarose beads directly on top of protein A agarose beads followed by the addition of cell lysates (at least 60 µg of proteins) to the whole sample (*see* **Note 11**). Make sure that all the steps are performed on ice or at 4°C.

5. Mix gently by inverting 5–6 times and incubate at 4°C for 3–4 h or overnight at 4°C on a rocking platform (*see* **Note 12**).

6. After incubation, wash the immune complexes three times with freshly prepared lysis buffer as described below:
   (a) First, centrifuge the immune complexes at $1,000 \times g$ for 1 min at 4°C

   (b) Aspirate the supernatant gently using a small syringe needle connected to the vacuum (*see* **Note 13**)

   (c) Add 500 µL of IP lysis buffer and mixed gently by inverting 5–6 times

   (d) Repeat **steps 6a–6c** three times

7. After the final washing, carefully aspirate out the supernatant completely (*see* **Note 13**). A schematic representation of this method is shown in **Fig. 1**.

3.4. SDS-PAGE

1. Wash the glass plates to be used to prepare the gels with a detergent and rinse extensively with distilled water. Let the plates air-dry in a plastic rack and keep them clean until use.

2. Prepare 10% sodium dodecyl sulfate-polyacrylamide gel electrophoresis (SDS-PAGE) gel by mixing 3.75 mL of water, 1.875 mL of 4× separating buffer, 1.875 mL of 40% acrylamide/bis solution, 37.5 µL of 10% APS, and 7.5 µL of TEMED (1.5-mm thick Bio-Rad mini blots) (*see* **Note 14**).

3. Pour the gel in the glass plate assembly and overlay with 0.1% SDS to avoid drying. Leave space for the stacking gel.

4. The gel should polymerize in about 20–30 min. Pour off 0.1% SDS by tilting the apparatus and absorbing the excess with a tissue or filter paper.

5. Prepare the stacking gel by mixing 1.625 mL of water, 0.625 mL of 4× stacking buffer, 0.25 mL of 40% acrylamide/*bis*

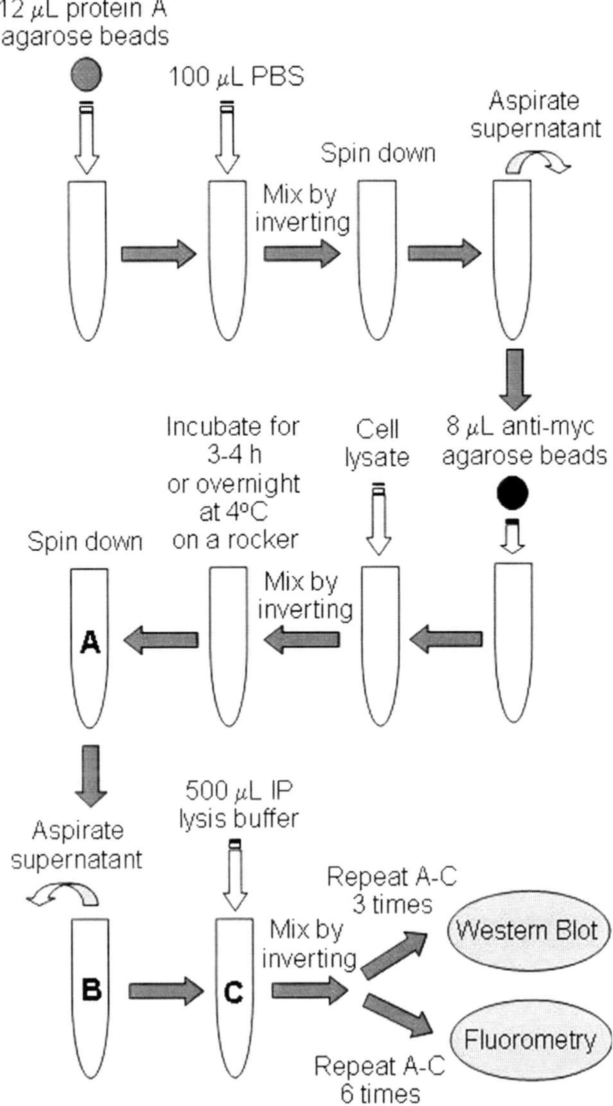

Fig. 1. Schematic representation of the various steps involved in immunoprecipitation.

solution, 12.5 µL of APS solution, and 2.5 µL of TEMED (*see* **Note 14**).

6. Pour the stacking gel on top of the separating gel. Immediately insert the comb avoiding air bubbles and wait for 20–30 min for the stacking gel to polymerize (*see* **Note 15**).

7. Remove the comb and rinse the gel with water or 1× running buffer.

8. Prepare the running buffer by diluting 100 mL of the 10× running buffer with 900 mL of water in a measuring cylinder. Invert to mix and add running buffer in the Mini Trans-Blot cell (Bio-Rad) holding the gel cassette (*see* **Note 16**).

9. Resuspend the immune complexes in 30–40 µL of 2× Laemmli sample buffer and boil at 95°C on a heat block for 5 min (*see* **Note 17**).

10. Briefly vortex the samples and spin down at high speed for about 10–20 s to precipitate the beads.

11. Load 10 µL of the molecular weight marker in the first well and 30–40 µL of each sample in the remaining wells (*see* **Note 18**).

12. Complete the assembly of the gel unit and connect to a power supply. Run the gel at 100 V till the ladder is about to run out of the separating gel (about 2 h) (*see* **Note 19**).

3.5. Western Blotting for S-Nitrosylated Protein

1. Transfer the samples separated by SDS-PAGE onto a 0.45-µm nitrocellulose membrane electrophoretically.

2. Prepare 1× transfer buffer: Dilute 80 mL of 10× transfer buffer with 720 mL of distilled water in a measuring cylinder. Add 200 mL of methanol to this solution and mix well (*see* **Note 20**).

3. Soak two pieces of sponge and two sheets of 3MM filter paper in 1× transfer buffer. In a separate tray containing 1× transfer buffer, wet a sheet of nitrocellulose membrane cut a little larger than the size of the separating gel.

4. Disconnect the gel unit from the power supply and discard the stacking gel. Carefully lay the membrane on the separating gel and cut one corner from the separating gel and the membrane to mark its orientation (*see* **Note 21**).

5. Assemble the sponges, filter paper, gel, and membrane on the transfer cassette in the following order. Cathode < sponge – filter paper – gel – membrane – filter paper – sponge < anode (*see* **Note 22**).

6. Place the cassette with the transfer assembly into the transfer tank (Bio-Rad) and fill it with 1× transfer buffer.

7. Put on the lid of the tank and turn on the power supply. Transfer the gel at 25 V for 2 h (*see* **Note 23**).

8. After transfer take out the cassette and remove the sponges, discard the filter papers and the gel. If the transfer is successful, then the molecular weight marker should be clearly visible on the membrane.

9. Block the membrane in 5% nonfat dry milk in TBST for 1 h at room temperature or overnight at 4°C on a rocker.

10. Discard the blocking buffer and rinse the membrane twice with TBST.

11. Incubate the membrane with anti-*S*-nitroso-cysteine antibody at 1:5,000 dilution in 5% nonfat dry milk in TBST for 1 h at room temperature or overnight at 4°C on a rocker.

12. Remove the primary antibody and wash the membrane three times for 5 min each with TBST (*see* **Note 24**).

13. Incubate the membrane with anti-rabbit secondary antibody at 1:5,000 dilution in 5% nonfat dry milk in TBST for 1 h at room temperature or overnight at 4°C on a rocker.

14. Discard the secondary antibody and wash the membrane three times for 15 min each with TBST.

15. After the final wash, remove the blot and place it on a plastic wrap. Mix 1 mL of Supersignal® West Pico stable peroxide solution and 1 mL of Supersignal® West Pico Luminol/Enhancer solution provided in the kit and immediately pour it on the blot. Make sure that the blot is completely covered with the solution and incubate it at room temperature for 5 min.

16. After 5 min, pick up the blot with forceps and drain the excessive solution by placing it on a tissue paper. Place the blot on a plastic wrap that is cut appropriately.

17. Wrap the membrane in plastic and place it in an X-ray film cassette.

18. Place a film on the blot in a dark room and expose for 10–60 s. Develop the blot in an X-ray processor. An example of the results produced is shown in **Fig. 2**.

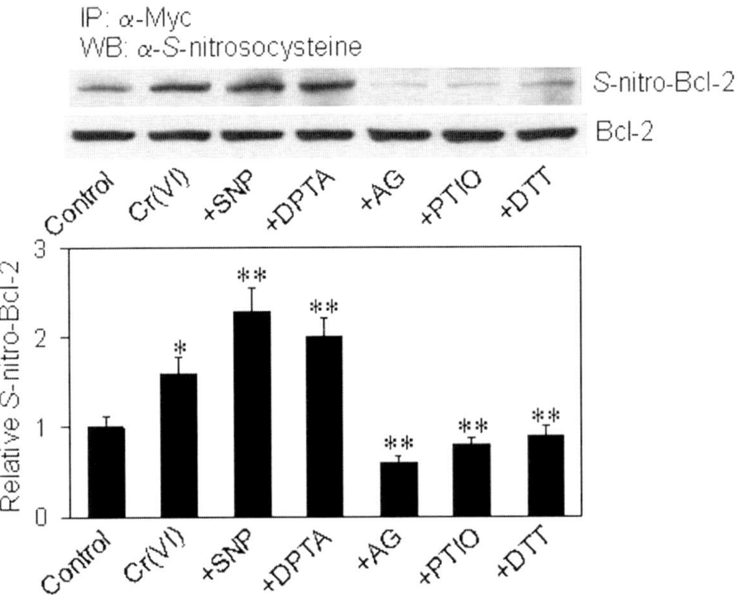

Fig. 2. *S*-nitrosylation of Bcl-2 analyzed by Western blot analysis. Subconfluent monolayers of H-460 cells overexpressing Bcl-2-myc were pretreated with SNP (500 µg/mL), DPTA NONOate (400 µM), AG (300 µM) and PTIO (300 µM), or DTT (10 mM) for 1 h. The cells were then treated with Cr(VI) (20 µM) for 3 h and cell lysates were prepared for immunoprecipitation using anti-myc antibody. The resulting immune complexes were analyzed for *S*-nitrosocysteine by Western blotting. Densitometry was performed to determine the relative *S*-nitrosocysteine levels after reprobing the membranes with anti-Bcl-2 antibody. Data are mean ±SD ($n = 4$). *Asterisk, $p < 0.05$ vs. nontreated control. Double asterisk, $p < 0.05$ vs.* Cr(VI)-treated controls (reproduced from (*6*) with permission from the *Journal of Biological Chemistry*).

3.6. Stripping and Reprobing Blots for Total Bcl-2

1. After a satisfactory blot for *S*-nitroso-Bcl-2 protein is obtained, strip the membrane and reprobe with an antibody that recognizes total Bcl-2. This provides a loading control and confirms that the observed changes in the level of *S*-nitroso-Bcl-2 are accurate.

2. Warm stripping buffer (10–15 mL per blot) to 55°C. Incubate the blot in the stripping buffer for 30 min on a rocking platform (*see* **Note 25**).

3. Wash with TBST (three times for 10 min each).

4. Block again in blocking buffer for 1 h at room temperature or overnight at 4°C on a rocker.

5. Discard the blocking buffer and rinse the membrane twice with TBST.

6. Incubate the membrane with anti-Bcl-2 antibody at 1:200 dilution in 5% nonfat dry milk in TBST for 1 h at room temperature or overnight at 4°C on a rocker.

7. Remove the primary antibody and wash the membrane three times for 5 min each with TBST.

8. Incubate the membrane with anti-mouse secondary antibody at 1:5,000 dilution in 5% nonfat dry milk in TBST for 1 h at room temperature or overnight at 4°C on a rocker.

9. Discard the secondary antibody and wash the membrane three times for 15 min each with TBST.

10. Finally, develop the blot as mentioned above.

11. Blots were quantified by imaging densitometry using UN-SCAN-IT® automated digitizing software (Silk Scientific). Mean densitometry data from independent experiments were normalized to the control. An example of the result obtained is shown in **Fig. 2**.

3.7. Fluorometric Measurement of S-Nitrosylation

1. After IP (**Subheading 3.3, step 5**), rinse the immunopellets four times with 500 μL of IP lysis buffer and twice with 500 μL of PBS.

    (a) First, centrifuge the immune complexes at $1,000 \times g$ for 1 min at 4°C

    (b) Aspirate the supernatant gently using a small syringe needle connected to the vacuum

    (c) Add 500 μL of IP lysis buffer and mix gently by inverting 5–6 times

    (d) Repeat **steps 1a** and **1b**

2. After the final step, aspirate out the supernatant completely and resuspend the immunopellets in 500 μL of PBS.

3. Add $HgCl_2$ (200 μM) and DAN (200 μM) and incubate in dark at room temperature for 0.5 h (*see* **Note 26**).

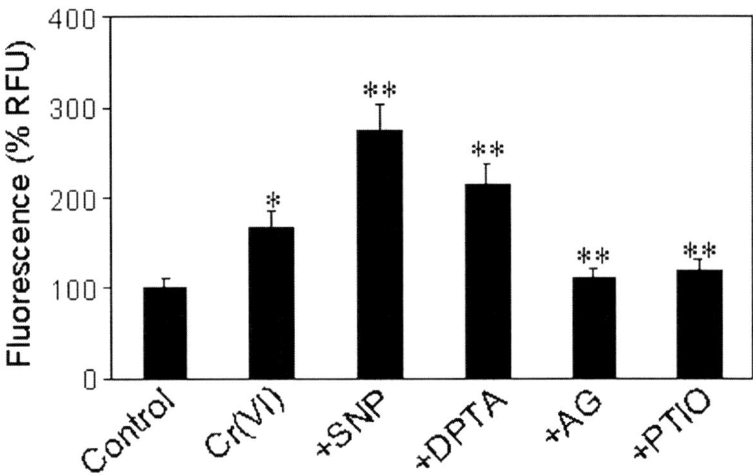

Fig. 3. *S*-nitrosylation of Bcl-2 analyzed by spectrofluorometric analysis. Subconfluent monolayers of H-460 cells overexpressing Bcl-2-myc were pretreated with SNP (500 µg/mL), DPTA NONOate (400 µM), AG (300 µM), or PTIO (300 µM) for 1 h. The cells were then treated with Cr(VI) (20 µM) for 3 h and cell lysates were prepared for immunoprecipitation using anti-myc antibody. The resulting immune complexes were incubated with 200 µM HgCl$_2$ and 200 µM DAN. NO released from nitrosylated Bcl-2 was quantified at 375/450 nm. Data are mean ±SD ($n = 4$). *Asterisk, $p < 0.05$ vs. nontreated control. Double asterisk, $p < 0.05$ vs. Cr(VI)-treated controls (reproduced from (6) with permission from the Journal of Biological Chemistry).*

4. Add 1 M NaOH and vortex each sample briefly.

5. Quantify the fluorescent triazole generated from the reaction between NO released from nitrosylated protein and DAN using a spectrofluorometer at the excitation and emission wavelengths of 375 and 450 nm, respectively. An example of the result obtained is shown in **Fig.** 3.

4. Notes

1. IP lysis buffer should be freshly prepared before use. Tris–HCl, NaCl, glycerol, and NP-40 can be mixed into a solution and stored at 4°C. However, PMSF and protease inhibitor should be added just before using the buffer for each experiment.

2. The standard referred to as "water" in this text is deionized water with a resistivity of 18 MΩ cm.

3. Running buffer (10×) and transfer buffer (10×) can also be bought commercially from Invitrogen.

4. We added LAC to the medium before treatment so as to prevent proteasomal degradation of the protein of interest (Bcl-2), as LAC is a well-established proteasome inhibitor. Since Bcl-2 undergoes proteasomal degradation, we added LAC so as to clearly detect nitrosylation of the protein.

5. Please check under the microscope if the cells have lysed completely before scraping. If necessary, add more lysis buffer and incubate for a longer time.

6. For each sample, 200 μL of the WR is required.

7. Always prepare extra volume of the WR. The WR can be stored in a closed container at room temperature for several days.

8. Before aspirating the beads, slightly cut the pipette tip or use a large orifice pipette tip so as to load equal volume of the beads in all microcentrifuge tubes. Make sure the beads are always kept on ice and vortex it every time before adding into a new tube.

9. The speed and the time can change depending on the centrifuge used. Adjust speed and time such that the beads settle at the bottom of the tube but do not form solid precipitate that is difficult to resuspend in the solution.

10. Adjust the vacuum pressure such that the beads are not aspirated out. Use the smallest syringe needles.

11. Add the cell lysate immediately so as to avoid drying of the beads. Always use the entire volume of cell lysates for better results.

12. Keep the samples on 180° rocker to allow proper mixing and binding of the protein to the beads.

13. Aspirate out the supernatant completely without affecting the beads. This is especially important after the first and the last step so as to get rid of the nonspecific proteins before analyzing the immune complexes.

14. Add TEMED just before pouring the gel as it is the main polymerizing agent. Add TEMED in the tube, shake it well, and immediately pour the gel.

15. Before inserting the comb, try to get rid of the bubbles with a pipette tip. After inserting the comb, there should not be any bubbles at the bottom of the well as it will hinder the protein from running on to the gel.

16. Running buffer can be reused 5–6 times.

17. For some samples, 1× or 1.5× Laemmli buffer works better. Therefore, optimize and dilute it accordingly in IP lysis buffer from the 6× Laemmli buffer stock.

18. Load the sample carefully avoiding beads. Press the tip gently against the bottom of the well and release it slowly to aspirate the solution only without the beads. If the wells do not retain the whole sample, load the volume that will comfortably fill the well without spilling over to next well. Then run the gel for about 5 min so that the loaded sample runs out of the well but not yet into the separating gel. Turn off the power supply, load the remaining sample, and run it again.

19. Gels can be run at 80–120 V at room temperature or at 40–50 V overnight. To get better protein resolution, run the samples at 40 V till they reach the separating gel and then increase it to 90 V.

20. Transfer buffer can be reused 2–3 times.

21. The transferred protein marker is a good indication of the orientation. But to avoid confusion it is better to cut the bottom right-hand corner of both the gel and the membrane together to minimize error.

22. Prepare the assembly such that the nitrocellulose membrane is always between the gel and the anode. The orientation is very important so that the gel is not lost into the buffer and is properly transferred on to the membrane.

23. Gels can also be transferred overnight at 10–15 V.

24. Primary antibodies diluted in 5% nonfat milk/TBST solution can be stored at –20°C for at least a month and reused a couple of times depending on their sensitivity.

25. Maintain the temperature while incubating the membrane with the buffer to ensure proper stripping. Do not leave it in the buffer for a long time otherwise the protein will be lost completely.

26. $HgCl_2$ facilitates the release of NO from the nitrosylated site, which then forms a fluorescent triazole with DAN. Therefore, first add $HgCl_2$ into the immune complex solution followed by the addition of DAN.

Acknowledgment

This work was supported by the NIH grant R01 HL763401.

References

1. Bredt, D. S., and Snyder, S. H. (1994). Nitric oxide: a physiologic messenger molecule. *Annu. Rev. Biochem.* **63**, 175–95.

2. Ignarro, L. J. (1991). Signal transduction mechanisms involving nitric oxide. *Biochem. Pharmacol.* **41**, 485–90.

3. Nathan, C., and Xie, Q. W. (1994). Nitric oxide synthases: roles, tolls, and controls. *Cell* **78**, 915–8.

4. Stamler, J. S., Simon, D. I., Osborne, J. A., Mullins, M. E., Jaraki, O., Michel, T., Singel, D. J., and Loscalzo, J. (1992). S-nitrosylation of proteins with nitric oxide: synthesis and characterization of biologically active compounds. *Proc. Natl Acad. Sci. USA* **89**, 444–8.

5. Stamler, J. S., Lamas, S., and Fang, F. C. (2001). Nitrosylation. the prototypic redox-based signaling mechanism. *Cell* **106**, 675–83.

6. Azad, N., Vallyathan, V., Wang, L., Tant-ishaiyakul, V., Stehlik, C., Leonard, S. S., and Rojanasakul, Y. (2006). S-nitrosylation of Bcl-2 inhibits its ubiquitin-proteasomal degradation. A novel antiapoptotic mechanism that suppresses apoptosis. *J. Biol. Chem.* **281**, 34124–34.

7. Chanvorachote, P., Nimmannit, U., Wang, L., Stehlik, C., Lu, B., Azad, N., and Rojanasakul, Y. (2005). Nitric oxide negatively regulates Fas CD95-induced apoptosis through inhibition of ubiquitin-proteasome-mediated degradation of FLICE inhibitory protein. *J. Biol. Chem.* **280**, 42044–50.

8. Kim, Y. M., Kim, T. H., Chung, H. T., Talanian, R. V., Yin, X. M., and Billiar, T. R. (2000). Nitric oxide prevents tumor necrosis factor alpha-induced rat hepatocyte apoptosis by the interruption of mitochondrial apoptotic signaling through S-nitrosylation of caspase-8. *Hepatology* **32**, 770–8.

9. Kim, Y. M., Talanian, R. V., and Billiar, T. R. (1997). Nitric oxide inhibits apoptosis by preventing increases in caspase-3-like activity via two distinct mechanisms. *J. Biol. Chem.* **272**, 31138–48.

10. Mannick, J. B., Schonhoff, C., Papeta, N., Ghafourifar, P., Szibor, M., Fang, K., and Gaston, B. (2001). S-Nitrosylation of mitochondrial caspases. *J. Cell. Biol.* **154**, 1111–6.

11. Wang, Y., Yun, B. W., Kwon, E., Hong, J. K., Yoon, J., and Loake, G. J. (2006). S-nitrosylation: an emerging redox-based post-translational modification in plants. *J. Exp. Bot.* **57**, 1777–84.

12. Stamler, J. S. (2004) S-nitrosothiols in the blood: roles, amounts, and methods of analysis. *Circ. Res.* **94**, 414–7.

13. Eu, J. P., Liu, L., Zeng, M., and Stamler, J. S. (2000). An apoptotic model for nitrosative stress. *Biochemistry* **39**, 1040–7.

14. McMahon, T. J., Moon, R. E., Luschinger, B. P., Carraway, M. S., Stone, A. E., Stolp, B. W., Gow, A. J., Pawloski, J. R., Watke, P., Singel, D. J., Piantadosi, C. A., and Stamler, J. S. (2002). Nitric oxide in the human respiratory cycle. *Nat. Med.* **8**, 711–7.

15. Stamler, J. S., Jaraki, O., Osborne, J., Simon, D. I., Keaney, J., Vita, J., Singel, D., Valeri, C. R., and Loscalzo, J. (1992). Nitric oxide circulates in mammalian plasma primarily as an S-nitroso adduct of serum albumin. *Proc. Natl Acad. Sci. USA* **89**, 7674–7.

16. Foster, M. W., and Stamler, J. S. (2004). New insights into protein S-nitrosylation. Mitochondria as a model system. *J. Biol. Chem.* **279**, 25891–7.

17. Jaffrey, S. R., Erdjument-Bromage, H., Ferris, C. D., Tempst, P., and Snyder, S. H. (2001). Protein S-nitrosylation: a physiological signal for neuronal nitric oxide. *Nat. Cell Biol.* **3**, 193–7.

Chapter 10

Application of In Vivo EPR for Tissue pO_2 and Redox Measurements

Nadeem Khan and Dipak K. Das

Summary

The technique of electron paramagnetic resonance (EPR) spectroscopy is more than 50 years old, but only recently it has been used for in vivo studies. Its limited application in the past was due to the problem of high nonresonant dielectric loss of the exciting frequency because of high water content in biological samples. However, with the development of spectrometers working at lower frequencies (1,200 MHz and below) during the last 15 years, it is now possible to conduct in vivo measurements on a variety of animals and isolated organs. This is further facilitated by the development of new resonators with high sensitivity and appropriate stability for in vivo applications. It now has become feasible to obtain new insights into the complex aspects of physiology and pathophysiology using in vivo EPR. Among several important applications of this technique, the in vivo tissue pO_2 (partial pressure of oxygen) and redox measurements seem to be the most extensive use of this technique. In this chapter, we describe the procedure for in vivo pO_2 and redox measurements in animal models.

Key words: EPR, ESR, Oximetry, pO_2, Redox, Nitroxide, Heart, In vivo

1. Introduction

The amount of oxygen in tissue plays an important role in many physiological and pathological processes, especially those associated with reactive intermediates (*1–3*). Over the last several years, considerable effort has been invested to develop techniques that can provide noninvasive and reliable measure of tissue oxygen (reported as the concentration $[O_2]$ or the partial pressure pO_2). The methods, such as the polarographic electrode, fluorescence quenching, O_2 binding to myoglobin, chemiluminescence, phosphorescence quenching, and spin label oximetry are useful but

Peter Erhardt and Ambrus Toth (eds.), *Apoptosis*, Methods in Molecular Biology, vol. 559
DOI 10.1007/978-1-60327-017-5_10
© Humana Press, a part of Springer Science + Business Media, LLC 2004, 2009

have significant limitations, especially when used in vivo. Methods which assess perfusion, such as magnetic resonance imaging, do not provide a direct measure of tissue pO_2. These methods have been reviewed recently (*4–7*).

EPR oximetry is a new emerging technique which has the capability to provide repeated, direct, and accurate measurement of tissue oxygen and the redox status of the tissue. The other important applications include the measurement of biophysical parameters such as macromolecular motion, membrane fluidity, viscosity, membrane potential, paramagnetic metal ions, pH, thiols, and detection and identification of free radicals by using the method of spin trapping.

1.1. In vivo EPR Oximetry for pO_2 Measurement

The basis of EPR oximetry is the paramagnetic nature of molecular oxygen, which therefore affects the EPR spectra of other paramagnetic species in its vicinity by altering their relaxation rates. The magnitude of this effect is directly related to the amount of oxygen that is present in the environment of the paramagnetic materials. Therefore, the line width of the EPR spectra of the paramagnetic materials, when injected into tissue, provides direct measurement of tissue pO_2. The placement of the paramagnetic material in the tissue of interest is minimally invasive (it usually requires an insertion via a 25 or 23 gauge needle, or direct implantation during a surgical procedure required for other reasons), but all subsequent measurements are entirely noninvasive (*8, 9*). Two types of oxygen sensitive paramagnetic materials are used: particulates (such as lithium phthalocyanine (LiPc) crystals) and soluble probes (such as nitroxides and trityl radicals). The unpaired electrons in particulates are distributed over complex arrays of atoms and crystalline materials, such as in LiPc (*10, 11*). LiPc is so far the most commonly used probe for pO_2 measurements. These are metabolically inert, and their EPR spectra have good response to the presence of oxygen. Once introduced into the tissue of interest, they allow repeated measurement of pO_2 at the same site for up to years after implantation (*12, 13*). The region that is measured directly is that immediately surrounding the paramagnetic material. If the paramagnetic material is a macroscopic particle (such as LiPc), then it reflects the pO_2 in the tissue that is in contact with the surface of the particle. If the paramagnetic material is a slurry of small particles (such as ink), then it reports the average pO_2 of the sum of the surfaces that are in contact with the individual components of the slurry. Under appropriate conditions, EPR oximetry using particulates can measure pO_2 levels in tissues over a wide range from the extremes of very low (<0.1%) to very high (up to 100%) oxygen levels. Extensive reviews of EPR oximetry with more information on the principles, methodology, and technical aspects are available in recent reports (*14, 15*). An important recent development in

EPR oximetry is the capability to measure pO$_2$ simultaneously from several sites (*16–18*). The High Spatial Resolution Multi-Site (HSR-MS) EPR oximetry technique can be used to provide pO$_2$ estimates from multiple LiPc implants with greatly increased spatial resolution (*16*). This technique has been recently refined to achieve improved oximetric sensitivity by the application of overmodulation during EPR measurements (*18*).

EPR oximetry with particulates has features that often compliment those available with other methods. These include the capability to make repeated noninvasive measurements with high sensitivity and accuracy, and the measurements can be made quickly (in as little as a few minutes) with a high degree of stability and inertness. The existing EPR oximetry techniques already are very useful for small animal studies where the depth of measurements is not an overriding issue. In large animals and potentially in human subjects, this technique seems to be immediately applicable for up to a distance of 10 mm from the surface. For depths greater than 10 mm, approaches such as implantable resonators can be successfully used for tissue pO$_2$ measurements; however, this involves some degree of invasiveness. The clinical use of EPR oximetry seems especially promising and is likely to be used for long-term monitoring of the status and response to treatments of peripheral vascular diseases and also to optimize radiotherapy by enabling it to be modified on the basis of the pO$_2$ measured in the tumor.

EPR oximetry has been used to study the tissue pO$_2$ in a wide range of experimental systems, including heart (*19, 20*), muscle (*21*), brain (*22, 23*), kidney (*24, 25*), liver (*26*), skin (*27*), and tumors (*28, 29*). In order to measure pO$_2$ using EPR oximetry, a detector (also referred to as a resonator) is placed on the surface of the skin above the tissue injected with the paramagnetic material. Through the application of an appropriate combination of an electromagnetic field (excitation frequency near 1,200 MHz) and a magnetic field (around 400 G), the scanning of the magnetic field produces a characteristic resonance signal (*14, 15*). By using an appropriate calibration curve, the line width of the EPR signal provides a sensitive measurement of tissue oxygen (*19–29*).

1.2. Measurement of Redox State of the Isolated Perfused Heart

Nitroxides, molecules with a stabilized unpaired electron on N–O bond, have been widely utilized in biophysical studies. The discovery that the reduction of nitroxides is dependent on the oxygen concentration has raised additional possibilities for measuring oxygen concentrations and related redox metabolism. Recently, EPR has been widely used for noninvasive studies of the pharmacokinetics of nitroxides. This provides an effective approach to understand the fundamental aspects of the metabolism and distribution of nitroxides in vivo.

Nitroxides can act as electron acceptors, forming hydroxylamines and, as electron donors, giving oxoammonium cations. Nitroxides may also react with other free radicals to give radical adducts, or disproportionate under strong acidic conditions to the hydroxylamine and oxoammonium salt. Oxygen-dependent changes in the rate of reduction of nitroxides occur when the oxygen concentration is relatively low (the apparent K_m for oxygen is approximately 1 μM (*30*)). The potential applications of EPR using nitroxides include biophysical and biochemical studies, such as oximetry, analysis of membrane fluidity, and polarity; detection of free radicals; and measurement of redox interactions with antioxidants and oxidants. Some of the potential biological uses of nitroxides have been summarized by Kocherginsky and Swartz (*31*).

2. Materials

2.1. In vivo pO₂ Measurements in Heart

1. Paramagnetic oximetry probe (such as LiPc).
2. 23 gauge needle/plunger.
3. Suitable anesthetic and inhaled oxygen (we use 1.5–1.8% isoflurane with 30% FiO_2 in our experiments).
4. 1.2 GHz (L-band) EPR spectrometer.

2.2. Redox Status Measurement in Isolated Perfused Heart

1. Nitroxides such as TEMPO (2,2,5,5-tetramethyl-4-piperidine-1oxyl; Sigma-Aldrich) or PCA (2,2,5,5-tetramethyl-3-carboxylpyrrolidine-*N*-oxyl; Sigma-Aldrich) could be used for this purpose.
2. 1.2 GHz (L-band) EPR spectrometer.
3. Sodium pentobarbital 80 mg/kg, i.p. (or other appropriate anesthetic).
4. Anticoagulant (heparin sodium, 500 IU/kg, i.v.).
5. Langendorff isolated heart perfusion apparatus.
6. Modified Krebs–Henseleit bicarbonate buffer (KHB): 118 mM NaCl, 4.7 mM KCl, 1.7 mM $CaCl_2$, 25 mM Na_2HPO_4, 0.36 mM KH_2PO_4, 1.2 mM $MgSO_4$ and 10 mM glucose.

3. Methods

3.1. In vivo pO₂ Measurements in Heart

1. Paramagnetic oximetry probe (such as LiPc) should be sterilized (autoclaved) prior to implantation in the tissue.

2. A response of the EPR signals of the LiPc crystals to different concentrations of perfused oxygen should be measured, either using a 1.2 GHz (or a ~9 GHz) EPR spectrometer to obtain the calibration plot. The dependence of the line width of EPR spectrum on pO$_2$ is determined by measuring the line width (LW) as a function of pO$_2$ in the perfused gas. The LW is defined as the difference in the magnetic field between the maximum and the minimum of the first-derivative EPR signal. The resulting calibration is fitted to a first-order regression equation, which then is used to convert the values of LW measured in the heart into appropriate values of pO$_2$.

3. LiPc crystals (40–60 µg) should be injected into the myocardial tissue of interest using a 23 gauge needle/plunger. This will involve minor surgery to open the chest and access the heart for LiPc implantation. This procedure also can be done during other surgeries, for example during the implantation of a pneumatic occluder for ischemia-reperfusion studies. The animals should be allowed to recover for 5–7 days. For studies in isolated perfused heart, the LiPc crystals can be injected after the heart is excised; the experiment set up is described by Grinberg et al. (*19, 20*).

4. For pO$_2$ measurements, the animals should be anesthetized using a suitable anesthetic and inhaled oxygen and gently placed in between the EPR magnets (we use 1.5–1.8% isoflurane with 30% FiO$_2$ in our experiments). This anesthetic maintains reasonably good SpO$_2$, heart beat, and blood pressure. The body temperature of the animals should be maintained at 37°C using a warm water pad and warm air blower.

5. The EPR resonator should be positioned on the skin above the myocardial tissue with the LiPc implant and the spectrometer tuned. The spectrometer parameters should be optimized and EPR data collected as desired.

6. The EPR spectra should be averaged to obtain better signal to noise ratio which will provide accurate estimates of tissue pO$_2$.

7. The fit of the EPR spectrum of the implanted LiPc crystals will provide line width, which is then converted to pO$_2$ using the calibration plot as described above (*see* **Note 1**).

This procedure will allow continuous noninvasive measurement of tissue pO$_2$, and these measurements can be repeated over days or weeks as desired. If the study is acute, a small incision in the chest could be used to gently insert the resonator for its placement directly on the myocardial tissue to achieve better signal to noise ratio of the EPR spectra. Currently, implantable resonators are being developed, which can provide much better signal to noise ratio, and myocardial tissue pO$_2$ can be measured repetitively for several weeks. The initial implantation procedure will be invasive, but the rest of the oximetry measurements will be entirely noninvasive.

3.2. Redox Status Measurement in Isolated Perfused Heart

A 1.2-GHz EPR spectrometer is used to determine the redox status of the isolated perfused heart by evaluating the reduction of a nitroxide during the ischemic challenge. The experiment setup is similar to that reported by Grinberg et al. (*19, 20*), Khan et al. (*32*), and references cited therein.

1. An appropriate nitroxide should be selected and its concentration should be determined for use at 1.2-GHz EPR spectrometer. TEMPO (2,2,5,5-tetramethyl-4-piperidine-1oxyl) or PCA (2,2,5,5-tetramethyl-3-carboxylpyrrolidine-Noxyl) are the most commonly used nitroxide for this purpose. A concentration of 0.2 mM or less is suggested for isolated perfused heart experiments. For direct in vivo measurements, 150 mg/kg of the nitroxide could be used.

2. The rat should be anesthetized (sodium pentobarbital 80 mg/kg, i.p. or other appropriate anesthetic), and an anticoagulant injected (heparin sodium, 500 IU/kg, i.v.).

3. After ensuring sufficient depth of anesthesia, a thoracotomy is performed, and the heart is perfused in the retrograde Langendorff mode at 37°C at a constant perfusion pressure of 100 cm of water (10 kPa) for a 5-min washout period. The perfusion buffer slightly varies from lab to lab. We use a modified Krebs–Henseleit bicarbonate buffer (KHB) as described in **Subheading 2**.

4. The heart is perfused with the KHB buffer containing the nitroxide for 15 min, and global ischemia is performed for 30 min. The reduction of the nitroxide by the myocardial tissue during ischemia will result in a decrease in EPR signal intensity with time.

5. The change in signal intensity with time should be plotted, and the data should be fitted depending on the nature of the reduction (first or second order exponential decay) to determine the reduction rate of the nitroxide. This provides the redox status of the heart which can be compared with other experimental conditions such as treatment of the heart with other drugs, etc. Details of the procedure can be obtained from a recent publication by Das et al. (*33*) (*see* **Note 1**).

The oxygen-dependent metabolism of nitroxides can be used with NMR to provide images that reflect these processes (*34*). In vivo EPR spectroscopy of nitroxides also provides a noninvasive method to measure the presence of reactive free radicals by their effects on the concentration of the nitroxides (*35–37*).

4. Notes

1. For tissue pO$_2$ measurements, the EPR spectrometer parameters especially the microwave power, modulation amplitude, and time constant should be carefully selected to avoid any EPR signal distortion. The power saturation should be checked by recording the EPR spectra at different microwave powers and then by plotting the EPR signal intensity vs. the square root of power. The EPR signal intensity should grow as the square root of the microwave power; however, at saturation the EPR signal intensity will become weaker (saturate). One can measure either the signal intensity or the line width to check power saturation (at saturation, the line width will broaden) and then select the microwave power which does not lead to saturation. To avoid line broadening due to field modulation, the "rule of thumb" is to keep the modulation amplitude at least half of the EPR line width. The time constant filters out the noise but, on the other hand, a high time constant will distort the signal. Generally, it is best to keep the time constant nearly ten times less than the scan time used to acquire the EPR signal. The oximetry probe can be requested from Professor Harold M. Swartz, Director, EPR Center (http://www.dartmouth.edu/~eprctr/), Dartmouth Medical School, Hanover, NH 03755, USA. The readers are encouraged to contact the authors for any query or for a potential collaboration to make these measurements.

Acknowledgements

NIH grants CA118069 and CA120919.

References

1. Gutteridge JM. (1993). Free radicals in disease processes: a compilation of cause and consequence. *Free Radical Res Commun* **19**, 141–158.

2. Ceconi C, Boraso A, Cargnoni A and Ferrari R. (2003). Oxidative stress in cardiovascular disease: myth or fact? *Arch Biochem Biophys* **420**, 217–221.

3. Warnholtz A, Wendt M, August M and Munzel T. (2004). Clinical aspects of reactive oxygen and nitrogen species. *Biochem Soc Symp* **71**, 121–133.

4. Nozue M, Lee I, Yuan F, Teicher BA, Brizel DM, Dewhirst MW, Milross CG, Milas L, Song CW, Thomas CD, Guichard M, Evans SM, Koch CJ, Lord EM, Jain RK and Suit HD. (1997). Interlaboratory variation in oxygen tension measurement by Eppendorf "Histograph" and comparison with hypoxic marker. *J Surg Oncol* **66**, 30–38.

5. Hunjan S, Zhao D, Constantinescu A, Hahn EW, Antich PP and Mason RP. (2001). Tumor oximetry: demonstration of an enhanced dynamic mapping procedure using fluorine-19

echo planar magnetic resonance imaging in the Dunning prostate R3327-AT1 rat tumor. *Int J Radiat Oncol Biol Phys* **49**, 1097–1108.

6. Young WK, Vojnovic B and Wardman P. (1996). Measurement of oxygen tension in tumors by time-resolved fluorescence. *Br J Cancer Suppl* **27**, S256–S259.

7. Swartz HM, Dunn JF (eds). (2002). Measurements of Oxygen in Tissues: Overview and Perspectives on Methods to Make the Measurements. *Oxygen Transport to Tissue XXII*. Pabst Science, Lengerich.

8. Dunn JF and Swartz HM. (2003). In vivo electron paramagnetic resonance oximetry with particulate materials. *Methods* **30**, 159–166.

9. Swartz HM. (2003). The measurement of oxygen in vivo using EPR techniques. *In Vivo EPR (ESR): Theory and Applications*, LJ Berliner (Ed.). Plenum, New York, NY.

10. Liu KJ, Gast P, Moussavi M, Norby SW, Vahidi N, Walczak T, Wu M and Swartz HM. (1993). Lithium phthalocyanine: a probe for electron paramagnetic resonance oximetry in viable biological systems. *Proc Natl Acad Sci USA* **90**, 5438–5442.

11. Afeworki M, Miller NR, Devasahayam N, Cook J, Mitchell JB, Subramanian S and Krishna MC. (1998). Preparation and EPR studies of lithium phthalocyanine radical as an oxymetric probe. *Free Radical Biol Med* **25**, 72–78.

12. Khan N, Williams BB and Swartz HM. (2006). Clinical applications of in vivo EPR: rationale and initial results. *Appl Magn Reson* **30**, 185–199.

13. Swartz HM, Khan N, Buckey J, Comi R, Gould L, Grinberg O, Hartford A, Hopf H, Hou H, Hug E, Iwasaki A, Lesniewski P, Salikhov I and Walczak T. (2004). Clinical applications of EPR: overview and perspectives. *NMR Biomed* **17**, 335–351.

14. Swartz HM. (2002). The measurement of oxygen in vivo using EPR techniques. Biological Magnetic *Resonance – Volume 20*: In vivo EPR (ESR): Theory and Applications, LJ Berliner (Ed.). Plenum, New York, NY.

15. Swartz HM and Halpern H. (1998). EPR studies of living animals and related model systems (in vivo EPR). *Biol Magn Reson* **14**, 367–404.

16. Grinberg OY, Smirnov AI and Swartz HM. (2001). High spatial resolution multi-site EPR oximetry. The use of convolution-based fitting method. *J Magn Reson* **152**, 247–258.

17. Smirnov AI, Norby SW, Clarkson RB, Walczak T and Swartz HM. (1993). Simul-taneous multi-site EPR spectroscopy *in vivo*. *Magn Reson Med* **30**, 213–220.

18. Williams BB, Hou H, Grinberg O, Demidenko E and Swartz H. (2007). High spatial resolution multi-site EPR oximetry of transient focal cerebral ischemia in the rat. *Antioxid Redox Signal* **9**, 1691–1698.

19. Grinberg OY, Friedman BJ and Swartz HM. (1997). Intramyocardial pO_2 measured by EPR. *Adv Exp Med Biol* **428**, 261–268.

20. Grinberg OY, Grinberg SA, Friedman BJ and Swartz HM. (1997). Myocardial oxygen tension and capillary density in the isolated perfused rat heart during pharmacological intervention. *Adv Exp Med Biol* **411**, 171–181.

21. Helisch A, Wagner S, Khan N, Drinane M, Wolfram S, Heil M, Ziegelhoeffer T, Brandt U, Pearlman JD, Swartz HM and Schaper W. (2006). Impact of mouse strain differences in innate hind limb collateral vasculature. *Arterioscler Thromb Vasc Biol* **26**, 520–526.

22. Hou H, Grinberg O, Grinberg S and Swartz H. (2005). Cerebral tissue oxygenation in reversible focal ischemia in rats: multi-site EPR oximetry measurements. *Physiol Meas* **26**, 131–141.

23. Swartz HM, Taie S, Miyake M, Grinberg OY, Hou H, el-Kadi H and Dunn JF. (2003). The effects of anesthesia on cerebral tissue oxygen tension: use of EPR oximetry to make repeated measurements. *Adv Exp Med Biol* **530**, 569–575.

24. James PE, Bacic G, Grinberg OY, Goda F, Dunn JF, Jackson SK and Swartz HM. (1996). Endotoxin-induced changes in intrarenal pO_2, measured by in vivo electron paramagnetic resonance oximetry and magnetic resonance imaging. *Free Radical Biol Med* **21**, 25–34.

25. James PE, Goda F, Grinberg OY, Szybinski KG and Swartz HM. (1997). Intrarenal pO_2 measured by EPR oximetry and the effects of bacterial endotoxin. *Adv Exp Med Biol* **411**, 557–568.

26. Towner RA, Sturgeon SA, Khan N, Hou H and Swartz HM. (2002). In vivo assessment of nodularin-induced hepatotoxicity in the rat using magnetic resonance techniques (MRI, MRS and EPR oximetry). *Chem-Biol Interact* **139**, 231–250.

27. Abramovic Z, Sentjurc M, Kristl J, Khan N, Hou H and Swartz HM. (2006). Influence of different anesthetics on skin oxygenation studied by electron paramagnetic resonance in vivo. *Skin Pharmacol Physiol* **20**, 77–84.

28. Hou H, Khan N, O'Hara JA, Grinberg OY, Dunn JF, Abajian MA, Wilmot CM, Demidenko E, Lu S, Steffen RP and Swartz HM. (2005). Increased oxygenation of intracranial tumors by

efaproxyn (efaproxiral), an allosteric hemoglobin modifier: In vivo EPR oximetry study. *Int J Radiat Oncol Biol Phys* **61**, 1503–1509.

29. O'Hara JA, Khan N, Hou H, Wilmo CM, Demidenko E, Dunn JF and Swartz HM. (2004). Comparison of EPR oximetry and Eppendorf polarographic electrode assessments of rat brain PtO$_2$. *Physiol Meas* **25**, 1413–1423.

30. Chen K, Glockner JF, Morse PD 2nd and Swartz HM. (1989). Effects of oxygen on the metabolism of nitroxide spin labels in cells. *Biochemistry* **28**, 2496–2501.

31. Kocherginsky N and Swartz HM. (1995). Nitroxide Spin Labels, Reactions in Biology and Chemistry. CRC, Boca Raton, FL.

32. Khan M, Kutala VK, Vikram DS, Wisel S, Chacko SM, Kuppusamy ML, Mohan IK, Zweier JL, Kwiatkowski P and Kuppusamy P. (2007). Skeletal myoblasts transplanted in the ischemic myocardium enhance in situ oxygenation and recovery of contractile function. *Am J Physiol Heart Circ Physiol* **293**, H2129–H2139.

33. Das S, Khan N, Mukherjee S, Bagchi D, Gurusamy N, Swartz H, Das DK. (2008). Redox regulation of resveratrol-mediated switching of death signal into survival signal. *Free Radical Biol Med* **44**, 82–90.

34. Swartz HM, Chen K, Pals M, Sentjurc M and Morse PD II. (1986). Hypoxia-sensitive NMR contrast agents. *Mag Reson Med* **3**, 169–174.

35. Miura Y, Utsumi H and Hamada A. (1992). Effects of inspired oxygen concentration on in vivo redox reaction of nitroxide radicals in whole mice. *Biochem Biophys Res Commun* **182**, 1108–1114.

36. Gomi F, Utsumi H, Hamada A and Matsuo M. (1993). Aging retards spin clearance from mouse brain and food restriction prevents its age-dependent retardation. *Life Sci* **52**, 2027–2033.

37. Utsumi H, Takeshita K, Miura Y, Masuda S and Hamada A. (1993). In vivo EPR measurement of radical reaction in whole mice: Influence of inspired oxygen and ischemia-reperfusion injury on nitroxide reduction. *Free Radical Res Commun* **19**, S219–S225.

Part IV

Analysis of the Function of Major Regulators of Apoptosis/Cell Survival

Chapter 11

Assays to Measure p53-Dependent and -Independent Apoptosis

Darren C. Phillips, Sean P. Garrison, John R. Jeffers, and Gerard P. Zambetti

Summary

Paramount to the maintenance of normal tissue homeostasis is the induction of programmed cell death, otherwise known as apoptosis. Several disease states, including cancer, are characterized by an inability to remove unwanted cells due to a failure to commit to apoptosis. What is more, apoptosis is the central functional response behind many agents utilized in the treatment of cancer. Many of these antitumorigenic agents rely on the activation of the tumor suppressor p53. As the physiological "guardian of the genome," p53's normal function is to sense stressed or damaged cells and arrest proliferation, allowing time for cellular repair. However, if the damage is excessive, cells are removed prior to the onset of malignancy through apoptosis. Current chemotherapeutic strategies manipulate this property by damaging cells and turning on p53's transcriptional function, which consequently upregulates the expression of proapoptotic proteins such as Puma. We have also demonstrated that Puma is capable of inducing apoptosis independent of p53. In this regard, defects in the apoptotic machinery or in p53 function itself lead to a resistant phenotype that in cancer results in chemotherapeutic failure, and more often than not, poor prognosis. This chapter describes protocols for the determination of p53-dependent and -independent apoptosis utilizing primary cells from genetically altered mice.

Key words: Apoptosis, p53, Puma, Bcl-2, Bone marrow-derived myeloid progenitors, Thymocytes, Knockout mice

1. Introduction

The ability of p53 to dictate and modulate cell growth, cell cycle progression, apoptosis and senescence, as well as respond to DNA damage, underscores p53's tumor suppressor function *(1)*. Armed

Peter Erhardt and Ambrus Toth (eds.), *Apoptosis*, Methods in Molecular Biology, vol. 559
DOI 10.1007/978-1-60327-017-5_11
© Humana Press, a part of Springer Science + Business Media, LLC 2004, 2009

with so many functional weapons, p53 is therefore capable of preventing the abnormal growth of stressed cells or cells with mutated DNA that may otherwise result in tumorigenesis. Dysregulation of any number of these cellular processes as a result of inhibition of normal p53 function through deletion, mutation, or the interaction with other proteins (e.g., MDM2, or viral proteins), eventually results in cellular transformation and the development of malignancy *(1–3)*. Mice null for p53 are viable, but succumb to a tumor phenotype that is also observed upon loss of a single p53 allele *(4, 5)*. Further, loss of p53 cooperates with other tumor suppressors or oncogenes to exacerbate the tumor incidence and type *(6–8)*. Given the tumor prone phenotype of mice defective in p53 function, it is of no surprise that p53 is one of the most frequently mutated genes in human cancers *(9, 10)*.

At the heart of the cellular responses to p53 is its ability to transcriptionally activate target genes. Enhanced expression of the p53 responsive genes, such as *p21$^{cip1/waf1}$*, *GADD45*, and *14-3-3*, results in growth arrest, a quiescent-like state that allows for the repair of damaged DNA prior to replication in mitosis. However, should cellular damage be too great or the arrest response compromised, p53, through the transcription of pro-apoptotic genes, initiates a sequence of events that eventually leads to deletion of the cell through programmed cell death. The p53-inducible genes that are responsible for the ensuing apoptotic response can be subdivided into two classes *(11)*. The first group belongs to the "extrinsic" pathway that is triggered by the activation of "death receptors" by their corresponding ligands. These include the receptors CD95 and DR4/DR5, and their respective ligands CD95L and TRAIL. The second group concerns components of the "intrinsic" pathway and numerous proapoptotic members of the Bcl-2 family including Bax, Noxa, and *p53-u*pregulated *mod*ulator of *a*poptosis (Puma) *(11)*. While the precise mechanism of how the proapoptotic Bcl-2 family members interact with antipoptotic Bcl-2 family members is contentious and several models have been proposed, Bax and Bak oligomerization are required to initiate the release of mitochondrial factors into the cytosol that propagate the apoptotic signal from the mitochondria *(12, 13)*. Upregulation of Puma expression is essential for p53-dependent apoptosis that is mediated by numerous chemotherapeutic agents or oncogenic stress. However, Puma is also required for apoptosis induced in a p53-independent fashion *(14, 15)*.

Many current chemotherapeutic strategies utilized clinically in the management of cancer rely on the activation of p53. Given the frequency of p53 mutations that are inherent or acquired, understanding the necessity for p53 in the initiation and regulation of apoptosis is of paramount importance. The vast majority of studies that investigate the mechanisms of tumoricidal action utilize cell lines that have been transformed or were obtained from cancers. Although informative from a biochemical or disease-specific perspective, many of these cell lines contain aberrant signaling networks that may interfere directly or indirectly with p53 function and hence can obscure the true requirement for p53 in the induction of apoptosis. With this in mind, we describe herein the use of primary cells obtained from mice treated in vivo or ex vivo that allows the resolution of p53-dependent or -independent apoptosis without interference from other mutated tumor suppressors or hyperexpressed oncogenes. What is more, these protocols can be applied to experiments where additional known genetic manipulations can be introduced in the mouse itself or through in vitro transfection of isolated cells to further dissect apoptotic signal transduction.

The following protocols describing the isolation of primary mouse cells from either the thymus or the bone marrow can be applied for immediate analysis of programmed cell death in vivo or for subsequent ex vivo manipulation prior to the determination of apoptosis in vitro.

2. Materials

2.1. Thymocyte/Bone Marrow-Derived Myeloid Progenitor Isolation and Culture

1. Bone marrow culture media: RPMI 1640 (Gibco) supplemented with 15% fetal bovine serum (FBS; Gibco), 1% penicillin/streptomycin (P/S; Gibco), 10 mM glutamine (BioWhittaker), 10 ng/mL mouse recombinant Stem Cell Factor (SCF; R&D), 10 ng/mL mouse recombinant IL-6 (R&D), and 40 U/mL mouse recombinant IL-3 (commercially available from R&D). Filter, store at 4°C, and warm before use. Use under aseptic conditions.

2. Cytokine withdrawal media: RPMI 1640 supplemented with 15% FBS (Gibco), 1% P/S (Gibco), 10 mM L-glutamine (Gibco). Store at 4°C and warm before use. Store under aseptic conditions.

3. RPMI1640: store at 4°C and warm before use. Store under aseptic conditions.

4. Thymocyte media: Dulbecco's Modified Eagle's Medium (DMEM; BioWhittaker) containing 5% FBS (Gibco), 1% L-Glutamine, 1% P/S, and 25 mM HEPES (Gibco). Store at 4°C and warm before use. Store under aseptic conditions.

5. Phosphate buffered saline (PBS): 137 mM NaCl, 2.7 mM KCl, 10 mM Na_2HPO_4, 2 mM KH_2PO_4. Adjust to pH 7.4.

6. Dulbecco's PBS (DPBS): 137 mM NaCl, 2.7 mM KCl, 1.5 mM KH_2PO_4, 8.1 mM $NaHPO_4$, 0.9 mM $CaCl_2$, 0.5 mM $MgCl_2$. Adjust to pH 7.4.

7. Culture flasks: T25, T75, and T225 (Sarstedt, Inc.).

8. 6- and 12-well plates (Corning, Inc.).

9. Red Blood Cell (RBC) Lysis solution: 150 mM NH_4Cl, 10 mM $KHCO_3$, 100 mM Na_2-EDTA in water at pH 7.2. Store at 4°C and warm before use. Store under aseptic conditions.

10. 1% acetic acid in water.

11. 10-mL syringes (Becton Dickinson).

12. 23 and 27.5 G needles (BD).

13. Dissection kit.

14. Foil.

15. Razor blades.

16. Petri dishes (10 cm²; BD).

17. 50-mL and 15-mL conical tubes (BD).

2.2. Flow Cytometry Reagents

1. Annexin-V-binding buffer: 10 mM HEPES, 0.9% NaCl, 2.5 μM $CaCl_2$, 1% BSA (Sigma).

2. Annexin-V-FIUOS (Roche/Boehringer Mannheim). Alternatively, Annexin-V-FITC, Annexin-V-PE, or -APC can all be obtained from BD-Pharmingen.

3. Propidium iodide (PI) (Sigma, *see* **Note 1**). Protective gloves and clothing should be used when handling this agent. PI and other DNA intercalating dyes are potentially mutagenic and may be absorbed through the skin, by inhalation or by ingestion.

4. Hypotonic fluorochrome solutions: 50 μg/mL PI, 0.1% sodium citrate (Fisher), 0.1% Triton X-100 (BioRad) in water.

5. Antibody staining solution: PBS + 5% FBS.

2.3. Antibodies

CD4-FITC (Clone GK1.5, BD); CD8-PE (Clone 53.6.7, BD); B220-PECy7 (Clone RA-6B, BD); TER119 (Clone TER119, BD); Mac-1-FITC (Clone M1/70, BD); F4-80 (Clone BM-8, *e*Biosciences); Gr1 (Clone RB6-8c5, BD), Sca-1 (Ly6E/A; Clone E13-161.7, BD); Lineage-PECy7/APC (BD), ckit (CD117; Clone 2B8, BD).

---

3. Methods

We have used mice genetically deficient in either *p53* or *Puma* for the resolution of p53-dependent and -independent apoptosis, and compared their responses to those obtained from their wild-type (WT) littermate controls (*see* **Notes 2** and **3**). Mice were left untreated or exposed to genotoxic stress. Subsequently, we determined the apoptotic status of cells isolated from the thymus or bone marrow in vivo, or following drug treatment ex vivo.

3.1. Bone Marrow Isolation

1. Euthanize mice by exposure to CO_2 and cervical dislocation or according to institutional policies.

2. Position the mouse on its back and saturate with 70% ethanol.

3. Using sharp dissecting scissors, cut an incision in the fur below the peritoneal membrane and peel back the skin to expose the hind limbs.

4. Remove hind legs from the hip joint with scissors, cut off feet at the ankles, and remove flesh.

5. Separate femur and tibula at knee joint with scissors (total of four bones per mouse). Using fingers, wrap bones with dry Kimwipes® (Kimtech Sciences) to strip away excess flesh (*see* **Note 4**). It is imperative to remove all flesh to minimize contamination of bone marrow cultures. Place bones in a 50-mL tube containing 15 mL of culture media (one tube per mouse).

6. Store on ice until all mice have been processed.

7. Define workspace with aluminum foil.

8. Pour media and bones into a 10-cm² Petri dish.

9. Using tweezers, place bones onto aluminum foil and cut away ends with a razor blade to expose interior marrow shaft. To minimize contamination, use separated areas for bones of different genotypes.

10. Draw 5–10 mL of media into a 10-mL syringe and cap with 23 G needle.

11. Insert bevel of the needle into the bone marrow shaft, and flush bone marrow from each end of the bone with media into the Petri dish. Once bone marrow has been removed, bones will appear clear and can consequently be discarded in the appropriate manner according to institutional protocol for the handling of biohazardous waste. The same media is used to remove the bone marrow from all four bones from the same mouse.

12. Gently draw and expel the bone marrow containing media into a 10-mL syringe/pipette to produce a single cell suspension. Repeat this process several times before transferring the cell suspension to a fresh 15-mL tube (*see* **Note 5**).

13. Store on ice until all samples have been processed prior to culture and expansion of bone marrow-derived myeloid progenitors (*see* **Subheading 3.2**). For immediate flow

cytometric analysis of immunophenotype (*see* **Subheadings 3.3** and **3.6**) or Annexin-V (*see* **Subheading 3.8**) positivity, RBC lysis is not necessary. However, nucleated cell number should be determined through lysis of RBC by diluting bone marrow suspension 1:1 with 1% acetic acid in water (*see* **Notes 6** and **7**). Typically, we harvest approximately 60×10^6 nucleated bone marrow cells from the hind legs of a single mouse.

14. Take care to ensure minimal cross-contamination of samples. Clean isolation equipment with 70% ethanol prior to handling samples of a different genetic or treatment background. Use fresh disposable plastic ware for each sample.

3.2. Culture of Bone Marrow–Derived Myeloid Progenitors

The following procedure should be performed under aseptic conditions:

1. Centrifuge bone marrow cell suspensions at $1,500 \times g$ for 5 min.

2. Resuspend bone marrow cells in 1 mL of "RBC lysis buffer" and incubate at room temperature for 5 min.

3. Increase volume fivefold and centrifuge at $1,500 \times g$ for 5 min at 4°C. Wash bone marrow twice with RPMI and determine cell number.

4. Culture bone marrow cells at 1×10^6/mL and incubate at 37°C in a humidified 95% air, 5% CO_2 atmosphere.

5. After 24 h of culture, transfer cell suspension to fresh culture flasks. Periodically thereafter, change media and flasks of the cells in suspension. Continue to culture until no adherent cells remain in the culture vessel.

6. Passage bone marrow-derived myeloid progenitors, maintaining cell number between 1×10^6 and 2×10^6 per mL. In our hands, these cells double in number every 4 days.

7. Prior to further biochemical or pharmacological manipulation, determine immunophenotype as described in **Subheadings 3.3** and **3.6** (*see* **Notes 6** and **7**).

3.3. Immunophenotype of Myeloid Progenitors by Flow Cytometry

Established homogeneous cultures of bone marrow-derived myeloid progenitors are lineage⁻ Sca-1⁺ cKit⁺. Cultures are negative for F4/80, Gr-1/Ly6G, CD4, CD8, B220, and TER-119. Freshly isolated bone marrow is used as a positive control to establish regions of lineage negativity and lineage-specific populations. The majority of cells analyzed from freshly isolated bone marrow will be lineage positive (*see* **Fig. 1**).

1. As a positive control, a WT mouse should be sacrificed and the bone marrow prepared as described in **Subheading 3.1** without culture.

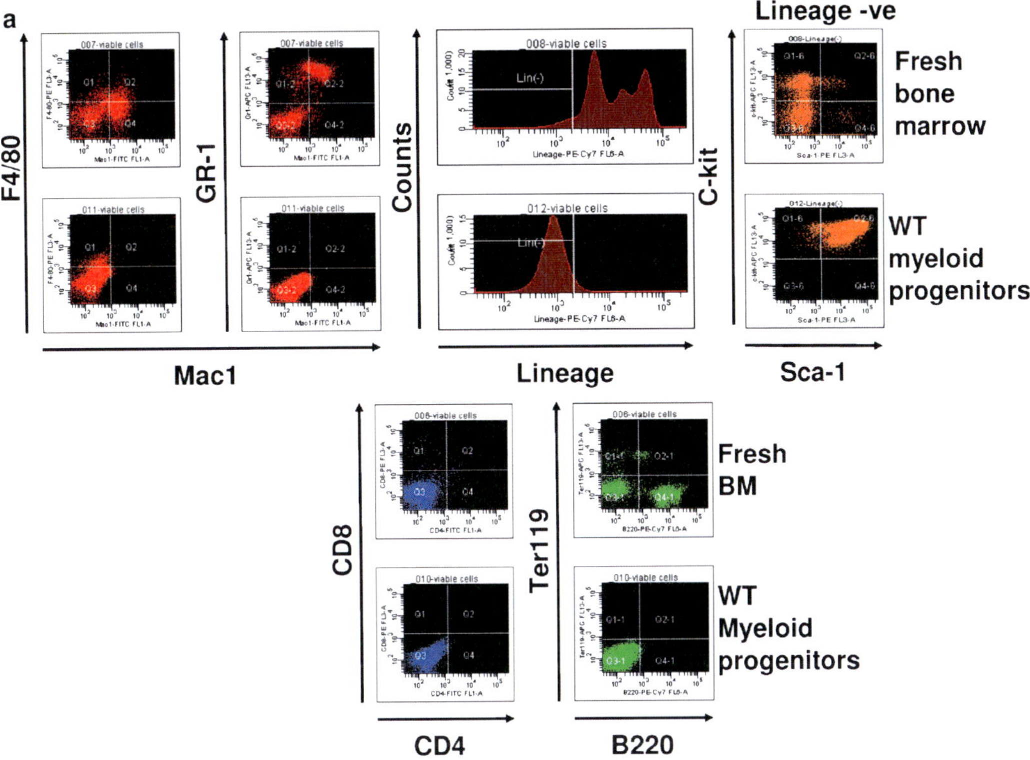

Fig. 1. Immunophenotype of bone marrow derived myeloid progenitors. Bone marrow was isolated from the hind legs of wild-type (WT) mice and mice null for *p53* and *Puma*. Cell suspensions were cultured in RPMI 1640 supplemented with 15% FBS, IL-6, IL-3, SCF, 1% P/S, and 10 mM glutamine. This media supports the culture and expansion of bone marrow-derived myeloid progenitors. Cultures were immunophenotyped by flow cytometry for expression of lineage[+], Mac-1, GR1, F4/80, CD8, CD4, B220, Ter119, and lineage[-] c[-]kit[+] Sca[+] populations and compared to that obtained from freshly isolated bone marrow. Shown are representative flow cytometry histograms obtained from WT bone marrow-derived myeloid progenitors. Identical results were obtained from cultures derived from mice null for *p53* and *Puma*.

2. Determine cell number and resuspend cell suspensions to 2×10^6 per mL.

3. Approximately 10×10^6 of freshly isolated bone marrow cells and 7×10^6 cultured myeloid progenitors are required.

4. *See* **Subheading 3.6** for staining procedure.

3.4. Treatment of Cultured Bone Marrow Cells

Cytokines are capable of efficiently abrogating p53-mediated apoptosis *(16)*. We have observed that apoptosis induced by DNA damage through ionizing radiation (IR) or etoposide is protected by the cytokine content of the bone marrow culture media. To resolve the DNA-damaging effects, bone marrow-derived myeloid progenitors should be cultured in cytokine-free media. The effects of p53-dependent and -independent apoptosis can be deciphered in this scenario. Although p53-deficient bone marrow-derived myeloid progenitors succumb to apoptosis via cytokine withdrawal to a similar extent as the WT cells (p53-independent death), unlike the WT cells, apoptosis is not further exacerbated by the further exposure to IR (*see* **Fig. 2a**). Deletion of Puma protects bone marrow-derived myeloid progenitors from both p53-dependent (IR) and -independent (cytokine withdrawal) apoptosis (*see* **Fig. 2b**). It should be appreciated that the ability of antitumorigenic drugs to induce apoptosis in bone marrow-derived myeloid progenitors in the presence of cytokines is agent specific. In our hands, we have observed that the platinum containing DNA alkylating agent carboplatin induces apoptosis of WT myeloid progenitors in culture despite the presence of cytokines (SCF, IL-3, and IL-6) and is attenuated by *Puma* deletion

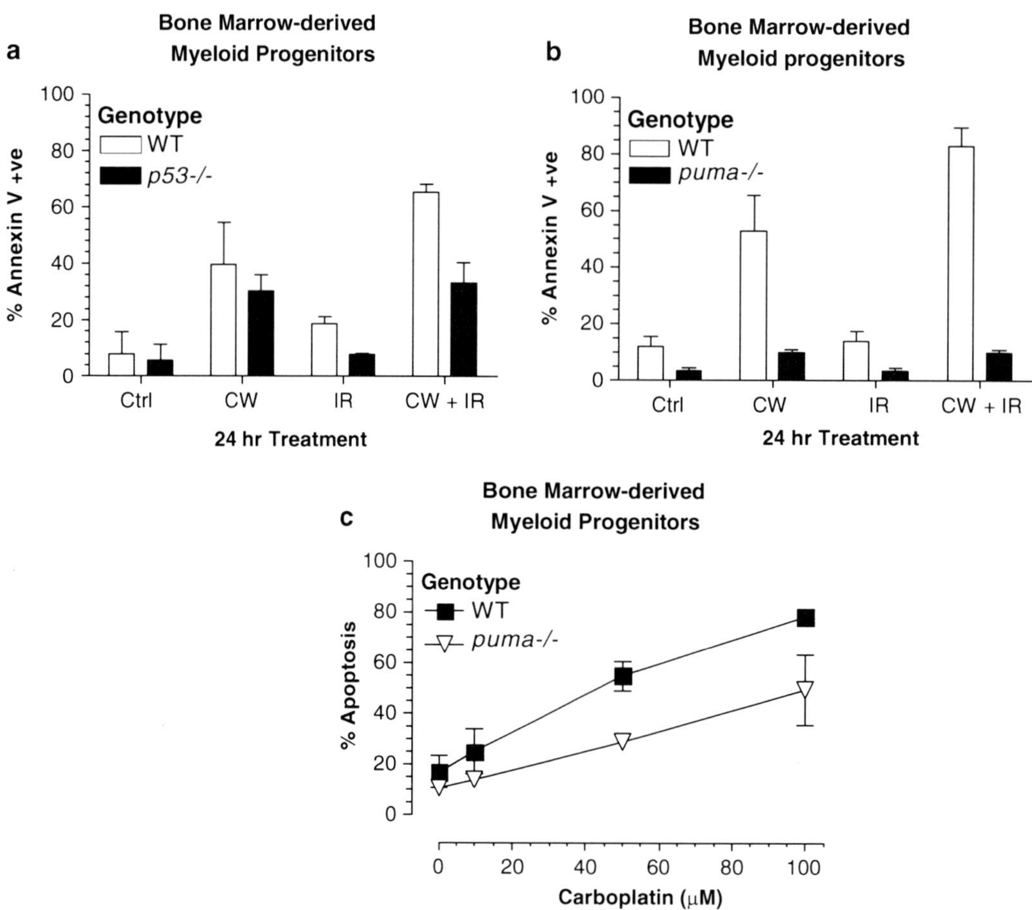

Fig. 2. Resolution of p53-dependent and -independent apoptotic responses in bone marrow-derived myeloid progenitors. Wild-type (WT), *Puma*[−/−] or *p53*[−/−] bone marrow-derived myeloid progenitors (0.25×10^6/mL) were cultured in fully supplemented bone marrow media or in media lacking IL-3, IL-6, and SCF (cytokine withdrawal, CW), and then exposed to ionizing radiation (IR) (5 Gy; **a** and **b**). Alternatively, cells were treated with carboplatin (0–100 μM) in fully supplemented media (**c**). Samples were cultured for 24 h in a humidified 95% air, 5% CO_2 atmosphere at 37°C. Cells were then washed and stained with Annexin-V-APC and PI, and the percentage of Annexin-V positive cells determined by flow cytometry. Data represent the mean ± SEM of at least three independent experiments from three different mice.

(*see* **Fig. 2c**). Additionally, dose and kinetics of apoptosis should also be considered when determining the effects of antitumorigenic agents on this cell type in vitro.

The following procedure should be performed under aseptic conditions:

1. Pre-warm all media.

2. Determine bone marrow cell number and wash twice with RPMI.

3. Resuspend cells at 0.25×10^6 to 1×10^6 per mL in either cytokine-free media RPMI (cytokine withdrawal) or in normal media.

4. Treat with agent of choice and culture at 37°C in a humidified 95% air, 5% CO_2 atmosphere for specified times.

5. Harvest cells and wash twice with PBS containing 1% BSA.

6. Evaluate degree of apoptosis by Annexin-V staining utilizing **Subheading 3.8**.

3.5. Isolation of Mouse Thymocytes

This method describes the isolation of thymocytes from mice with subsequent ex vivo manipulation biochemically or pharmacologically (*see* **Notes 2** and **3**). Alternatively, thymocytes from mice treated in vivo can be harvested utilizing the same technique prior to further processing. We have observed the effects of spontaneous apoptosis in culture ex vivo to be p53-independent but to be somewhat dependent upon Puma expression. Spontaneous apoptosis of WT thymocytes in vitro is enhanced by the additional exposure to IR. However, this potentiation is abrogated by loss of either *p53* or *Puma* (*see* **Fig. 3b**). Both genes are required for apoptosis of thymocytes in vivo following a DNA damaging insult *(14)* or as shown here, following a myelosuppressive regimen of carboplatin plus total body ionizing (TBI) radiation *(17)* (*see* **Fig. 3c**).

1. Euthanize mice by exposure to CO_2 and cervical dislocation.

2. Position the mouse on its back and saturate with 70% ethanol

3. Dissect mouse, locate and isolate thymus *(18)*.

4. Trim thymus of associated connective tissue and store on ice in 2–3 mL of media in 50-mL tubes until all mice have been processed.

5. Mince the thymus using scissors.

6. Remove the plunger from a 5-mL syringe. Place 100 μM sterile nylon mesh/screen above a 50-mL tube containing 3 mL of thymocyte media. Using the syringe plunger, press the tissue though a 100 μM strainer/mesh.

7. Draw and expel strained thymocytes through a 5-mL syringe with 23 G needle to disrupt aggregates.

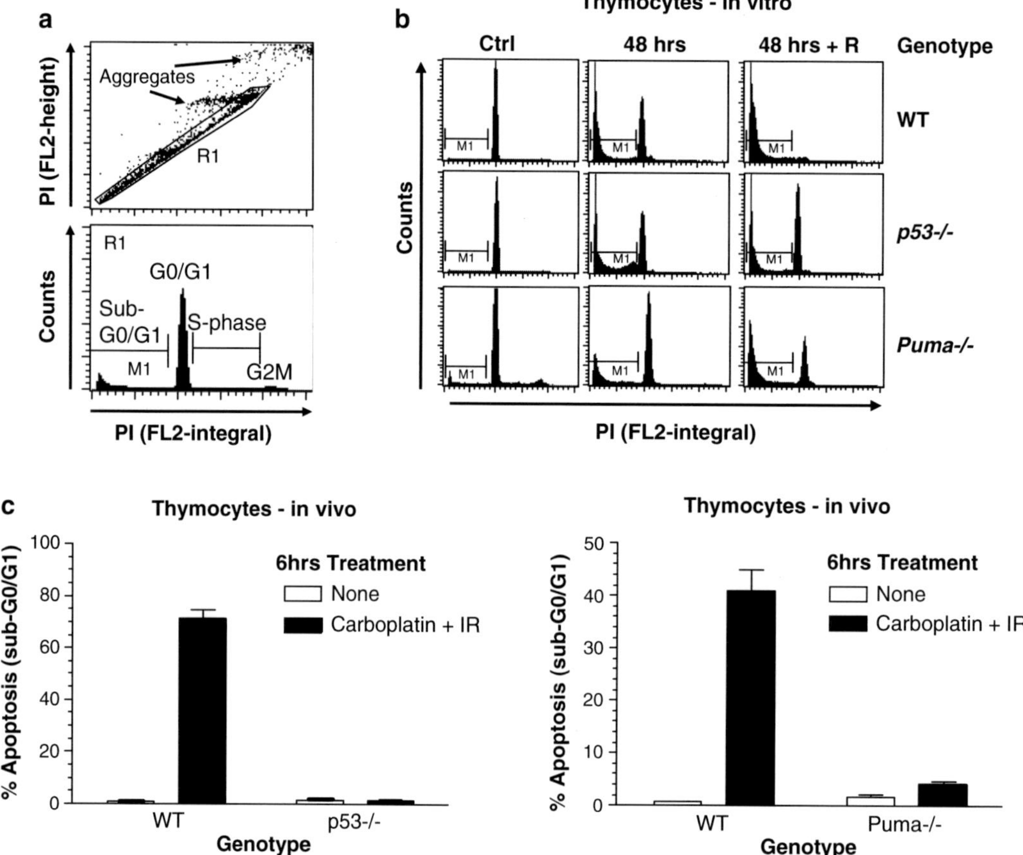

Fig. 3. Differentiation of p53-dependent and -independent apoptotic responses in thymocytes in vitro and in vivo. For evaluation of spontaneous apoptosis in vitro, thymocytes were isolated from untreated wild-type (WT) mice and mice null for *p53* or *Puma*. Untreated thymocytes (0.5×10^6/mL) were cultured for a further 48 h with or without exposure to ionizing radiation (IR; 5 Gy). Cells were then harvested, resuspended in 500 microliter (µL) of hypotonic fluorochrome solution and incubated in the dark at 4°C for a minimum of 30 min. DNA cell cycle profiles were determined by flow cytometry utilizing a dual parameter histogram of FL2 integral vs. FL2-height to gate out aggregates (as shown in (**a**); gate R1). Shown in (**b**) are representative histograms of the gated DNA cell cycle profiles of thymocytes obtained from WT, *p53*$^{-/-}$, or *Puma*$^{-/-}$ mice cultured for 48 h with or without IR. The percentage apoptosis was determined from the sub-G0/G1 fraction placed to the left of the G0/G1 peak and labeled "M1" (**b**). Alternatively, thymocytes were harvested from WT, *p53*$^{-/-}$, or *Puma*$^{-/-}$ mice treated with carboplatin (10 mg/kg) plus TBI (7.5 Gy) for 6 h in vivo. Thymocytes were washed and immediately resuspended in hypotonic fluorochrome solution prior to DNA cell cycle analysis as described (**c**). Data represent the mean ± SEM of 2–4 individual experiments.

8. Allow sample to stand for 5 min to allow aggregates to settle before transferring cell suspension to a fresh 15-mL conical tube.

9. Centrifuge at $1{,}200 \times g$ for 5 min.

10. Resuspend thymocytes in 1 mL RBC lysis media and leave at room temperature for 5 min.

11. Increase volume fivefold with thymocyte media and wash twice with DPBS containing 1% BSA.

12. Resuspend thymocytes to 1×10^6 per mL in thymocyte media.

13. For determination of apoptosis in vivo, proceed to **Subheading 3.7** immediately following isolation (for further biochemical manipulation, *see* **Note 6**). Alternatively, culture with or without chemotherapeutic agents at 37°C in a humidified 95% air, 5% CO_2 atmosphere.

3.6. Immunofluorescence Staining Protocol for Flow Cytometry

1. Prepare 1×10^6 of cultured or freshly isolated bone marrow cells in 100 µL of staining medium containing mouse Fc Block.

2. Vortex-mix and incubate on ice for 30 min.

3. Increase volume by adding 100 µL of staining medium, centrifuge at $1,500 \times g$ for 5 min at 4°C and remove supernatant. Wash cells once in staining solution.

4. Resuspend cells in 50 µL of antibody mix. Vortex-mix and incubate on ice for 30 min. We utilize the following antibody combinations for immunophenotyping of bone marrow-derived myeloid progenitors; CD4-FITC/CD8-PE/ B220-PECy7/TER-119-APC, Mac-1/F4-80/GR1-APC, Sca-1-PE, lineage-PECy7, ckit-APC. These antibodies were used at a dilution of 1:50 (*see* **Notes 7–9**). For evaluation of cell death of lineage negative/positive bone marrow cells by Annexin-V, samples were prestained with lineage-APC.

5. Increase volume by adding 100 µL of staining solution, centrifuge at $1,500 \times g$ for 5 min at 4°C and remove supernatant. Wash cells once in staining solution.

6. If apoptosis of the lineage negative or positive population is to be determined, begin Annexin-V protocol (*see* **Subheading 3.8**). Otherwise, filter suspension through a 40 µM nylon mesh to prevent blocking of flow cytometer and determine immunophenotype of the viable population. Immunophenotyping was performed on a BD LSR-II flow cytometer or similar instrumentation.

3.7. Flow Cytometric DNA Cell Cycle Analysis

Determination of nuclear DNA content reveals information on the cell cycle and apoptosis. The DNA intercalating fluorescent dye PI binds specifically and stoichiometrically to nucleic acids, where fluorescence is enhanced on binding that can be quantified by flow cytometry. However, PI is not cell permeable and hence, prior to DNA staining, cells are lysed to isolate nucleoids *(19)*. Flow cytometric evaluation of nucleoid-associated PI fluorescence permits the quantification of both stages of the cell cycle and apoptosis. New synthesis of DNA marks the entry of the cell into the synthetic phase (S-phase) where the DNA content of the cell increases until it doubles that of G0/G1 cells. The cell is now considered to be in the G2 phase of the cell cycle and DNA synthesis is terminated. Finally the cell enters the mitotic phase (M), to divide into two daughter cells, which revert back to G1 phase for sustained cell division, or G0 phase. PI-stained nucleoids in the S phase contain DNA and hence fluorescence that is intermediate of G0/G1 and G2/M (*see* **Fig. 3a**). DNA fragmentation represents a biochemical and morphological feature of apoptosis *(20, 21)*. Fragmented DNA binds less fluorochrome due to its smaller size producing decreased fluorescence in comparison to aneuploid DNA, appearing to the left of the G0/G1 peak, and is consequently termed sub-G0/G1, subdiploid, or hypoploid. The amount of sub-G0/G1 DNA increases as the degree of fragmentation increases due to an apoptotic insult *(22)*.

1. Harvest 5×10^5 cells and wash twice in PBS containing 1% BSA.

2. Resuspend cell pellet in 250 µL of hypotonic fluorochrome solution, agitate to mix and incubate in the dark at 4°C for a minimum of 30 min and a maximum of 24 h.

3. Analyze the DNA cell cycle profiles by flow cytometry (BD FACScalibur). Doublets are removed by applying the appropriate gating ("R1") on a dual parameter histogram of linear height FL3 fluorescence vs. linear integral FL3 fluorescence (*see* **Fig. 3a**, upper panel).

4. A minimum of 20,000 events per sample are collected on a gated ("R1") histogram of FL3 integral vs. count. In order to optimally resolve sub-G0/G1 DNA content (apoptosis), the G0/G1 peak is positioned at approximately 400 arbitrary units on the *x*-axis of FL3 integral (*see* **Fig. 3a**, lower panel).

5. The percentage apoptosis is determined from the sub-G0/G1 population. A marker (M1) is placed to the left of the G0/G1 peak, spanning the sub-G0/G1 data, but not overlapping the *y*-axis to prevent interference by machine noise (*see* **Fig. 3a**). Effects on the cell cycle profile should be evaluated using appropriate software such as ModFit LT 3.1 (Verity Software House) or Multi-Cycle (Phoenix Flow Systems).

3.8. Annexin-V Staining Protocol

Preceding the condensation of chromatin and fragmentation of intranucleasomal DNA following an apoptotic insult is the translocation of phosphatidylserine (PS) from the inner to the outer leaflet of the plasma membrane. This early apoptotic event can be detected by flow cytometry using fluorescently tagged Annexin-V, a high affinity Ca^{2+}-dependent PS binding protein *(23, 24)*. Loss of membrane impermeability characterizes late apoptotic death. In conjunction with the membrane impermeable DNA intercalating dye PI (*see* **Notes 7** and **8**), early (Annexin-V positive,

PI negative) and late (Annexin-V positive, PI positive) apoptotic cells can be determined by flow cytometry.

1. Prepare 3×10^5 bone marrow cells in 1 mL of PBS and wash.

2. Resuspend in 100 µL of Annexin-V staining buffer containing 2 µL of Annexin-V-FLUOS (if using other fluorescent probes, Annexin V-FLUOS can be replaced with 10 µL of Annexin-V-FITC, Annexin-V-PE or Annexin V-APC).

3. Gently vortex and incubate in the dark at room temperature for 15 min.

4. Add 300 µL of cold Annexin-V binding buffer containing 10 µg/mL PI (*see* **Note 8**).

5. Filter suspension through a 40 µM nylon mesh to prevent blocking of flow cytometer.

6. Samples were analyzed on a FACScalibur (BD) or LSR-II (BD). The percentage of cell death is quantified on a dual parameter histogram of Annexin-V fluorescence vs. PI fluorescence, applying appropriate gating if sample has been prestained for a particular immunophenotype (e.g., lineage; *see* **Subheading 3.6**). The percentage of live cells is determined from quadrant Q3. Annexin-V positive regions are found in Q2 and Q4. Percentage apoptosis was determined from the summation of events found in Q2 plus Q4 (*see* **Fig. 4b**).

We have treated *p53*$^{-/-}$ mice and their littermate WT controls with the myelosuppressive regimen of carboplatin (10 mg/kg, tail vein injection) in combination with TBI (7.5 Gy) *(17)*. Subsequently, bone marrow was extracted according to **Subheading 3.1** and apoptosis in vivo determined by Annexin-V staining (*see* **Subheading 3.7**) of the lineage negative populations (*see* **Subheading 3.6**). The lineage negative region was defined by flow cytometric analysis of bone marrow samples stained with an APC isotype negative control, and termed P4 (*see* **Fig. 4a**). The degree of Annexin-V vs. DAPI staining of the lineage negative (P4) was subsequently determined (*see* **Notes 8** and **9**). Genetic deletion of *p53* abrogates apoptosis (Annexin-V positive) of the lineage negative bone marrow population induced by a myelosuppressive regimen in vivo (*see* **Fig. 4c**).

4. Notes

1. As an alternative to PI, 7-AAD (7-amino-actinomycin D; BD) or DAPI (4′,6-diamidino-2-phenylindole; Molecular Probes) may also be used. However, use of DAPI requires samples to

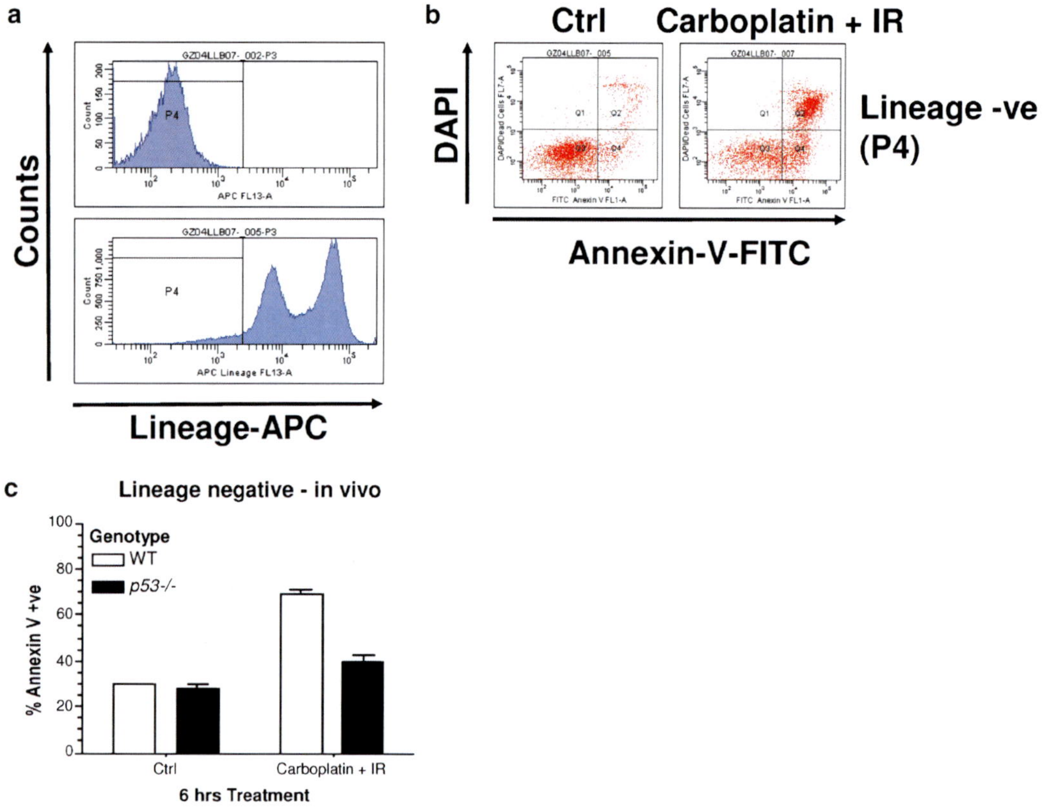

Fig. 4. Deletion of *p53* protects bone marrow progenitors against apoptosis mediated by a myelosuppresive treatment regimen in vivo. Mice null for *p53* and their wild-type (WT) littermates were treated with a myelosuppressive regimen (carboplatin, 10 mg/kg plus 7.5 Gy TBI) for 6 h. Mice were sacrificed, and bone marrow isolated from the hind legs. Bone marrow suspensions were stained with lineage specific antibodies prior to Annexin-V-APC and DAPI. The lineage negative bone marrow population was determined utilizing the isotype negative control for APC and the region defined as "P4" ((**a**), *upper panel*). The majority of bone marrow cells are lineage positive ((**a**), *lower panel*). The percentage apoptosis of the lineage negative populations was determined from dual parameter histograms of Annexin-V-APC vs. DAPI. Flow cytometry histograms shown are representative of results obtained from two different WT mice (**b**). Graphs depict the percentage apoptosis (Annexin-V positive) in vivo of lineage negative bone marrow populations obtained from control (Ctrl) and carboplatin plus IR-treated WT and *p53*$^{-/-}$ mice (**c**). Data represent the mean ± SEM of samples obtained from two mice.

be analyzed on a flow cytometer with either a violet or an UV excitation source. Use of DAPI will reduce fluorescence overspill into the FITC and APC channels.

2. All animal-related procedures must be approved by an institutional animal care and use committee as required by the Animal Welfare Act and NIH policy.

3. If littermate controls are not available, mice should be age matched. This is imperative when comparing thymocyte number since the mouse thymus involutes at 4–6 weeks of age *(25)*. Furthermore, recent data describes the tumor suppressor

function and transcriptional activity of p53 to be progressively muted in aging mice *(26)*. It is also of significant importance to compare animals that are of an identical background breed. For example, C57BL/6 mice are more susceptible to fatality as a result of IR than those on a S129 background. In this regard, thymocytes isolated from C57BL/6 mice exposed to TBI possess a higher level of in vivo apoptosis than those obtained from mice on a C57BL/6.S129 background treated with TBI (*see* **Fig. 3**).

4. We have found that Kimwipes® adhere to the flesh surrounding the bones of the hind legs and enable more efficient processing than with scissors alone.

5. Repeated drawing and expulsion of cells through a needle attached to a syringe may induce sheer damage and skew the incidence of death.

6. Although beyond the methodological scope of this chapter, cell samples taken directly from the mouse or following ex vivo manipulation may be utilized for other purposes. For example, extraction of RNA or protein for real-time PCR and Western Blot analysis, respectively.

7. Prior to acquiring flow cytometric information on immunophenotype or Annexin-V positivity ensure that the appropriate color compensation has been performed. Additionally, isotype negative controls for each fluorochrome should be used to determine regions of positive and negative fluorescence. Choice of fluorescently tagged antibodies is dictated by the specifications of the flow cytometer available.

8. It is recommended that antibodies be carefully titrated for optimum performance.

9. The degree of in vivo apoptosis determined by Annexin-V staining can be applied to cells of differing phenotypes by simply substituting in, or adding other antibodies to that of lineage described herein (*see* **Subheadings 3.3**, **3.6**, and **3.8**).

Acknowledgments

We thank the Animal Resources Center and Drs. Ashmun & Hamilton-Easton of the Flow Cytometry and Cell Sorting Facility at St. Jude Children's Research Hospital for their technical assistance. We also thank Drs. Kirsteen H. Maclean and Richard Cross for their critical review.

References

1. Vogelstein, B., Lane, D., and Levine, A. J. (2000). Surfing the p53 network. *Nature* **408,** 307–310.

2. Momand, J., Zambetti, G. P., Olson, D. C., George, D., and Levine, A. J. (1992). The mdm-2 oncogene product forms a complex with the p53 protein and inhibits p53-mediated transactivation. *Cell* **69,** 1237–1245.

3. Zambetti, G. P. (2007). The p53 mutation "gradient effect" and its clinical implications. *J Cell Physiol* **213,** 370–373.

4. Donehower, L. A., Harvey, M., Slagle, B. L., McArthur, M. J., Montgomery, C. A., Jr., Butel, J. S., et al. (1992). Mice deficient for p53 are developmentally normal but susceptible to spontaneous tumours. *Nature* **356,** 215–221.

5. Attardi, L. D., and Jacks, T. (1999). The role of p53 in tumour suppression: lessons from mouse models. *Cell Mol Life Sci* **55,** 48–63.

6. Eischen, C. M., Weber, J. D., Roussel, M. F., Sherr, C. J., and Cleveland, J. L. (1999). Disruption of the ARF-Mdm2-p53 tumor suppressor pathway in Myc-induced lymphomagenesis. *Genes Dev* **13,** 2658–2669.

7. Schmitt, C. A., McCurrach, M. E., de Stanchina, E., Wallace-Brodeur, R. R., and Lowe, S. W. (1999). INK4a/ARF mutations accelerate lymphomagenesis and promote chemoresistance by disabling p53. *Genes Dev* **13,** 2670–2677.

8. Johnson, L., Mercer, K., Greenbaum, D., Bronson, R. T., Crowley, D., Tuveson, D. A., et al. (2001). Somatic activation of the K-ras oncogene causes early onset lung cancer in mice. *Nature* **410,** 1111–1116.

9. Olivier, M., Eeles, R., Hollstein, M., Khan, M. A., Harris, C. C., and Hainaut, P. (2002). The IARC TP53 database: new online mutation analysis and recommendations to users. *Hum Mutat* **19,** 607–614.

10. Soussi, T., Ishioka, C., Claustres, M., and Beroud, C. (2006). Locus-specific mutation databases: pitfalls and good practice based on the p53 experience. *Nat Rev Cancer* **6,** 83–90.

11. Olsson, A., Manzl, C., Strasser, A., and Villunger, A. (2007). How important are post-translational modifications in p53 for selectivity in target-gene transcription and tumour suppression? *Cell Death Differ* **14,** 1561–1575.

12. Willis, S. N., and Adams, J. M. (2005). Life in the balance: how BH3-only proteins induce apoptosis. *Curr Opin Cell Biol* **17,** 617–625.

13. Youle, R. J., and Strasser, A. (2008). The BCL-2 protein family: opposing activities that mediate cell death. *Nat Rev Mol Cell Biol* **9,** 47–59.

14. Jeffers, J. R., Parganas, E., Lee, Y., Yang, C., Wang, J., Brennan, J., et al. (2003). Puma is an essential mediator of p53-dependent and -independent apoptotic pathways. *Cancer Cell* **4,** 321–328.

15. Villunger, A., Michalak, E. M., Coultas, L., Mullauer, F., Bock, G., Ausserlechner, M. J., et al. (2003). p53- and drug-induced apoptotic responses mediated by BH3-only proteins puma and noxa. *Science* **302,** 1036–1038.

16. Weber, J. D., and Zambetti, G. P. (2003). Renewing the debate over the p53 apoptotic response. *Cell Death Differ* **10,** 409–412.

17. Pestina, T. I., Cleveland, J. L., Yang, C., Zambetti, G. P., and Jackson, C. W. (2001). Mpl ligand prevents lethal myelosuppression by inhibiting p53-dependent apoptosis. *Blood* **98,** 2084–2090.

18. Fredrickson, T.N., and Harris, A.W. (eds) (2000). *Atlas of Mouse Hematopathology.* Harwood Academic, London.

19. Nicoletti, I., Migliorati, G., Pagliacci, M. C., Grignani, F., and Riccardi, C. (1991). A rapid and simple method for measuring thymocyte apoptosis by propidium iodide staining and flow cytometry. *J Immunol Methods* **139,** 271–279.

20. Kerr, J. F., Wyllie, A. H., and Currie, A. R. (1972). Apoptosis: a basic biological phenomenon with wide-ranging implications in tissue kinetics. *Br J Cancer* **26,** 239–257.

21. Wyllie, A. H. (1980). Glucocorticoid-induced thymocyte apoptosis is associated with endogenous endonuclease activation. *Nature* **284,** 555–556.

22. Ormerod, M. (ed) (1999). *Flow Cytometry,* 2nd edition. Bios, Oxford.

23. Koopman, G., Reutelingsperger, C. P., Kuijten, G. A., Keehnen, R. M., Pals, S. T., and van Oers, M. H. (1994). Annexin V for flow cytometric detection of phosphatidylserine expression on B cells undergoing apoptosis. *Blood* **84,** 1415–1420.

24. Raynal, P., and Pollard, H. B. (1994). Annexins: the problem of assessing the biological role for a gene family of multifunctional calcium- and phospholipid-binding proteins. *Biochim Biophys Acta* **1197,** 63–93.

25. Chaplin, D. D. (2003). Lymphoid tissues and organs, in *Fundamental Immunology,*

5th edition (Paul, W.E., ed.). Lippincott, Williams and Wilkins, Philadelphia, PA.

26. Feng, Z., Hu, W., Teresky, A. K., Hernando, E., Cordon-Cardo, C., and Levine, A. J. (2007). Declining p53 function in the aging process: a possible mechanism for the increased tumor incidence in older populations. *Proc Natl Acad Sci USA* **104,** 16633–16638.

Chapter 12

Measurement of Changes in Cdk2 and Cyclin O-Associated Kinase Activity in Apoptosis

Ramon Roset and Gabriel Gil-Gómez

Summary

Many cell cycle regulatory proteins have been shown to be able to regulate cell death. Activation of Cdk2 has been shown to be necessary for apoptosis of quiescent cells such as thymocytes, neurons, and endothelial cells. This activation is stimulus-specific because it occurs in glucocorticoid and DNA damage but not in CD95-induced apoptosis in thymocytes. Apoptotic Cdk2 activation in lymphoid cells is controlled by a recently identified protein, cyclin O, and its activity is modulated by p53 and members of the Bcl-2 protein family. In this chapter, we describe methods for measuring changes in Cdk2 activity during apoptosis. In addition, we also show the details of the generation of an antibody able to immunoprecipitate the cyclin O complexes from apoptotic cells in native conditions and its use to measure the kinase activity associated with this proapoptotic cyclin.

Key words: Cdk2, Apoptosis, Thymocytes, Cell cycle, Kinase, Intrinsic stimuli

1. Introduction

Living cells have developed conserved mechanisms that ensure the correct replication of their genetic material and its distribution into the two daughter cells. In addition, pluricellular organisms have developed biochemical pathways that allow damaged or useless cells to be killed and their corpses removed. Each of these mechanisms is executed and regulated by an exclusive biochemical machinery, reflecting a high degree of specialization acquired during evolution. Nevertheless, apparently opposite mechanisms of cell division and cell death have been shown to share a number of proteins. Among these, activation of some of

Peter Erhardt and Ambrus Toth (eds.), *Apoptosis*, Methods in Molecular Biology, vol. 559
DOI 10.1007/978-1-60327-017-5_12
© Humana Press, a part of Springer Science+Business Media, LLC 2004, 2009

the Cyclin-Dependent Kinases (Cdks) has been shown to be a crucial step in the regulation of both mitosis and apoptosis (*1*). Cdk members are a family of serine/threonine-dependent protein kinases that share a high degree of structural homology and although their function is highly heterogeneous, all of them share a common mechanism that involves binding to an activating subunit. This positive regulatory subunit either belongs to a family of conserved proteins called cyclins or it is nonhomologous to cyclins, exemplified by the p35 protein needed to activate Cdk5 (*2*). The first members of the cyclin family were proteins whose abundance was cell cycle regulated by means of transcriptional and/or postranscriptional mechanisms (*3*). Nevertheless, as the family grew, other members were shown to share some degree of structural homology while their abundance was constant along the cell division cycle.

In addition to the cyclins, negative regulatory Cdk subunits were isolated and named Inhibitors of the Cdks (CDKIs) (*4*). The kinase activity of the complex is also regulated by phosphorylation/dephosphorylation mechanisms, which ensure a quick and complete switching on and off of Cdk activity (*5*).

Proteins regulating the apoptotic pathways were first isolated by genetic approaches involving the nematode *C. elegans*. We know now that the cell death pathways are finely tuned by a growing number of regulators. They detect cell damage, transmit this information to repair systems, decide if the damage can be repaired and, if this is not possible, execute the cell death program (*6*). Removal of dead cells is also regulated by a number of recently isolated genes (*7*), completing a program that surveys the homeostasis of a pluricellular organism. The detection of damaged genetic material is accomplished by a number of proteins among which the tumour suppressor p53 is the best characterized. Lesions in the DNA trigger activation of p53 and its function as a transcription factor, promoting the expression of DNA repair and apoptosis-related target genes whose function is not completely understood (*8*). Other apoptotic stimuli do not depend on the action of p53, although they need de novo transcription and translation, like glucocorticoids (*9*). Both groups of stimuli constitute the so-called intrinsic stimuli. A third group of stimuli (extrinsic stimuli) do not require de novo transcription and translation, exemplified by the cytotoxic cytokine TNFα and the CD95/CD95L system (*10*). The Bcl2 family members constitute another group of apoptosis-related proteins. They regulate cell survival positively (Bcl2 subfamily) or negatively (Bax subfamily) and are thought to participate in determining the decision point of life and death (*11*). Finally, the caspase family represents the main proteins involved in the execution phase of apoptosis. The members of this family participate in several regulatory steps of the apoptosis signal transduction pathway, from sensing the cellular damage ("apical" caspases) to the dismantling ("executioner" caspases) of nuclear and cytoskeletal architectures (*12*) and are believed to be activated by all the apoptotic stimuli and hence their activation is a marker of apoptotic death fate.

Many cell cycle regulatory proteins have been shown to regulate cell death. Rather than apoptosis being regarded as an abortive attempt to re-enter cell cycle, we have shown that activation of the cell cycle machinery in apoptotic quiescent cells does not follow the orderly fashion characteristic of the mitotic cell cycle but reflects common biochemical machinery shared between both processes. Activation of Cdk2 has been shown to be necessary for apoptosis of quiescent cells like thymocytes (*13*), neurons (Brady et al., personal communication) and endothelial cells undergoing apoptosis induced by the withdrawal of trophic factors (*14*). Apoptotic Cdk2 activation has been shown to be controlled by apoptosis regulatory proteins like p53 and Bcl2 family members and it is specific since other members of the Cdk family like Cdk1 are not activated under the same conditions (*13*) (*see* **Fig. 1**). Cdk2 can be negatively regulated mainly by p21^{Cip1} and p27^{Kip1} CDKIs. In postmitotic thymocytes undergoing apoptosis, p27^{Kip1} levels decline correlating with Cdk2 activation (*13*). In apoptosing human umbilical vein endothelial cells, both p21^{Cip1} and p27^{Kip1} are degraded by activated caspases, again correlating with Cdk2 activation (*14*). Nevertheless, their function in regulating Cdk2 activity during apoptosis is still unclear. Although the exact biochemical function of Cdk2 activation during apoptosis is not known, we have positioned Cdk2 upstream from mitochondrial apoptotic changes, phosphatidylserine exposure, and apical caspase activation (*15*). We have recently described a new cyclin (cyclin O), which is required to activate Cdk2 during apoptosis induced by intrinsic stimuli in lymphoid cells. Cyclin O is expressed upon apoptosis triggering preceding cell death, and its downregulation abrogates γ-radiation and glucocorticoid but not CD95-induced apoptosis (*16*).

Excellent reviews of methods to determine Cdk2, Cdk4 and Cdk6 activities in cycling cells are available (*17, 18*). The objective of this chapter is to describe the techniques that we are using in our laboratory to measure Cdk2 kinase activity during apoptosis. We will describe the modifications we have introduced into

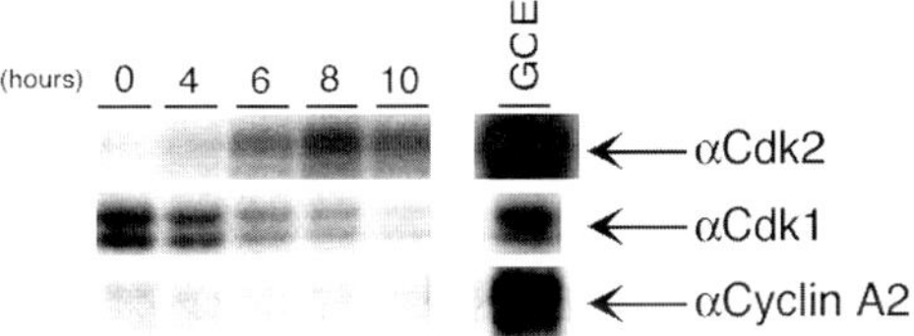

Fig. 1. Cdk2, Cdk1 and cyclin A-associated kinase activities in CD2Bax transgenic thymocytes after γ-radiation induced apoptosis. *GCE* Growing cells extract.

the general method to adapt it to our model of mouse apoptotic thymocytes. We will also show the details of the generation of an antibody able to immunoprecipitate the cyclin O complexes from apoptotic cells in native conditions and its use to measure the kinase activity associated to this proapoptotic cyclin.

The measurement of Cdk2 kinase activity is based on its isolation together with associated proteins by immunoprecipitation using specific antibodies. The immunopurified protein complexes are depleted of spurious proteins by washing them with lysis buffer, and finally the kinase activity is revealed by incorporation of ^{32}P from labelled [γ-^{32}P]ATP into non-specific exogenous substrates like Histone H1. Variations of the method include pull down with purified cyclin fusion proteins, use of Cdk2 interacting proteins like p9[CKShs1] coupled to Sepharose beads to recover the kinase complexes (*19*) or using recombinant bona fide Cdk2 substrates like Rb, Cdc6 or NPAT (*20*).

The antibodies used should be able to recognize and immunoprecipitate the native kinase complexes without impairing their activity by steric hindrance or disruption of the associated regulatory proteins. The crystal structure of Cdk2-cyclin A complexes predicts that the C-terminal 16 amino acid residues are pointing out of the protein complex (*21*), which suggests that antibodies directed against this domain would be able to recognize the complexes without disrupting their structure. Antibodies directed against similar protein domains have been used to measure kinase activities of other members of the Cdk family like Cdk4 and Cdk6 (*18*) or Cdk5 (*22*). Assays of Cdk2 kinase activity are complicated given the number of mechanisms that affect its activity. Monomeric Cdk2 protein is inactive as a kinase. Binding of a positive regulatory subunit, namely a cyclin, activates the complex. A number of activating and inactivating phosphorylations regulate the activity of preformed cyclin–Cdk complexes (*5*). In order to preserve the phosphorylation status of the Cdk2 complexes in cells, phosphatase inhibitors are added to the lysis buffer to avoid changes due to the action of protein phosphatases present in the crude extract during the immunoprecipitation procedure. Nevertheless, no apparent changes in the Cdk2 phosphorylation status have been observed by Western Blot during thymocyte apoptosis (*13*) and similar conclusions have been reached by more accurate methods in human umbilical vein endothelial cells undergoing apoptosis (*14*).

The fact that Cdk2 needs to bind to a cyclin in order to become active opens the possibility of using anti-cyclin antibodies to immunoprecipitate the complexes. Cdk2-cyclin A active complexes can be easily recovered from extracts obtained from growing human or mouse cell lines by immunoprecipitation with anti-cyclin A antibodies. From freshly isolated thymocytes, very low but measurable amounts of cyclin A associated kinase activity is recovered but this initial activity decays as they undergo apoptosis, arguing against cyclin A being the cyclin responsible for Cdk2 activation during thymocyte apoptosis (*13*) (*see* **Fig. 1**). The same holds true for the other "cell cycle" Cdk2 activating cyclins, cyclin E1 (*13*), and cyclin E2 (*23*).

Given the number of Cdk2 regulatory mechanisms, careful attention must be paid to the interpretation of the immunoprecipitation/kinase assay results. Interaction of the Cdk–cyclin complexes with regulatory proteins may be lost if their interaction is not tight enough. Even more confusing results may arise from loss of cellular compartmentalization of regulatory proteins at the time of preparing the cell lysate. Proteins separated in different cell compartments may then interact and mislead the kinase assay results.

2. Materials

1. RPMI Complete Medium: RPMI 1640 medium (Gibco), 10% fetal bovine serum, streptomycin (100 μg/mL), penicillin (100 UI/mL), β-mercaptoethanol (50 μM from a 25 mM stock in mQ water sterilized by filtration through a 0.22 μm filter and kept at –20°C). Store at 4°C in the dark.

2. Cdk2 lysis buffer: 50 mM Tris–HCl pH 7.5, 150 mM NaCl, 20 mM EDTA, 0.5% NP40 stored at 4°C. For complete Cdk2 lysis buffer, prior to use add 1 mM DTT; protease inhibitors: 2 μg/mL aprotinin, 2 μg/mL leupeptin, 2 μg/mL antipain, 20 μg/mL soybean trypsin inhibitor and 2 mM Pefabloc SC; phosphatase inhibitors:1 mM sodium fluoride, 1 mM glycerophosphate, 1 mM sodium pyrophosphate, 0.2 mM activated sodium orthovanadate; and stocks: aprotinin, leupeptin and antipain: 0.2 mg/mL each in water, soybean trypsin inhibitor (Roche) 2 mg/mL in water, Pefabloc SC (Roche) 2 M in water, DTT 1 M in water. Aliquot and keep frozen. The following are made as stock solutions: NaF: 1 M in water. Keep at 4°C. β-glycerophosphate 0.5 M in water. Adjust to pH 7.5. Sodium pyrophosphate 0.1 M in water. Keep at room temperature.

3. Activated sodium orthovanadate 20 mM stock (*see* **Note 1**). Prepare a 20 mM sodium orthovanadate (Sigma-Aldrich) solution in mQ water. Adjust the pH to 10 with HCl. Should turn yellow-orange due to the formation of decavanadate. Boil for 15 min. Decavanadate depolymerises and solution turns colourless. Aliquot and keep frozen.

4. Protein A/G beads: Protein A or G covalently bound to Sepharose CL-4B beads (GE Healthcare) should be washed four times with plain Cdk2 lysis buffer prior to use in order to remove the ethanol used as preservative. Five to twenty microlitres of the slurry were used per sample in the immunoprecipitation experiments (*see* **Note 2**).

5. Antibodies

   (a) Cdk1 (Cdc2): Monoclonal antibody A17 (BD Pharmingen, NJ) raised against a 15 amino acid sequence from the C-terminus of human Cdk1 was used in the Cdk1 immunoprecipitation/kinase assays.

   (b) Cdk2: We routinely used an affinity purified polyclonal antibody directed against the 16 C-terminal amino acids of human Cdk2 (M2 antibody from Santa-Cruz) for Cdk2 immunoprecipitation/kinase assays. The antibody is also excellent for Western Blotting.

   (c) Cyclin A: We use two different antibodies raised either against the full length human cyclin A (Santa-Cruz anti-Cyclin A H-432) or against a peptide corresponding to the 19 C-terminal amino acids of the mouse cyclin A sequence (Santa-Cruz anti-Cyclin A C-19). Both antibodies work similarly for immunoprecipitation/kinase assay experiments. For Western Blotting we have better results with the H-432 antibody.

   (d) Cyclin O: We raised a rabbit antiserum (C2) against the peptide H-SSLPRILPPQIWERC-NH2 present at the C-terminus of mouse cyclin O sequence (*see* **Note 3**). The C2 serum was used to detect the cyclin O associated kinase activity by immunoprecipitation + kinase technique.

6. Cdk2 kinase buffer: 50 mM Tris–HCl pH 7.5, 10 mM $MgCl_2$. Keep at 4°C.

7. Cdk2 hot mix (per sample): 20 µM ATP, 10 µCi γ-^{32}P]ATP, 1 mM DTT, and 2 µg histone H1. Stocks: ATP, 1 mM in Cdk2 kinase buffer; DTT, 0.2 mM in Cdk2 kinase Buffer; and histone H1 (from Roche), 1 µg/µL in Cdk2 kinase buffer. All of them should be aliquoted and kept frozen. To prepare the hot mix, mix the necessary amounts and complete up to 20 µL per sample with Cdk2 kinase buffer. Follow at all times the Recommended Laboratory Safety Procedures for manipulation of radioactive material.

8. Laemmli Buffer (2×): 125 mM Tris–HCl, pH 6.8, 20% glycerol, 4% SDS; 1.4 M β-mercaptoethanol, 5% saturated bromophenol blue solution. Aliquot and keep at 4°C.

9. Coomassie blue staining solution: 40% methanol, 10% acetic acid and 0.1% Coomassie brilliant blue R-250. Keep at room temperature.

10. Destaining solution: 10% acetic acid and 40% methanol.

3. Methods

3.1. Mouse Thymocyte Isolation

1. Sacrifice the mice by CO_2 asphyxiation. We usually use 5–10-weeks-old FVB/N mice. Older mice can be used but the size of the thymus decreases with age.

2. Remove the thymus with the aid of a pair of tweezers and put them in cold plain RPMI medium kept on ice.

3. Place a 70-µm cell strainer (Falcon) inside a 35-mm Petri dish and transfer the thymuses onto the cell strainer. Add 2.5 mL of complete RPMI medium and disrupt the thymus with the aid of a syringe plunger.

4. Collect single cells, transfer them to a sterile tube and wash the strainer and the plunger with 2.5 mL of complete RPMI medium. Transfer the cell suspension to the tube.

5. Pellet the thymocytes by spinning at $1{,}000 \times g$ for 5 min.

6. Aspirate the medium.

7. Resuspend the cell pellet in 5 mL of Red Blood Lysis Buffer (Sigma-Aldrich).

8. Incubated for 5 min at room temperature in order to remove the red blood cells.

9. Pellet the cells by spinning at $1{,}000 \times g$ for 5 min.

10. Resuspend the cell pellet in 10 mL of RPMI complete medium.

11. Pellet the cells by spinning at $1{,}000 \times g$ for 5 min.

12. Resuspend the cells in 5 mL of medium.

13. Determine cell yield by counting Trypan Blue excluding cells with a haemocytometer.

3.2. Apoptosis Induction in Thymocytes

1. Dilute thymocyte suspension to 1×10^6 to 5×10^6 cells/mL with RPMI complete medium.

2. Cell suspensions are put in culture at 37°C, 5% CO_2 for the indicated time either with or without apoptosis induction.

3. Apoptotic inducers are added prior to culture. The stimuli used were γ-radiation: from a ^{131}Cs source (5 Gy). Dexamethasone: 1 µM from a 1 mM stock in absolute ethanol.

4. After incubation, thymocytes were aliquoted into 0.5-mL tubes, spun at 600 × *g* for 5 min, supernatant was aspirated and the dry pellets (0.5×10^6 to 2×10^6 cells/pellet) were immediately frozen at –80°C until analysis. Deep freezing of the cell pellets is necessary to help extract the Cdk2 complexes.

3.3. Culture of EL-4 Derived Cell Clones and Apoptosis Induction

1. The mouse lymphoma cell line EL-4 derived clones Luc2 and L3-2 (expressing an shRNA against GFP or against mouse cyclin O, respectively) (*16*), were cultured in standard conditions, never allowing the culture to reach cell densities higher than 10^6 cells/mL.

2. Apoptosis induction: cell suspensions (5×10^5 cells/mL) were put in culture at 37°C, 5% CO_2 after treatment with γ-radiation from a ^{131}Cs source (5 Gy) and at the indicated times, aliquots were taken, spun at 600 × *g* for 5 min, supernatant was aspirated and the dry pellets (about 3×10^6 cells/pellet) were immediately frozen at –80°C until analysis.

3.4. Cell Extract Preparation

1. Resuspend the thymocyte pellets (*see* **Note 4**) in 100 μL of Cdk2 lysis buffer plus protease/phosphatase inhibitors and DTT by pipetting up and down or vortexing. In the case of Luc2 and L3-2 cells, frozen pellets were resuspended in 250 μL of the buffer, the protein concentration was determined by the Bradford method (Bio-Rad) and 100 μg of the cell lysate were diluted up to 500 μL of final volume with complete Cdk2 lysis buffer and used per immunoprecipitation (*see* **Note 4**). Use 0.5-mL eppendorf tubes during the whole procedure since it is easier to see the protein A/G beads and remove thoroughly the supernatants.

2. Incubate 20 min on ice.

3. Clarify the extract by spinning in the microcentrifuge at 18,000 × *g* for 20 min at 4°C.

4. Transfer the supernatant to a fresh chilled 0.5-mL tube. Be very careful of not taking along any traces of the pellet since this will result in high background (*see* **Note 5**).

5. Proceed immediately with the immunoprecipitation procedure.

3.5. Immunoprecipitation

1. Prepare the antibody mix (per sample): *x* μL of antibody (*see* **Note 6**), 5–20 μL of protein A/G beads (washed), and 100 μL of lysis buffer/inhibitors, and DTT.

2. Mix: 100 μL thymocyte cell extract, 120 μL antibody mix and 180 μL complete Cdk2 lysis buffer.

3. Rotate for 2–3 h at 4°C.

4. Spin the beads down with a very short pulse (not higher than $850 \times g$).

5. Remove the supernatant thoroughly.

6. Resuspend the beads in 0.5 mL of plain Cdk2 lysis buffer plus 1 mM DTT.

7. Repeat **steps 4–6** two more times.

8. Wash once with 0.5 mL of kinase buffer plus 1 mM DTT and remove the supernatant thoroughly to get rid of all traces of detergent coming from the Cdk2 lysis buffer.

9. Proceed immediately with the kinase reaction.

3.6. Kinase Reaction

1. Resuspend the beads in 20 µL of Cdk2 hot mix.

2. Incubate for 30 min at 30°C.

3. Stop the reaction with 30 µL of 2× Laemmli Buffer. At this point, samples can be kept overnight at –80°C. Otherwise proceed with the SDS-PAGE.

3.7. SDS-PAGE and Detection

1. Boil the samples 5–10 min.

2. Load the samples on a 12% SDS-PAGE gel (20-cm long). Stop the run when the bromophenol blue just run off the gel.

3. Stain the gel with Coomassie blue staining solution for 30 min gently rocking. Staining solution may be reused several times.

4. Destain the gel with destaining solution until you can visualize the histone H1 bands (*see* **Note 7**).

5. Dry the gel and expose it to an X-ray film.

4. Notes

1. Vanadium anions interconvert in solution. Activation of the sodium orthovanadate solution is thought to yield higher concentration of the monovanadate anion that is the most potent protein phosphotyrosine-phosphatase inhibitor form (*24*).

2. Protein A/G beads can be mixed with washed, empty Sepharose beads in order to increase their bulk to facilitate recovery of the pellet during the washing steps.

3. The C2 peptide was coupled with 3-maleimidobenzoic acid *N*-hydroxysuccinimide ester (MBS, Sigma-Aldrich) to keyhole limpet haemocyanine (KLH, Sigma-Aldrich). Production of the specific antibody in the different batches of sera was confirmed and titrated by ELISA. Given the fact that the

3-maleimidobenzoic acid *N*-hydroxysuccinimide ester (MBS) used to cross-link C2 peptide to the KLH can elicit a potent antibody response in rabbits, the antigen used to coat the ELISA plates was cross-linked to the carrier ovoalbumin using glutaraldehyde. The C2 anti-cyclin O antibody fails to immunoprecipitate significant kinase activity from clone L3-2 cell extracts where cyclin O has been downregulated (*see* **Fig. 2**) after apoptosis induction, indicating that it specifically immunoprecipitates the cyclin O-containing kinase complexes. All the antibody techniques were done as described (*25*). The amino acid sequence of the C2 peptide is poorly conserved in human, so the antibody does not recognise the human protein. Attempts to raise antibodies directed against the whole mouse protein or against an N-terminal peptide or a peptide predicted to be located in a loop of the protein failed to yield sera useful in immunoprecipitation experiments.

4. The number of thymocytes to be used should be determined in each case. Routinely we can detect Cdk2 kinase activity increases in 5×10^5 wild type thymocytes from FVB/N mice treated with 1 μM dexamethasone or 5 Gy or γ-radiation after 5 h of incubation. In the case Cdk2 activity is assayed from lysates coming from growing cells, much less sample is required as Cdk2 activity can be almost two orders of magnitude higher than in apoptotic quiescent cells. Detection of cyclin O associated kinase activity is less sensitive, at least using the C2 antiserum. Our attempts to detect it in apoptotic thymocytes have been negative up to now, and in the case of the EL-4 derived clones, only using high amounts of protein extract we succeeded in its detection after apoptosis induction (*see* **Fig. 2**). This is likely to be a consequence of the low abundance of the cyclin O protein in the extracts (*16*).

5. When assaying cyclin O associated kinase activity in extracts from EL-4 derived clones, prior to the immunoprecipitation, a preclearing step helps to reduce the background. For this

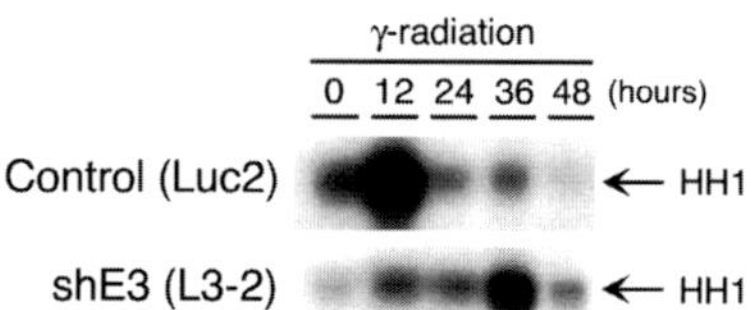

Fig. 2. Cyclin O downregulation in EL-4 lymphoid cells. EL-4 mouse lymphoma cells stably transfected with a plasmid encoding an shRNA construct against the mouse cyclin O or a firefly Luciferase control shRNA were selected with puromycin and single cell clones isolated. Expression of cyclin O was detected in EL-4 cell clones by immunoprecipitation + kinase assay using the C2 antibody. Clones Luc2 (transfected with luciferase shRNA) and L3-2 (transfected with cyclin O shRNA) were harvested 0, 12, 24, 36 and 48 h after treatment with γ-radiation (5 Gy).

purpose, to 500 µg of cell extract we add Cdk2 lysis buffer/inhibitors/DTT up to 280 µL and 10 µL of washed protein A beads, rotate for 1 h, spin the beads down with a very short pulse and carefully transfer the supernatant into a fresh, chilled, 0.5-mL tube. Directly add the 120 µL of Antibody Mix (*see* **Subheading 3.5**, **step 2**) to the precleared 280 µL of cell extract and proceed with the indicated protocol for immuno-precipitation.

6. The amount of antibody to be used should be determined in each case. Use appropriate amounts of protein A/G beads for the amount of antibody used according to their antibody binding capacity. In the case of assaying cyclin O associated kinase activity in EL-4 derived clones, 2 µL of crude C2 antiserum was used per immunoprecipitation (100 µg of cell extract) and 10 µL of protein A beads.

7. In order to speed up the destaining procedure, change the destaining solution often and add a small piece of paper towel or Kimwipes. Cellulose binds the Coomassie Blue dye extracted from the gel accelerating the destaining procedure. Commercial histone H1 preparations run as two close bands of around 27 kDa and a third less intense band of around 23 kDa.

8. When assaying Cdk2 or Cdk1 kinase activities (*see* **Note 9**) from mouse strains with high number of cycling thymocytes like CD2Bax (*26*) or CD2Bad (*27*) transgenic mice, it can be found that Cdk2 activity is higher at time zero with respect to the wild type littermates. This activity goes down after some hours of culture, right before the rise in Cdk2 activity although both peaks may overlap if the first time point the sample is not taken shortly enough after apoptosis induction. We think that this first peak of Cdk2 activity comes from the proliferating compartment of thymocytes and then it is due to cyclin A and cyclin E–Cdk2 complexes. When thymocytes are isolated from their normal thymic microambient lose contact with stromal cells and locally produced growth factors, resulting in cell cycle exit and apoptosis. As the proliferating thymocytes go out of cycle, initial Cdk2 activity declines and later rises again. This second peak of Cdk2 activity is specific to the apoptotic cells and it is most likely attibutable to upregulation of cyclin O (*16*), since the kinase activity associated to cyclins E or A has fallen to background levels (*see* **Fig. 1**).

9. Cdk1 (Cdc2) activity can be measured using exactly the same protocol as described for Cdk2 using appropriate antibodies (*see* **Subheading 2**, **item 5**). In the case of thymocytes, Cdk1 initial activity is downregulated during the apoptosis time course (*see* **Fig. 1**), ruling out a causal role in apoptosis (*13, 28*). As in the case of Cdk2, we assume that Cdk1 initial activity is due to the proliferative fraction of thymocytes (*see* **Note 8**).

References

1. Golsteyn, R. M. (2005). Cdk1 and Cdk2 complexes (Cyclin dependent kinases) in apoptosis: a role beyond the cell cycle. *Cancer Lett.*, **217**, 129–138.

2. Nebreda, A. R. (2006). CDK activation by non-Cyclin proteins. *Curr. Opin. Cell Biol.*, **18**, 192–198.

3. Norbury, C. and Nurse, P. (1992). Animal cell cycles and their control. *Annu. Rev. Biochem.*, **61**, 441–470.

4. Besson, A., Dowdy, S. F., and Roberts, J. M. (2008). CDK inhibitors: cell cycle regulators and beyond. *Dev. Cell*, **14**, 159–169.

5. Morgan, D. O. (1995). Principles of CDK regulation. *Nature*, **374**, 131–134.

6. Harper, J. W. and Elledge, S. J. (2007). The DNA damage response: ten years after. *Mol. Cell*, **28**, 739–745.

7. Ravichandran, K. S. and Lorenz, U. (2007). Engulfment of apoptotic cells: signals for a good meal. *Nat. Rev. Immunol.*, **7**, 964–974.

8. Aylon, Y. and Oren, M. (2007). Living with p53, dying of p53. *Cell*, **130**, 597–600.

9. Herr, I., Gassler, N., Friess, H., and Buchler, M. W. (2007). Regulation of differential pro- and anti-apoptotic signaling by glucocorticoids. *Apoptosis*, **12**, 271–291.

10. Park, H. H., Lo, Y. C., Lin, S. C., Wang, L., Yang, J. K., and Wu, H. (2007). The death domain superfamily in intracellular signaling of apoptosis and inflammation. *Annu. Rev. Immunol.*, **25**, 561–586.

11. Roset, R., Ortet, L., and Gil-Gomez, G. (2007). Role of Bcl-2 family members on apoptosis: what we have learned from knock-out mice. *Front Biosci.*, **12**, 4722–4730.

12. Kumar, S. (2007). Caspase function in programmed cell death. *Cell Death Differ.*, **14**, 32–43.

13. Gil-Gomez, G., Berns, A., and Brady, H. J. (1998). A link between cell cycle and cell death: Bax and Bcl-2 modulate Cdk2 activation during thymocyte apoptosis. *EMBO J.*, **17**, 7209–7218.

14. Levkau, B., Koyama, H., Raines, E. W., Clurman, B. E., Herren, B., Orth, K., Roberts, J. M., and Ross, R. (1998). Cleavage of p21Cip1/Waf1 and p27^{Kip1} mediates apoptosis in endothelial cells through activation of Cdk2: role of a Caspase cascade. *Mol. Cell*, **1**, 553–563.

15. Granes, F., Roig, M. B., Brady, H. J. M., and Gil-Gomez, G. (2004). Cdk2 activation acts upstream of the mitochondrion during glucocorticoid induced thymocyte apoptosis. *Eur. J. Immunol.*, **34**, 2781–2790.

16. Roig, M. B.*, Roset, R.*, Ortet, L., Balsiger, N. A., Anfosso, A., Cabellos, L., Garrido, M., Alameda, F., Brady, H. J. M. and Gil-Gómez, G. *equal contribution (2009). Identification of a novel cyclin required for the intrinsic apoptosis pathway in lymphoid cells. *Cell Death Diff.*, **16**, 230–243.

17. Sheaff, R. J. (1997). Regulation of mammalian Cyclin-dependent kinase 2. *Methods Enzymol.*, **283**, 173–193.

18. Phelps, D. E. and Xiong, Y. (1997). Assay for activity of mammalian Cyclin D-dependent kinases CDK4 and CDK6. *Methods Enzymol.*, **283**, 194–205.

19. Azzi, L., Meijer, L., Ostvold, A. C., Lew, J., and Wang, J. H. (1994). Purification of a 15-kDa Cdk4- and Cdk5-binding protein. *J. Biol. Chem.*, **269**, 13279–13288.

20. Asada, A., Zhao, Y., Kondo, S., and Iwata, M. (1998). Induction of thymocyte apoptosis by Ca^{2+}-independent protein kinase C (nPKC) activation and its regulation by calcineurin activation. *J. Biol. Chem.*, **273**, 28392–28398.

21. Jeffrey, P. D., Russo, A. A., Polyak, K., Gibbs, E., Hurwitz, J., Massague, J., and Pavletich, N. P. (1995). Mechanism of CDK activation revealed by the structure of a Cyclin A-CDK2 complex. *Nature*, **376**, 313–320.

22. Tsai, L. H., Delalle, I., Caviness, V. S., Jr., Chae, T., and Harlow, E. (1994). p35 is a neural-specific regulatory subunit of Cyclin-dependent kinase 5. *Nature*, **371**, 419–423.

23. Gil-Gomez, G. (2004). Measurement of changes in apoptosis and cell cycle regulatory kinase Cdk2. *Methods Mol. Biol.*, **282**, 131–144.

24. Gordon, J. A. (1991). Use of vanadate as protein-phosphotyrosine phosphatase inhibitor. *Methods Enzymol.*, **201**, 477–482.

25. Harlow, E. and Lane, D. (1999) Using Antibodies. A Laboratory Manual, Cold Spring Laboratory, Cold Spring Harbor, NY.

26. Brady, H. J., Gil-Gomez, G., Kirberg, J., and Berns, A. J. (1996). Bax alpha perturbs T cell development and affects cell cycle entry of T cells. *EMBO J.*, **15**, 6991–7001.

27. Mok, C. L., Gil-Gomez, G., Williams, O., Coles, M., Taga, S., Tolaini, M., Norton, T., Kioussis, D., and Brady, H. J. (1999). Bad can act as a key regulator of T cell apoptosis and T cell development. *J. Exp. Med.*, **189**, 575–586.

28. Norbury, C., MacFarlane, M., Fearnhead, H., and Cohen, G. M. (1994). Cdc2 activation is not required for thymocyte apoptosis. *Biochem. Biophys. Res. Commun.*, **202**, 1400–1406.

Chapter 13

Fluorometric Methods for Detection of Mitochondrial Membrane Permeabilization in Apoptosis

Soumya Sinha Roy and György Hajnóczky

Summary

The mitochondrial regulation of cell death involves the release of proapoptotic factors, such as cytochrome c, Smac-DIABLO, AIF, OMI/HtrA2, by disruption of the outer mitochondrial membrane (OMM) permeability barrier that is controlled by pro- and antiapoptotic proteins of the Bcl-2 family. One of the mechanisms contributing to the OMM permeabilization is dependent on the interaction of proapoptotic Bcl-2 family proteins and other factors straight with the OMM. Another mechanism is initiated by the permeability transition of the inner mitochondrial membrane (IMM), leading to an increase in the matrix volume and reorganization of the IMM structure, which in turn, influence the OMM permeability barrier. The OMM also provides surface for the assembly of the apoptosome, where the mitochondria-derived proapoptotic factors induce caspase activation. Fluorescence measurements have been devised for evaluation of the barrier function of both OMM and IMM and of the downstream effectors of the factors released from the mitochondria to the cytosol. Many of these measurements are real-time, quantitative, and can be conveniently performed in a fluorometer cuvette containing suspensions of permeabilized cells or isolated mitochondria. This chapter provides a step-by-step manual for the measurements of the mitochondrial membrane potential, retention of Ca^{2+} and cytochrome c, matrix volume, and caspase activation and discusses protocols for discrimination between different mechanisms of the OMM permeabilization.

Key words: Apoptosis, Mitochondrial membrane permeability, Fluorometer, Mitochondrial membrane potential, Western blot calcium

1. Introduction

The number of studies concerned with mitochondria and apoptosis showed a progressive rise in the past 15 years. Mitochondria were recognized first as the main source of cellular ATP production, but over recent years their role has also been established in many aspects of

Peter Erhardt and Ambrus Toth (eds.), *Apoptosis*, Methods in Molecular Biology, vol. 559
DOI 10.1007/978-1-60327-017-5_13
© Humana Press, a part of Springer Science+Business Media, LLC 2004, 2009

cell physiology and pathophysiology *(1)*. For example, to regulate cell survival, mitochondria participate in calcium signaling by rapidly accumulating and releasing Ca^{2+} and retain proteins that induce execution of apoptosis upon release to the cytosol *(2)*. Initiation of the mitochondrial pathway of apoptosis signaling results in the permeabilization of the outer mitochondrial membrane (OMM) and the release of proapoptotic mitochondrial proteins which are mainly the residents of intermembrane space (IMS) *(3, 4)*. Presently, there are two recognized mechanisms for the OMM permeabilization (**Fig. 1**).

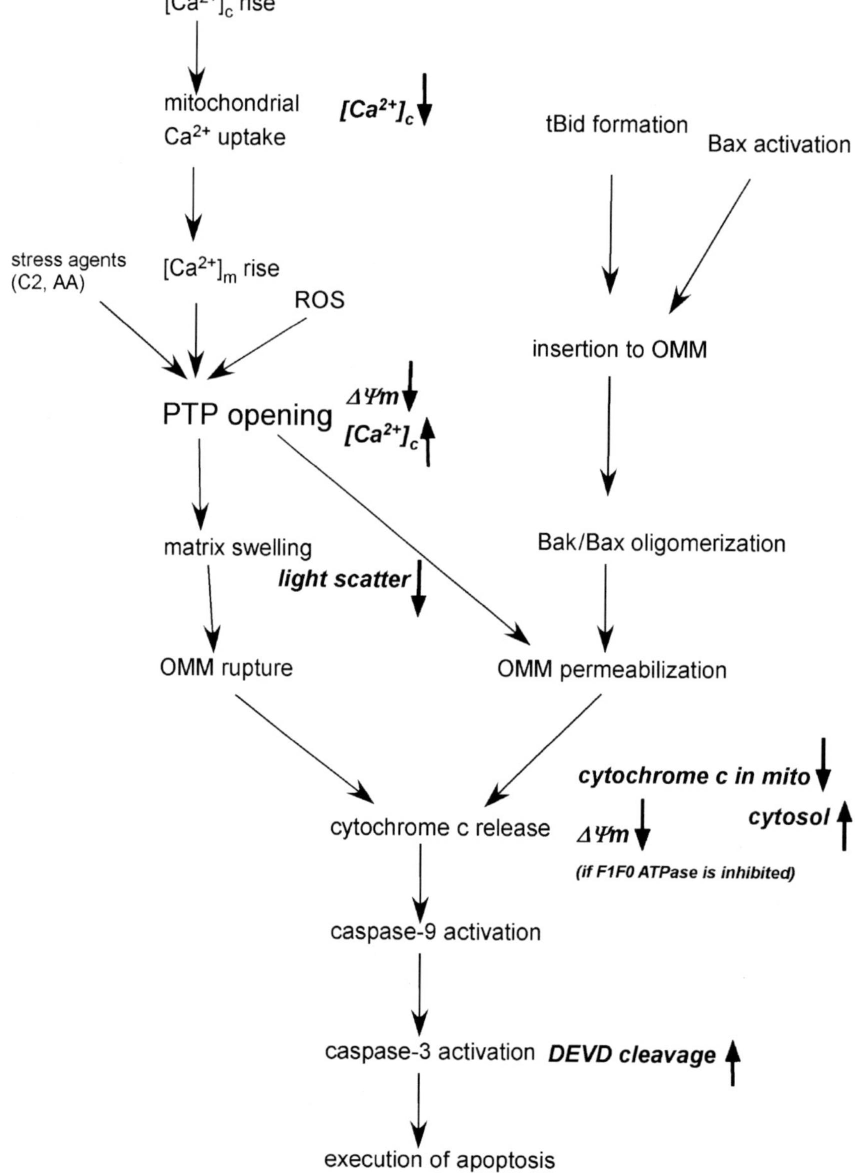

Fig. 1. Mitochondrial phase of apoptosis: mechanisms and measurements. Mechanisms of the mitochondrial phase of calcium and tBid/Bax-dependent apoptosis are depicted in *black*. The parameters measured by the presently described fluorometric approaches are shown in *italics*. *C2* C2-ceramide, *AA* Arachidonic acid.

The first involves the opening of the permeability transition pore (PTP) that seems to be a multiprotein complex localized at the contact sites of the OMM and inner mitochondrial membrane (IMM). Activation of the PTP permits the flux of ions and solutes between the matrix space and the extramitochondrial space, leading to dissipation of the membrane potential, loss of the ATP production, pH gradient, and gradual expansion of the matrix volume. Reorganization of the cristae allows matrix swelling without rupture of the IMM, but the OMM cannot accommodate to a large increase in volume and therefore prone to break. Notably, PTP opening may initiate cytochrome c release in the absence of large scale matrix swelling and permanent metabolic impairments in some conditions, raising the possibility that an alternative mechanism may also couple reversible PTP opening to the permeabilization of the OMM. Based on early biochemical studies, the PTP was envisioned as a complex formed by the voltage-dependent anion channel (VDAC), adenine nucleotide translocator (ANT), and mitochondrial cyclophilin D (CypD). Recent genetic studies have reported that the knockout of any ANT isoforms or VDAC isoforms or CypD does not affect or only evokes a quantitative change in the activation of the PTP, arguing for the involvement of some other factors in the formation of the pore *(5)*. The PTP activation is usually triggered by mitochondrial Ca^{2+} uptake and the ensuing rise in mitochondrial matrix $[Ca^{2+}]$ ($[Ca^{2+}]_m$) and/or by reactive oxygen species (ROS) *(6)*. However, other factors, including pH, $\Delta\Psi_m$, Bcl-2/xL, and adenine nucleotides also modulate the PTP opening. These factors synergize with each other in PTP activation and receive input from a variety of signaling molecules such as ceramide or arachidonic acid *(7, 8)*.

The second mechanism of OMM permeabilization is induced by a proapoptotic protein of the Bcl-2 family, such as Bax or Bid. Bcl-2 family proteins play a fundamental role in the integration of proapoptotic and antiapoptotic signals and many of those proteins are localized to and control the permeability properties of intracellular membranes *(9–11)*. Among the proapoptotic Bcl-2 family proteins, Bak and a small fraction of Bax are associated with the mitochondria, whereas several BH3-only proteins, e.g., Bid and the major part of Bax are in the cytoplasm of surviving cells, but exhibit redistribution to the mitochondria during apoptosis. Oligomerization of OMM-integrated Bax and Bak is thought to contribute to the formation of a pore upon apoptotic insult, which permits the release of proapoptotic mitochondrial proteins from IMS *(12)*. In response to engagement of the death receptors Bid is cleaved by activated caspase-8, and subsequently, the truncated C-terminus Bid (tBid) is translocated from cytoplasm to the mitochondria. After translocation to the OMM, tBid either directly activates oligomerization of Bax and Bak to form a pore or antagonizes their antiapoptotic counterparts (Bcl-2, Bcl-xl, etc.) and indirectly helps oligomerization of Bax and Bak *(13, 14)*.

Only a smaller fraction of the IMS is in the peripherial IMS and the majority of the proteins are compartmentalized in the cristae. Evidence has been presented that the release of the IMS proteins from the cristae requires opening and reorganization of the cristae *(15)*.

The following subheadings describe methods for the study of the mitochondrial phase of apoptosis using a fluorometer and different fluorescent dyes in suspensions of permeabilized cells. The evaluated parameters are marked in **Fig. 1** (italics).

2. Materials

2.1. Cell Culture

1. Human hepatoma cell line HepG2 (ATCC) and rat cardiac muscle cell line H9c2 (ATCC).
2. Dulbecco's Modified Essential Medium (DMEM) supplemented with 10% fetal bovine serum (Gibco).
3. Sodium pyruvate (Gibco), L-glutamine (Biowhittaker) and Penicillin/Streptomycin (10,000 U/mL and 10,000 µg/mL, respectively) (Biowhittaker).
4. Trypsin (0.25%) and EDTA (1 mM) (Gibco).
5. tBid (Serono Pharmaceutical or R&D Systems).

2.2. Assay Buffers

1. Ca^{2+} free extracellular buffer (Na-HEPES-EGTA): 120 mM NaCl, 5 mM KCl, 1 mM KH_2PO_4, 0.2 mM $MgCl_2$, 0.1 mM EGTA, 20 mM HEPES–NaOH pH 7.4.
2. Intracellular medium (ICM): 120 mM KCl, 10 mM NaCl, 1 mM KH_2PO_4, 5% Dextran, 20 mM HEPES–Tris pH 7.2 supplemented with 1 µg/mL of each of antipain, leupeptin, and pepstatin (Sigma). For Ca^{2+} measurement ICM was passed through a Chelex column (Chelex 100 resins) (Bio-Rad) prior to addition of protease inhibitors to lower the ambient $[Ca^{2+}]$.
3. Caspase assay buffer: 10% w/v sucrose, 0.1% w/v CHAPS, 5 mM dithiothreitol, 100 mM HEPES–Tris pH 7.2.
4. RIPA buffer (Sigma).

2.3. Western Blotting

1. Precast 15% SDS-PAGE gel (Bio-Rad).
2. Running buffer (5×): 125 mM Tris, 960 mM glycine, 0.5%(w/v) SDS.
3. Prestained molecular weight marker, Kaleidoscope markers (Bio-Rad).
4. Transfer buffer: 20 mM Tris, 150 mM glycine, 20%(v/v) methanol.

5. Nitrocellulose membrane and chromatography paper (both from Bio-Rad).

6. Tris-buffered saline with Tween(TBS-T) as powdered form (Sigma). Make a 1-L solution with distilled water and store at room temperature.

7. Blocking buffer: 5% (w/v) nonfat dry milk in TBS-T.

8. Antibody dilution buffer: 0.5% (w/v) nonfat dry milk in TBS-T.

9. Primary antibody: Monoclonal anticytochrome c (clone 7H8.2C12, BD Pharmingen)

10. Secondary antibody: Antimouse IgG conjugated to horse radish peroxidase (GE healthcare).

11. SuperSignal West Pico chemiluminescent substrate and CL-XPosure films (both from Pierce).

2.4. Fluorophores

1. Ratiometric calcium probes: fura/FA (K_d 224 nM, Teflabs or Molecular Probes) and fura2FF/FA (K_d~4 μM, Teflabs) (ex: 340 and 380 nm, em: 535 nm). Make 1 mM stock by dissolving in distilled water. Prepare aliquots and store at –20°C.

2. Ratiometric mitochondrial membrane potential probe JC1 (ex: 570 nm, 490 nm; em: 595 nm, 535 nm) (Molecular Probes). Make 1 mM stock by dissolving in DMSO. Prepare aliquots and store at –20°C.

3. Mitochondrial membrane potential probe TMRE (ex: 540 nm, em: 580 nm) (Molecular Probes). Make 1 mM stock by dissolving in DMSO. Store at –20°C.

4. Fluorogenic caspase substrate: Ac-DEVD-AMC (ex: 380 nm, em: 460 nm) (BD Pharmingen). Reconstitute a stock of 1 mg/mL in DMSO, make aliquots and store at –20°C.

3. Methods

3.1. Simultaneous Measurement of the Ca^{2+}-Induced Changes of $\Delta\Psi_m$ and $[Ca^{2+}]_c$ in Suspension of Permeabilized HepG2 Cells

The cytoplasmic free Ca^{2+} concentration ($[Ca^{2+}]_c$) is maintained at about 100 nM, a very low level relative to the extracellular fluid ($[Ca^{2+}]_{EC} \approx 1.2$ mM). Temporally and spatially organized increases in $[Ca^{2+}]_c$ and $[Ca^{2+}]_m$ represent one of the most commonly used intracellular signals. However, prolonged changes in Ca^{2+} distribution including an elevation in $[Ca^{2+}]_c$ and $[Ca^{2+}]_m$ trigger a variety of cascades that lead to cell death. The mitochondrial matrix is separated from the cytoplasm by two membranes and the IMM has a very limited permeability to ions. However, conditions of elevated $[Ca^{2+}]_c$ evoke activation of the Ca^{2+} uniporter that mediates a $\Delta\Psi_m$ driven Ca^{2+} uptake to the mitochondria. The uniporter-mediated Ca^{2+} uptake can be effectively inhibited by ruthenium red or by Ru360. Calcium uptake by the mitochondria lowers the $[Ca^{2+}]_c$ and is associated with some, typically transient depolarization. Mitochondrial Ca^{2+} uptake also results in a $[Ca^{2+}]_m$ elevation that is transient under physiological conditions and is central for the activation of the Ca^{2+}-dependent processes of mitochondrial ATP production. However, excess load of Ca^{2+} to the mitochondria leads to the opening of PTP, permeabilization of the IMM for solutes and ions, and the ensuing collapse of $\Delta\Psi_m$ and release of Ca^{2+} to the cytoplasm, referred as delayed Ca^{2+} dysregulation *(2, 16)*. The PTP activation can be blocked by cyclosporin A (CsA) and sanglifehrin A, which bind to CypD and by bongkrekic acid that binds to the ANT *(3)*. Simultaneous measurements of $\Delta\Psi_m$ and $[Ca^{2+}]_c$ allow real-time recording and quantitative analysis of these processes and therefore represent useful means to evaluate the role of mitochondria in Ca^{2+}-dependent cell death (**Fig. 2**) *(7)*. For monitoring of $\Delta\Psi_m$, a commonly used probe is JC-1, a lipophilic fluorescent cation. JC-1 is present as a green-fluorescent monomer in the cytoplasm. The highly negative $\Delta\Psi_m$ (–180 mV) facilitates the mitochondrial accumulation of JC-1 that in turn, incorporates into the mitochondrial membrane and forms aggregates (J aggregates). This aggregation changes the fluorescence properties of JC-1 leading to a shift from green to orange/red fluorescence. Thus, quantitation of both the mitochondrial red fluorescence and the cytoplasmic green fluorescence provides information on the $\Delta\Psi_m$. JC-1 data is commonly presented as the ratio of the red and green fluorescence, where the ratio is in a direct relationship with the $\Delta\Psi_m$. Calculation of the JC-1 ratio improves the signal and helps to eliminate artifacts that affect both green and red fluorescence but the user has to remember that the green and red fluorescence are originated from two different cellular compartments. For the measurement of $[Ca^{2+}]_c$ a ratiometric Ca^{2+} probe is practical: fura2/FA (if the $[Ca^{2+}]_c$ is <2 μM) or the lower affinity furaFF/FA. Fura-dyes are used in excitation ratio mode (near-UV excitation, ex1: 340 nm, ex2: 380 nm) and

of permeabilized cells using a membrane potential probe, JC1, and a Ca^{2+} tracer, fura2FF, respectively. Permeabilized cells were treated with C2 or DMSO 180 s prior to the addition of Ca^{2+} pulses. Representative analog traces (*black* – control and *gray* – C2) in the *left panel* shows the C2-induced enhancement of the $\Delta\Psi_m$ loss and $[Ca^{2+}]_c$ rise (referred as delayed Ca^{2+} dysregulation). *Right panel* shows the same measurements in the presence of ruthenium red (2 μM RuRed added 60 s prior to Ca^{2+}). As RuRed blocks the mitochondrial Ca^{2+} uptake, it protects the cells from the C2-induced enhancement of the $\Delta\Psi_m$ loss and $[Ca^{2+}]_c$ rise. (**b**) To establish the role of PTP, similar measurements were carried out in presence of CsA. Blocking of C2-induced enhancement of the $\Delta\Psi_m$ loss and $[Ca^{2+}]_c$ rise by CsA indicate the involvement of PTP in the process of C2-induced delayed Ca^{2+} dysregulation. (**c**) Simultaneous measurements of light scattering and $[Ca^{2+}]_c$ were carried out in suspensions of permeabilized HepG2 cells in presence of C2 only or C2 + CsA. In presence of C2, rise of $[Ca^{2+}]_c$ activates PTP opening and simultaneous swelling of mitochondria which is detected by the drop of the light scattering. CsA prevents PTP opening and protects mitochondria from C2 + Ca^{2+}-induced swelling. Panels (**a**) and (**b**) are reproduced from *(7)* with modification with permission from *The EMBO Journal*.

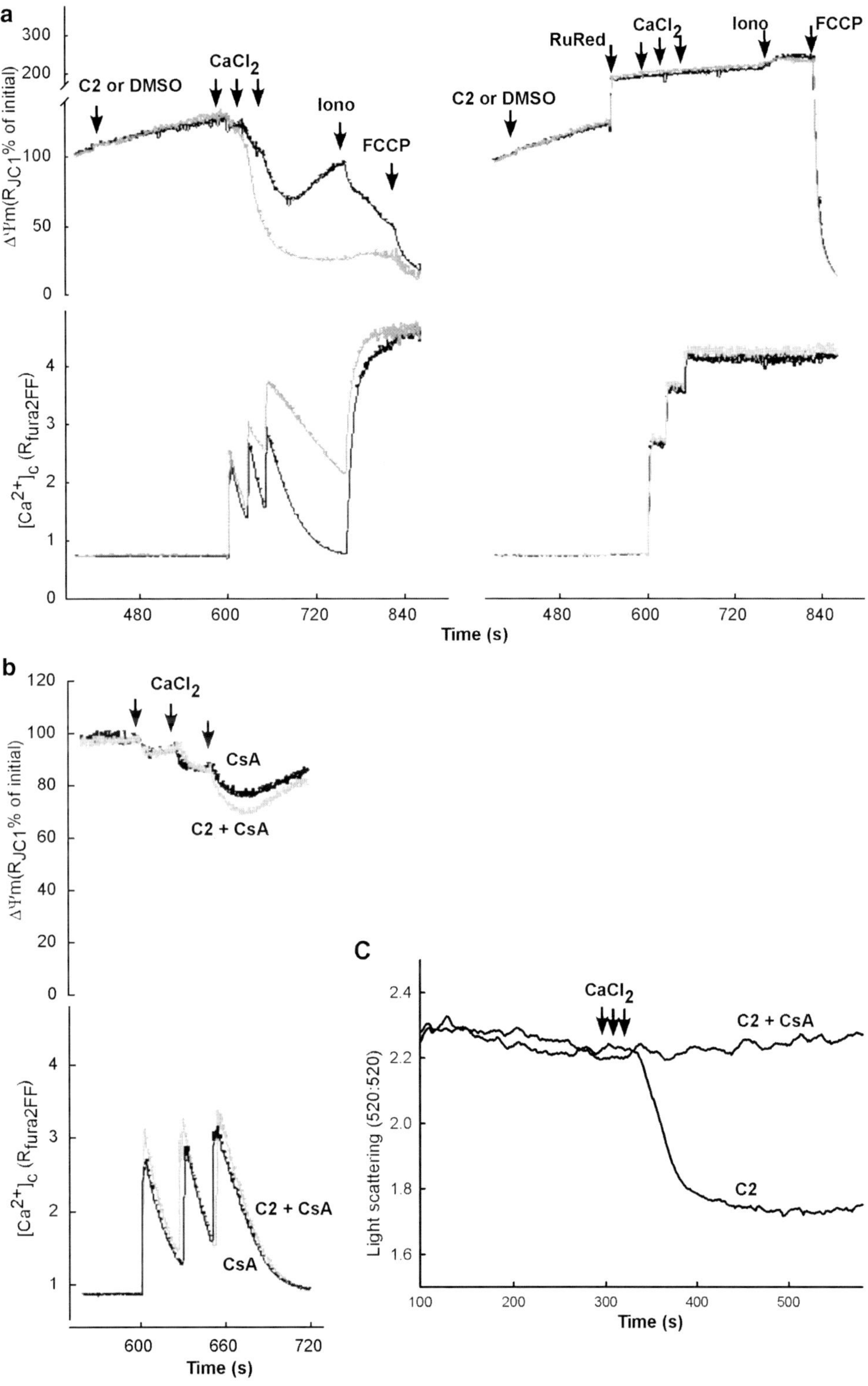

Fig. 2. Ca^{2+}-induced $\Delta\Psi_m$, $[Ca^{2+}]_c$ and mitochondrial swelling responses in permeabilized HepG2 cells exposed to proapoptotic stimuli. (**a**) Simultaneous measurements of $\Delta\Psi_m$ and $[Ca^{2+}]_c$ were carried out in suspensions

the emitted fluorescence can be conveniently measured in the blue-green range. The hydrophilic FA forms are confined to the cytoplasm. While calibration of the JC-1 fluorescence in terms of mV-s is very difficult, calibration of the fura ratio in terms of nM $[Ca^{2+}]_c$ can be done by a simple calibration procedure. Also, from changes in the fura ratio evoked by the addition of $CaCl_2$, one can determine the amount of calcium accumulated or released by the mitochondria.

1. Adherent cell cultures were grown to confluency in T75 flasks, then were washed with PBS and harvested by using 0.25% trypsin–EDTA.

2. Harvested cells were washed with ice cold Ca^{2+}-free extra-cellular buffer (*see* **Subheading 2.2**) and total cell proteins were estimated (by Lowry's method).

3. Equal aliquots of the cell suspension (2.4 mg protein each) were generated and were stored in Ca^{2+}-free extracellular buffer on ice (*see* **Note 1**).

4. Before using the cell aliquots were centrifuged ($150 \times g$ for 5 min) and the supernatant was quantitatively removed (*see* **Note 2**).

5. An aliquot of cells was resuspended in 1.5 mL of ICM (*see* **Subheading 2.2**) and permeabilized with 30–40 µg/mL digitonin in a spectrofluometric cuvette under magnetic stirring at 37°C (*see* **Note 3**).

6. In order to carry out simultaneous measurements of $[Ca^{2+}]_c$ and $\Delta\Psi_m$, the permeabilized cells were supplemented with 0.5 µM furaFF/FA and 800 nM JC-1 (*see* **Note 4**).

7. Fluorescence measurements were done in a multi-wavelength-excitation dual wavelength-emission fluorimeter (Delta RAM, PTI) using 340- and 380-nm excitation and 535-nm emission for fura2FF whereas 490-nm excitation/535-nm emission and 570-nm excitation/595-nm emission were used for JC1 (*see* **Note 5**).

8. After addition of digitonin, furaFF/FA and JC-1 start to record the fluorescence.

9. At 50 s, 2 mM succinate (complex II substrate) or 1 mM/ 5 mM malate/glutamate (complex I substrate) was added to energize the mitochondria.

10. At 100 s, 2 mM MgATP and ATP regenerating system composed of 5 mM phosphocreatine, 5 U/mL creatine kinase was added to provide energy for the ATP-dependent processes (*see* **Note 6**).

11. After reaching a steady state in the JC-1 ratio, three pulses of $CaCl_2$ (40 µM each) were added in 30-s intervals in the cuvette by a Hamilton syringe during continuous recording.

The amount of added $CaCl_2$ was chosen so that the last pulse evoked a reversible depolarization and $[Ca^{2+}]_c$ rise gradually recovering in 2–3 min in the control cells. Both the Ca^{2+}-induced depolarization and the decay of the $[Ca^{2+}]_c$ rise were eliminated if the permeabilized cells were pretreated with ruthenium red (2 μM), an inhibitor of the uniporter (**Fig.** 2a; *see* **Note 7**).

12. Stress agents like C2-ceramide (40 μM), which were tested for sensitization of the Ca^{2+}-induced activation of the PTP were added 180 s prior to start the Ca^{2+} pulsing (**Fig. 2a**) *(6, 7, 17)*.

13. Cyclosporin A (CsA, 5 μM) was added from the start of the recording to clarify the involvement of PTP in the stress agent and Ca^{2+}-induced changes in $\Delta\Psi_m$ and $[Ca^{2+}]_c$ (**Fig. 2b**; *see* **Note 8**).

14. After calcium pulsing the recording was continued for at least 5 min and then an uncoupler, FCCP (5 μM) was added to determine the effect of complete dissipation of the $\Delta\Psi_m$ on the JC-1 ratio. In some experiments, a Ca^{2+} ionophore was added prior to FCCP addition to mobilize the total Ca^{2+} that was retained in the mitochondria and ER.

15. Calibration of the furaFF signal was carried out at the end of each measurement adding 1.5 mM $CaCl_2$ and subsequently 10 mM EGTA–Tris pH 8.5. $[Ca^{2+}]_c$ was calculated by using a K_d of 4 μM. Increases of cytosolic $[Ca^{2+}]$ obtained by addition of $CaCl_2$ were calculated using constants obtained from *(18)*.

16. In most measurements, thapsigargin, an inhibitor of the SERCA Ca^{2+} pump (2 μM) was included to abolish the ER Ca^{2+} uptake so the added Ca^{2+} was accumulated only by the mitochondria. However, in some experiments thapsigargin was omitted and the ER was preloaded by mini pulses of Ca^{2+} (1 μM each) and then ER Ca^{2+} release was induced by IP3. Due to a local Ca^{2+} transfer between mitochondria and adjacent ER, Ca^{2+} release through the IP3 receptor was very effectively relayed to the mitochondria and caused PTP opening in cells exposed to some stress agents (e.g., C2).

3.2. Fluorometric Measurement of the tBid-Induced Loss of $\Delta\Psi_m$ in Suspension of Permeabilized HepG2 and H9c2 Cells

In response to engagement of the death receptors, procaspase-8 is activated to cleave Bid, a BH3-only Bcl-2 family protein, generating truncated C-terminus Bid (tBid). Recent data showed that the cleavage and the ensuing activation of Bid can also be catalyzed by several enzymes including calpain, caspases, cathepsins, and granzyme B in various cell death paradigms *(19)*. tBid interacts with the mitochondria to cause permeabilization of the OMM and subsequent release of proteins from the IMS, which mediate the execution of apoptosis. The tBid-induced OMM permeabilization depends on interplay between pro- and antiapoptotic Bcl-2 family proteins in the OMM (Bak, Bax, and Bcl-xL) and involves cardiolipin but the exact mechanism remains unclear. During tBid-induced release of apoptotic factors from the mitochondria, the inner IMM barrier function is maintained in several experimental models, although remodeling of the cristae seems to be necessary to allow the discharge of the IMS proteins. The IMS protein, cytochrome c is necessary for the mitochondrial electron transport chain function and upon release to the cytoplasm, induces caspase activation in the apoptosome. Thus, if $\Delta\Psi_m$ generation was due to the electron transport chain activity, release of cytochrome c from the mitochondria to the cytosol appears as mitochondrial depolarization *(20)*. This process was monitored by various potentiometric fluorophores (JC-1, tetramethylrhodamine ethyl esters (TMRE), tetramethylrhodamine methyl esters (TMRM), and Rhodamine123 (R123)) in permeabilized HepG2 and H9c2 cells. Fluorometric recording of $\Delta\Psi_m$ by JC-1 has been described (*see* **Subheading 3.1**). All the other probes are also cationic fluorophores that are accumulated in the mitochondria in proportion to the $\Delta\Psi_m$. These dyes do not form aggregates in the membranes and interact with membrane proteins minimally and thus are preferred probes for quantitative measurements of the membrane potential using Nernst equation. However, if these probes are present in the cytoplasm at micromolar concentration, accumulation in the mitochondria leads to quenching of their fluorescence. Mitochondrial depolarization due to cytochrome c release or to uncoupling appeared as dequenching of the TMRE, TMRM, or R123 fluorescence (*see* **Fig. 3**). However, the tBid-induced dequenching was dependent on the presence of oligomycin, an inhibitor of the F1F0 ATPase, indicating that the ATP synthase function in the reverse mode was competent to maintain the $\Delta\Psi_m$ despite the release of cytochrome c and other IMS proteins (*see* **Fig. 3a**). This result was an evidence that tBid does not interfere with the IMM barrier function. While the tBid effect on the $\Delta\Psi_m$ was recorded in the fluorometer, at any time points,

---

phosphokinase) in the absence of oligomycin showed relatively small tBid-induced depolarization. Addition of oligomycin augmented tBid-induced depolarization (**b**), suggesting that $\Delta\Psi_m$ was maintained after cytochrome c release utilizing extramitochondrial ATP in the reverse function of the F_1F_0 ATPase. *Middle panel* shows the comparative measurements of $\Delta\Psi_m$ using JC1 (**c**) and TMRE (**d**) in permeabilized HepG2 cells incubated in the presence of succinate, ATP, ATP regenerating system, and oligomycin. tBid causes a decreases of the JC1 ratio (**c**) and an increase in TMRE fluorescence, reflecting the release of TMRE from the depolarized mitochondria (**d**). The time course of TMRE dequenching is very similar to the decrease in JC1 ratio providing the evidence that both of these dyes could be used for efficient detection of tBid-induced $\Delta\Psi_m$ loss. In the *lower panel*, $\Delta\Psi_m$ was monitored using TMRE in permeabilized H9c2 cells treated with two different doses of tBid (**e**) for 5 min. The loss of $\Delta\Psi_m$ after treatment with tBid exhibited both a dose-dependent lag time of response and a lower rate of depolarization. Panels (**a**)–(**d**) are reproduced from *(20)* with modification with permission from *The Journal of Biological Chemistry*.

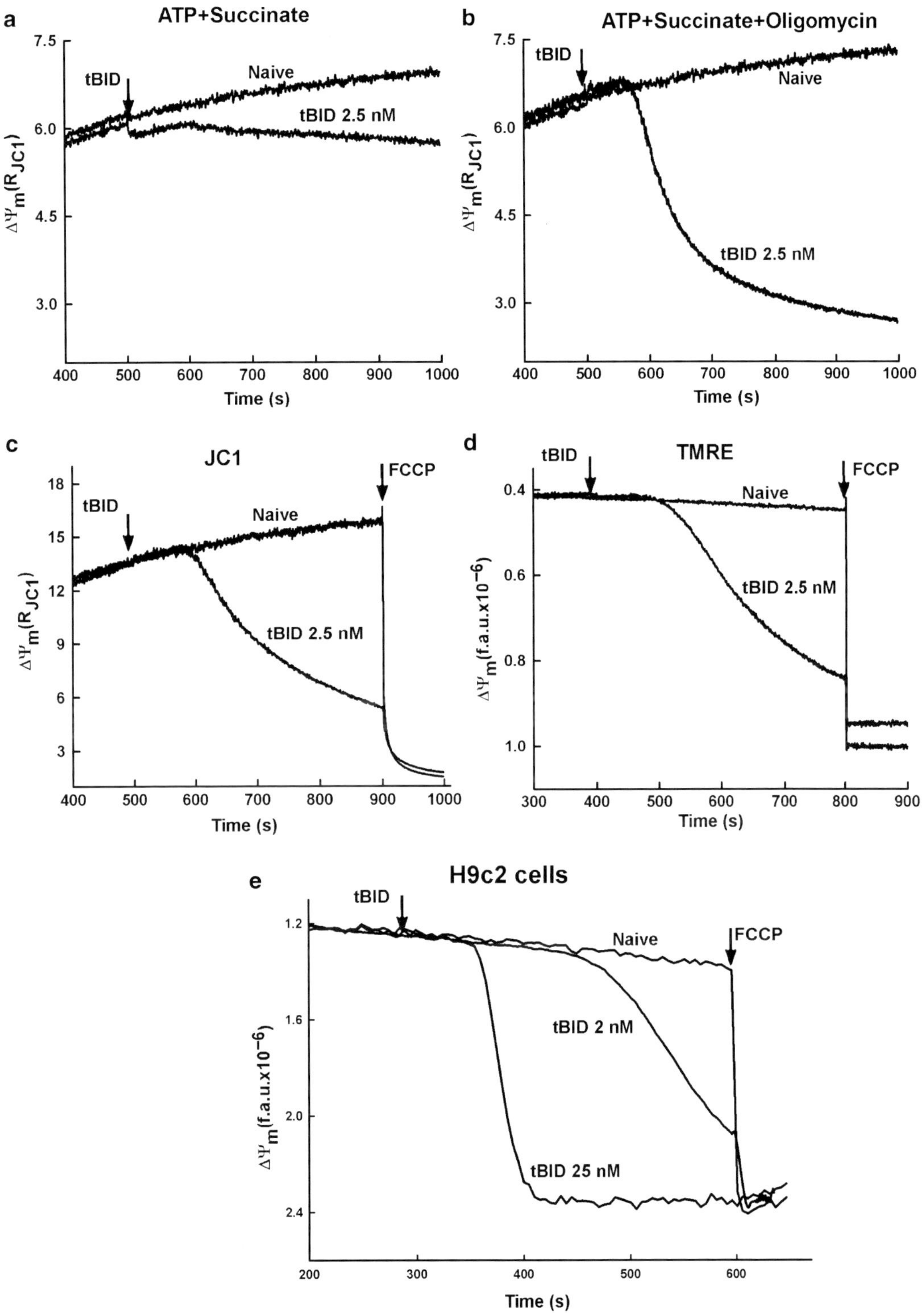

Fig. 3. tBid-induced OMM permeabilization and subsequent $\Delta\Psi_m$ loss in permeabilized cells. *Upper panel* shows the measurements of $\Delta\Psi_m$ using JC1 in permeabilized HepG2 cells incubated in the presence of succinate, ATP, and ATP regenerating system (**a**) succinate, ATP, ATP regenerating system, and oligomycin (**b**). As shown in (**a**), permeabilized cells supplied with succinate (complex II substrate) and ATP regenerating system (ATP, creatine phosphate, and creatine

aliquots of the permeabilized cell suspension can be obtained and rapidly processed for separation of cytosol and membrane fractions that can be used for biochemical or fluorometric detection of the distribution of the mitochondria-derived apoptotic factors (*see* **Subheading 3.4**) *(20)*. The protocol for the fluorometric measurement of TMRE is as follows:

1. **Steps 1–5** were the same as in **Subheading 3.1**.

2. For measurements of $\Delta\Psi_m$, the permeabilized cells were supplemented with 2 μM TMRE.

3. Fluorescence measurement was done in the same fluorometer described in **Subheading 3.1** using 540-nm excitation and 580-nm emission.

4. **Step 8–10** were the same as above.

5. At 150 s 2.5-μg/mL oligomycin was also added to prevent the function of the mitochondrial F1F0 ATPase.

6. At 300 s recombinant tBid was added. tBid could be added at different concentration to establish a dose–response relationship for the OMM-permeabilization.

7. FCCP (5 μM) was added at the end of each run to achieve complete depolarization of the mitochondria.

3.3. The Ca²⁺ and tBid-Induced Changes in Mitochondrial Volume with Time as Monitored by Light Scattering in Permeabilized HepG2 and H9c2 Cells

Under conditions of prolonged opening of the PTP (e.g., Ca^{2+} overload), the mitochondrial matrix volume increases. By contrast, selective permeabilization of the OMM by tBid results in no change or a decrease in matrix volume. Therefore, measurement of the mitochondrial volume may provide some clues to the mechanism of the mitochondrial membrane permeabilization. As mitochondria swell, their refractive index decreases, yielding a drop in light absorbance at higher wavelengths. Monitoring 90° light scatter in mitochondrial suspensions has been commonly used to evaluate changes in mitochondrial volume. Light scattering is usually measured at 520 nm, an isosbestic point for the mitochondrial cytochromes and thus insensitive to changes in redox state. Importantly, light scattering can be monitored simultaneously with the measurement of $[Ca^{2+}]_c$ or $\Delta\Psi_m$. The absolute values for A_{520} are dependent on several factors in addition to the matrix volume, including the concentration of the mitochondria and the optical geometry of the fluorometer. In addition, light scattering is sensitive to changes in the shape and the IMS of the mitochondria *(21, 22)*. Therefore, light scattering measurements can be used to assess the rate of change of matrix volume but do not allow determination of absolute values. Measurement of mitochondrial volume in permeabilized cells is further complicated by the contribution of other organelles to the light scatter. To test the contribution of mitochondrial volume to the changes in light scatter pharmacological means can be used. For example, CsA prevents the Ca^{2+}-induced decrease of absorbance in permeabilized cells (*see* **Fig. 2c**), providing an evidence for PTP-mediated mitochondrial swelling.

3.4. Separation of Cytosolic Fraction and Western Blot Analysis

Both the Ca^{2+}-induced PTP opening and the tBid-induced OMM permeabilization of mitochondria commit the cell to apoptosis by stimulating the release of apoptosis promoting factors to the cytoplasm. These are mitochondrial IMS proteins like cytochrome c, AIF, Smac/DIABLO, OMI/HtrA2, pro-caspases, etc. which are involved in the formation of apoptosome or initiate other apoptosis executing mechanisms *(3)*. To quantitate the release of the IMS proteins, the permeabilized cell suspensions have to be rapidly fractionated in the end of the fluorometric measurements. The mitochondria containing membrane fraction can be separated from the cytoplasm by a centrifugation protocol in 5 min. Furthermore, a rapid filtration-based protocol has also been set up, which allows one to obtain mitochondria-free cytoplasm in 5 s *(17)*. To assess the time course of the release of IMS proteins relative to changes in $\Delta\Psi_m$ and $[Ca^{2+}]_c$, cell samples can be obtained and filtered at multiple time points during the fluorometric measurement. Western blotting of the cytosolic and membrane fractions provides a simple approach to evaluate the release of each IMS protein (*see* **Fig. 4a–d**). The detailed method for separation of cytosolic and membrane fractions and Western blotting of cytochrome c is as follows:

1. After completion of the fluorometric measurement of $\Delta\Psi_m$ and $[Ca^{2+}]_c$ the cell suspension was rapidly centrifuged at $12,000 \times g$ for 5 min (*see* **Note 9**).

2. Alternatively, the cell suspension was rapidly passed through a syringe-less filter device (Autovial, 0.45-µm cellulose acetate membrane, Whatman) that retains the cellular membranes.

3. The supernatant (cytoplasm) and pellet (cellular membranes) produced by the centrifuge approach and the flow-trough generated by the filtration were put immediately in liquid nitrogen for rapid freezing. Subsequently, the samples were stored at –80°C.

4. Membrane fractions separated by centrifugation were lyzed by RIPA buffer. The membranes were incubated for 40 min in 4°C by rotating it and then centrifuged for 15 min in $12,000 \times g$. Supernatant were saved at –80°C for future use.

5. Proteins (25 µg) were separated by SDS-PAGE (15% gel) under reducing condition.

6. Separated proteins were transferred to nitrocellulose membrane.

7. Blocked the membrane in blocking buffer overnight in 4°C with gentle shaking.

8. Washed the membrane with TBS-T at room temperature with shaking for 5 min.

9. The membrane was incubated with the primary antibody (anticytochrome c) at 1:500 dilutions in antibody dilution buffer with gentle shaking for 5 h in room temperature.

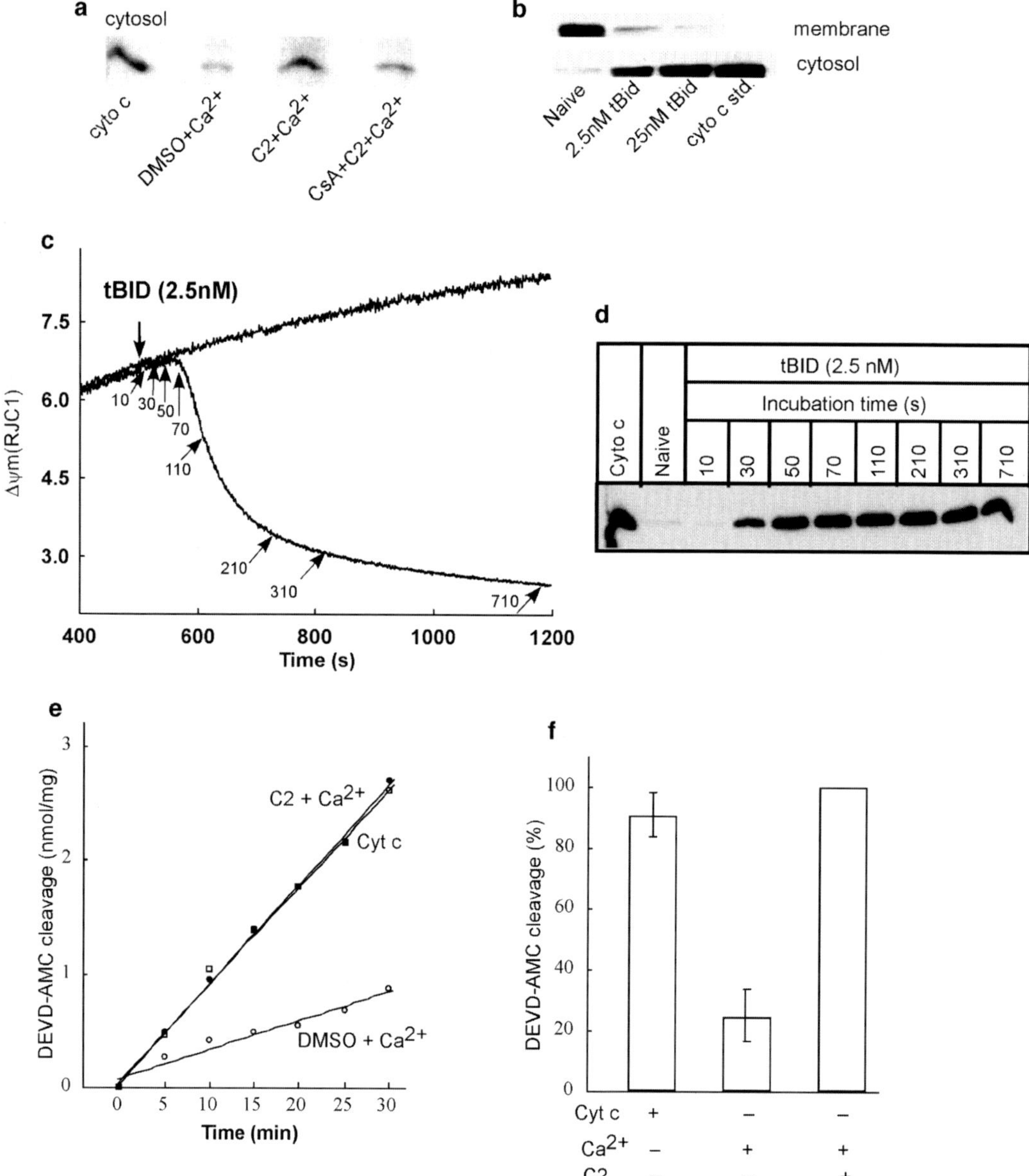

Fig. 4. Release of cytochrome *c* and activation of caspases after $\Delta\Psi_m$ loss in permeabilized HepG2 cells. *Upper panel* shows the immunoblots of cytochrome *c* in cytosolic and membrane fractions generated after the measurements of $\Delta\Psi_m$ in permeabilized cells. Cytosolic samples generated from the experiments described in **Fig.** 2a, b shows higher amount of cytochrome *c* in presence of C2 + Ca^{2+} compared to that of C2 + Ca^{2+} + CsA reflecting PTP opening (**a**). Membrane and cytosolic samples generated after $\Delta\Psi_m$ measurements in presence of two different concentrations of tBid shows dose-dependent release of cytochrome *c* in cytosol from the mitochondria (**b**). *Middle panel* shows that tBid-induced $\Delta\Psi_m$ loss is closely coupled to cytochrome *c* release. $\Delta\Psi_m$ was monitored in permeabilized cells treated with 2.5 nM tBid in the presence of succinate, ATP, ATP regenerating system, and oligomycin. *Arrows* indicate the time points at which cytosolic samples were generated by rapid filtration of the cells (**c**). Panel (**d**) shows time course of tBid-induced cytochrome *c* release as determined by Western blotting of the cytosolic samples. *Lower panel* shows caspase activation associated with Ca^{2+}-induced opening of PTP. Fluorometric assay of DEVD–AMC cleavage in cytosol extracts prepared at the end of measurements of $\Delta\Psi_m$ (*see* **Fig.** 2a) was carried out. Data in (**e**) shows the increase of Caspase 3 activity in presence of C2 + Ca^{2+} compared to DMSO control. Addition of exogenous cytochrome *c* in DMSO control cytosol increased its activity of Caspase 3 at the level of C2 + Ca^{2+} condition indicate that activation of Caspase 3 is downstream to the cytochrome *c* release. Data in (**f**) shows the DEVD-AMC cleavage normalized to the activity obtained with C2 + Ca^{2+}. Panels (**a**, **e**) and (**f**) are reproduced from *(7)* with modification with permission from *The EMBO Journal*, whereas panels (**c**) and (**d**) from *(20)* with modification with permission from *The Journal of Biological Chemistry*.

10. The membrane was washed with TBS-T in room temperature with shaking for six times with 5 min duration for each washing cycle.

11. The membrane was incubated in secondary antibody at 1:5,000 dilutions in antibody dilution buffer and was shaken gently for 1.5 h at room temperature.

12. The membrane was washed with TBS-T in room temperature with shaking for six times for 5 min for each washing cycle.

13. The blots were taken to the darkroom and were developed using an enhanced chemiluminescence kit.

3.5. Caspase Assay

The mitochondrial phase of apoptosis involves the release of apoptosis promoting proteins from the mitochondria and the assembly of the apoptosome. Here, cytochrome c binds Apaf-1 and in the presence dATP or ATP, recruits and activates procaspase-9 to form a complex that initiates activation of the effector caspases (e.g., caspase-3). The ICE family member Caspase-3 is involved in proteolysis of several important proteins including poly (ADP ribose) polymerase (PARP). The caspase-3-specific cleavage site is formed by DEVD that has also provided a sequence for generation of an artificial peptide substrate. Binding of the DEVD peptide to a fluorophore is utilized for an in vitro assay of caspase-3 activity. For example, when caspase-3 is incubated with Ac-DEVD-aminomethylcoumarin (AMC), the active caspase-3 cleaves the tetrapeptide between D and AMC, thus releasing fluorogenic AMC that can be measured in a fluorometer (ex: 380 nm, em: 460 nm). The detailed protocol of measuring caspase-3 activity in the cytoplasm fractions (*see* **Subheading 3.4**) is described below (*see* **Fig. 4e, f**).

1. 180-μL of cytoplasmic extract (protein concentration: 0.8–1.2 mg/mL) was added to 1,420 μL of caspase assay buffer in a cuvette.

2. The mixture was incubated for 10 min at 37°C and then the recording of fluorescence.

3. Add 12.5 μM substrate, Ac-DEVD-AMC, into the mixture and take another reading.

4. Assay mixture was incubated in the presence of substrate for 30 min at 37°C and the fluorescence of free AMC (ex: 380 nm, em: 460 nm) was monitored using the fluorometer described above.

5. Addition of exogenous cytochrome c (400 nM) to the assay mixture brought about a large increase in DEVD-AMC cleavage proved the requirement of cytochrome c for activation of Caspase 3.

6. Fluorescence was also measured in samples that did not contain any cytosol or were incubated in the presence of a caspase inhibitor. The results were used to make correction

for the noncaspase-dependent increase in AMC fluorescence. Furthermore, the fluorescence of various amounts of free AMC was also measured to establish the AMC vs. fluorescence relationship and to calculate the number of DEVD-AMC molecules cleaved by the effector caspase enzymes.

4. Notes

1. Cell aliquots of different cell types can be stored for 4–12 h without an apparent change in the mitochondrial and ER Ca^{2+} handling and $\Delta\Psi_m$ generation.

2. EGTA, a Ca^{2+} chelator was present in the cell storage buffer and affected the fura2 fluorescence changes if it was not removed from the cell suspension.

3. To achieve permeabilization of the plasma membrane in >99% of the cells and to avoid permeabilization of the mitochondrial and ER membranes, the digitonin concentration has to be carefully titrated for each cell type. To determine the lowest effective concentration of digitonin, we used Trypan Blue that only enters the cells upon the loss of the plasma membrane barrier and visualized the blue permeabilized cells under a microscope.

4. The relatively low affinity of fura2FF for Ca^{2+} ($K_d \sim 4$ μM for fura2FF) was favorable for avoiding saturation of the dye during large increases of $[Ca^{2+}]_c$.

5. The cuvette chamber was attached with a circulating water bath to maintain 37°C temperature continuously.

6. Mitochondria that have intact IMM but lack normal respiratory chain activity could generate a $\Delta\Psi_m$ using ATP utilizing the reversed operation of the F1F0-ATP synthase.

7. Usually less amount of calcium pulsing is required to evoke the similar effect when complex I substrates were used.

8. CsA targets multiple factors in intact cells but the cytoplasmic targets are largely diluted and effectively eliminated after cell permeabilization. As a negative control, FK506 (5 μM) was used that fails to stabilize the closed state of the PTP but inhibits the cytoplasmic targets of the CsA.

9. It is recommended that the samples are centrifuged at 4°C but the cytochrome c distribution did not seem to be altered when centrifuging was done at room temperature.

Acknowledgments

We would like to thank Drs. György Csordás, Gábor Szalai, Xuena Lin, Muniswamy Madesh, Cecilia Garcia, Ludivine Walter, and Mrs. Erika Davies for their dedicated efforts in setting up the approaches described here. This work was supported by R01-GM59419 from the National Institutes of Health to G.H.

References

1. Duchen, M.R. (2004). Mitochondria in health and disease: perspectives on a new mitochondrial biology. *Mol Aspects Med* **25**, 365–451.

2. Rizzuto, R., and Pozzan, T. (2006). Microdomains of intracellular Ca^{2+}: molecular determinants and functional consequences. *Physiol Rev* **86**, 369–408.

3. Kroemer, G., Galluzzi, L., and Brenner, C. (2007). Mitochondrial membrane permeabilization in cell death. *Physiol Rev* **87**, 99–163.

4. Chipuk, J.E., Bouchier-Hayes, L., and Green, D.R. (2006). Mitochondrial outer membrane permeabilization during apoptosis: the innocent bystander scenario. *Cell Death Differ* **13**, 1396–1402.

5. Bernardi, P., and Forte, M. (2007). The mitochondrial permeability transition pore. *Novartis Found Symp* **287**, 157–164; discussion 164–159.

6. Hajnoczky, G., Davies, E., and Madesh, M. (2003). Calcium signaling and apoptosis. *Biochem Biophys Res Commun* **304**, 445–454.

7. Szalai, G., Krishnamurthy, R., and Hajnoczky, G. (1999). Apoptosis driven by IP(3)-linked mitochondrial calcium signals. *EMBO J* **18**, 6349–6361.

8. Scorrano, L., Penzo, D., Petronilli, V., Pagano, F., and Bernardi, P. (2001). Arachidonic acid causes cell death through the mitochondrial permeability transition. Implications for tumor necrosis factor-alpha aopototic signaling. *J Biol Chem* **276**, 12035–12040.

9. Adams, J.M., and Cory, S. (2007). Bcl-2-regulated apoptosis: mechanism and therapeutic potential. *Curr Opin Immunol* **19**, 488–496.

10. Kuwana, T., and Newmeyer, D.D. (2003). Bcl-2-family proteins and the role of mitochondria in apoptosis. *Curr Opin Cell Biol* **15**, 691–699.

11. Danial, N.N., and Korsmeyer, S.J. (2004). Cell death: critical control points. *Cell* **116**, 205–219.

12. Antignani, A., and Youle, R.J. (2006). How do Bax and Bak lead to permeabilization of the outer mitochondrial membrane? *Curr Opin Cell Biol* **18**, 685–689.

13. Luo, X., Budihardjo, I., Zou, H., Slaughter, C., and Wang, X. (1998). Bid, a Bcl2 interacting protein, mediates cytochrome *c* release from mitochondria in response to activation of cell surface death receptors. *Cell* **94**, 481–490.

14. Wei, M.C., Lindsten, T., Mootha, V.K., Weiler, S., Gross, A., Ashiya, M., Thompson, C.B., and Korsmeyer, S.J. (2000). tBID, a membrane-targeted death ligand, oligomerizes BAK to release cytochrome *c*. *Genes Dev* **14**, 2060–2071.

15. Scorrano, L., Ashiya, M., Buttle, K., Weiler, S., Oakes, S.A., Mannella, C.A., and Korsmeyer, S.J. (2002). A distinct pathway remodels mitochondrial cristae and mobilizes cytochrome *c* during apoptosis. *Dev Cell* **2**, 55–67.

16. Hajnoczky, G., Csordas, G., Das, S., Garcia-Perez, C., Saotome, M., Sinha Roy, S., and Yi, M. (2006). Mitochondrial calcium signalling and cell death: approaches for assessing the role of mitochondrial Ca^{2+} uptake in apoptosis. *Cell Calcium* **40**, 553–560.

17. Madesh, M., and Hajnoczky, G. (2001). VDAC-dependent permeabilization of the outer mitochondrial membrane by superoxide induces rapid and massive cytochrome *c* release. *J Cell Biol* **155**, 1003–1015.

18. Bers, D.M., Patton, C.W., and Nuccitelli, R. (1994). A practical guide to the preparation of Ca^{2+} buffers. *Methods Cell Biol* **40**, 3–29.

19. Yin, X.M. (2006). Bid, a BH3-only multifunctional molecule, is at the cross road of life and death. *Gene* **369**, 7–19.

20. Madesh, M., Antonsson, B., Srinivasula, S.M., Alnemri, E.S., and Hajnoczky, G. (2002). Rapid kinetics of tBid-induced cytochrome *c* and Smac/DIABLO release and mitochondrial depolarization. *J Biol Chem* **277**, 5651–5659.

21. Das, M., Parker, J.E., and Halestrap, A.P. (2003). Matrix volume measurements challenge the existence of diazoxide/glibencamide-sensitive KATP channels in rat mitochondria. *J Physiol* **547**, 893–902.

22. Petronilli, V., Cola, C., Massari, S., Colonna, R., and Bernardi, P. (1993). Physiological effectors modify voltage sensing by the cyclosporin A-sensitive permeability transition pore of mitochondria. *J Biol Chem* **268**, 21939–21945.

Chapter 14

Regulation of Apoptosis by the Unfolded Protein Response

Andrew Fribley, Kezhong Zhang, and Randal J. Kaufman

Summary

In eukaryotic cells, the endoplasmic reticulum (ER) serves many specialized functions including bio-synthesis and assembly of membrane and secretory proteins, calcium storage and production of lipids and sterols. As a plant for protein folding and posttranslational modification, the ER provides stringent quality control systems to ensure that only correctly folded proteins exit the ER and unfolded or misfolded proteins are retained and ultimately degraded. Biochemical, physiological, and pathological stimuli that interfere with ER function can disrupt ER homeostasis, impose stress to the ER, and subsequently cause accumulation of unfolded or misfolded proteins in the ER lumen. To deal with accumulation of unfolded or misfolded proteins, the cell has evolved highly specific signaling pathways collectively called the "unfolded protein response" (UPR) to restore normal ER functions. However, if the overload of unfolded or misfolded proteins in the ER is not resolved, the prolonged UPR will induce ER stress-associated programmed cell death, apoptosis, to protect the organism by removing the stressed cells. In this chapter, we summarize our current understanding of UPR-induced apoptosis and various methods to detect ER stress and apoptosis in mammalian cells.

Key words: Apoptosis, Endoplasmic Reticulum Stress, Unfolded Protein Response

1. Introduction

1.1. The Unfolded Protein Response

As a unique protein-folding compartment and a dynamic calcium store, the ER is very sensitive to alterations in intracellular homeostasis. Perturbations that alter ER homeostasis disrupt folding and lead to the accumulation of unfolded proteins and protein aggregates which are detrimental to cell survival. Disturbances in intraluminal calcium or celluar redox status, increased demand for protein folding due to elevated synthesis of secretory proteins, the expression of mutant or misfolded proteins, nutrient/glucose deprivation or viral infection can all lead to stress in the ER. The ER has evolved highly specific signaling pathways collectively referred

Peter Erhardt and Ambrus Toth (eds.), *Apoptosis*, Methods in Molecular Biology, vol. 559
DOI 10.1007/978-1-60327-017-5_14
© Humana Press, a part of Springer Science + Business Media, LLC 2004, 2009

to as the unfolded protein response (UPR) that alter intracellular transcriptional and translational programs to deal with the accumulation of unfolded or misfolded proteins. Following the introduction of stress these pathways prevent continued accumulation of unfolded protein in the ER lumen by transiently attenuating general protein synthesis, increasing the lumenal folding capacity and the degradation of misfolded proteins through ER-associated protein degradation (ERAD) or autophagy *(1–3)*.

The complex network of the UPR to ER stress is mediated by only a few ER transmembrane proteins: PERK (PKR-like ER kinase), IRE1 (inositol-requiring enzyme 1), and ATF6 (activating transcription factor 6) *(3, 4)*. All these stress sensors have an ER lumenal domain that can sense the presence of unfolded or misfolded proteins, an ER transmembrane domain that targets the protein to the ER, and a functional cytosolic domain. Under resting conditions, all the three ER stress sensors are maintained in an inactive state through their association with the chaperone GRP78/BiP. When cells encounter ER stress, GRP78/BiP dissociates from the stress sensors and binds to underglycosylated, misfolded, or unassembled proteins, thus leading to activation of the UPR **(Fig. 1)**.

1.1.1. PERK/eIF2α/CHOP Signaling and Apoptosis

The most immediate response to the accumulation of unfolded or misfolded proteins is the activation of the ER transmembrane kinase PERK. Activated PERK phosphorylates the alpha subunit of eukaryotic translation initiation factor 2 (eIF2α) and attenuates general protein synthesis to reduce the lumenal client load in the stressed cells. Murine cells deleted in PERK or mutated at Ser51 in eIF2α to prevent phosphorylation did not attenuate protein synthesis upon ER stress and were significantly less able to survive stressful stimuli *(5, 6)*. PERK activation promotes cell survival by limiting the protein-folding load on the ER and by inducing the transcription of specific UPR survival genes downstream of eIF2α *(7)*. However, under prolonged or severe ER stress PERK-mediated activation of the transcription factor ATF4 has been demonstrated to induce CHOP/GADD153 *(7)* and also the proapoptotic BH3-only BCL2 family members NOXA and BIM *(8)*. CHOP (C/EBP homologous protein) is a bZIP-containing transcription factor that was identified as a member of the CCAAT/enhancer binding protein (C/EBP) family *(9)*. Overexpression of CHOP induces cell cycle arrest or apoptosis by regulating expression of multiple genes encoding proapoptotic factors including DR5 (death receptor 5), TRB3 (Tribbles homolog 3), CAVI (carbonic anhydrase VI) *(10–14)*; and CHOP-deficient cells are protected from ER stress-induced apoptosis *(15)*. CHOP also contributes to apoptosis through activating ERO1α, an ER oxidase that promotes hyperoxidization of the ER *(16)*, and through dimerization with cAMP-responsive

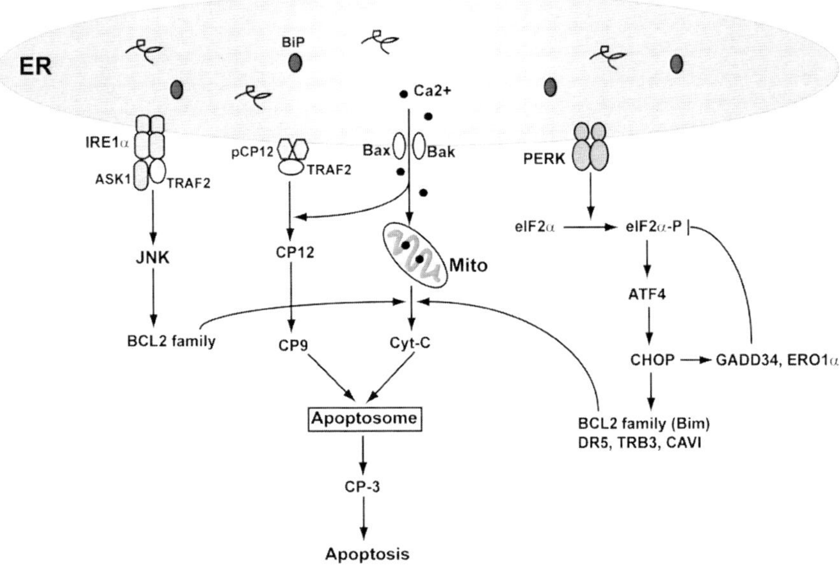

Fig. 1. The apoptotic pathways regulated by the UPR. In response to ER stress, PERK, and IRE1α are activated through their homodimerization and autophosphorylation. Activated PERK phosphorylates translation initiation factor eIF2α, which can selectively induce expression of proapoptotic transcription factors ATF4 and CHOP. CHOP induces expression of numerous proapoptotic factors including DR5, TRB3, CAVI, and BCL2 family proteins. CHOP can also induce expression of GADD34 and ERO1α, leading to apoptosis by increasing protein synthesis and oxidation in the ER of stressed cells. On activation of the UPR, IRE1α serves as a scaffold protein to form a complex with TRAF2 and ASK1, which subsequently activates the JNK-mediated apoptotic pathway. Furthermore, ER stress can induce BAX and BAK localization and oligomerization at the ER, which promotes calcium release from the ER to the cytosol. Increased cytosolic calcium concentration stimulates calcium uptake into the mitochondrial matrix, which can result in depolarization of the inner membrane and transition of the outer membrane permeability pore. This causes cytochrome *c* release and Apaf-1-dependent activation of the apoptosome leading to apoptosis. In addition, ER stress may also promote dissociation of TRAF2 from ER membrane-resident procaspase-12, allowing caspase-12 activation to mediate apoptosis. *Mito* Mitochondria, *Cyt-c* Cytochrome *c*, *pCP12* Procaspase-12, *CP12* Caspase-12, *CP9* Caspase-9, *CP-3* Caspase-3.

element binding protein (CREB) which can suppress the expression of the survival protein *Bcl2 (11)*.

1.1.2. The IRE1α Pathway

Upon activation of the UPR, IRE1 is activated through homodimerization and *trans*-autophosphorylation which can also lead to transcriptional activation of CHOP *(17)*. The mammalian genome contains two homologs of yeast *IRE1* designated as *IRE1α* and *IRE1β*. Whereas IRE1α is expressed in most cells and tissues, with high-level expression in the pancreas and placenta, IRE1β expression is prominent only in intestinal epithelial cells *(18, 19)*. Activated IRE1α can function as an endoribonuclease to initiate removal of a 26-nucleotide intron from X-box binding protein 1 (*Xbp1*) mRNA. Spliced *Xbp1* mRNA encodes a protein with a novel carboxy-terminus that acts as a potent transcriptional activator and spliced XBP1 induces expression of genes encoding enzymes that facilitate protein folding, secretion, or degradation *(20–23)*. In addition to promoting CHOP accumulation to induce apoptosis, under ER stress conditions, IRE1α can serve as a scaffold to recruit TNF receptor-associated factor 2 (TRAF2) and c-Jun-N-terminal inhibitory kinase (JIK) to the ER membrane *(24, 25)*. TRAF2 activates the apoptosis-signaling kinase 1 (ASK1), a mitogen-activated protein kinase kinase kinase (MAPKKK) *(26)*. Activated ASK1 leads to activation of the JNK protein kinase and mitochondria-dependent caspase activation *(26, 27)*. JNK activation is known to influence the cell-death machinery through phosphorylation of BCL2 family proteins including BCL2 and BIM, which subsequently promote cell death programs *(28)*; activated IRE1α can modulate apoptosis through interaction with BAX and BAK *(29, 30)*.

1.1.3. The ATF6 Pathway

ATF6 is a bZIP transcription factor of the CREB/ATF family. There are two forms of ATF6, ATF6α, and ATF6β (also known as CREB-RP). Upon UPR activation, ATF6 is transported to Golgi where it is cleaved by site-1 protease (S1P) and site-2 protease (S2P) to generate a 50-kDa cytosolic b-ZIP-containing fragment that migrates to the nucleus to activate transcription in coordination with the transcription factor NF-Y which is required for ATF6 docking to ER-associated binding elements present in the promoter regions of the UPR target genes. Although ATF6α facilitates adaptation to ER stress, and ATF6 null MEFs appear to be more sensitive to ER stress-induced cell death, it is dispensable for classical UPR signaling *(17, 31)*.

1.2. ER Stress-Induced Caspase Activation

Recent evidence suggests that under ER stress, proapoptotic Bcl2 proteins, Bak and Bax, undergo conformational alteration in the ER membrane to permit Ca^{2+} efflux into the cytoplasm *(32, 33)*. The increase in Ca^{2+} concentration in the cytoplasm activates the calcium-dependent cysteine protease m-Calpain, which subsequently cleaves and activates the ER-resident procaspase-12 *(34, 35)*. Activated caspase-12 cleaves and activates procaspase-9 and consequently leads to activation of the caspase cascade. Calpain-deficient MEFs have reduced ER stress-induced caspase-12 activation and are resistant to ER stress-associated apoptosis *(36)*. In addition to Calpain, caspase-7 translocates from the cytosol to the cytoplasmic side of the ER membrane under ER stress, and interacts with caspase-12, leading to its activation *(35)*. Activated caspase-12 activates caspase-9, which, in turn, forms the apoptosome with released Cytochrome *c* and Apaf-1 to activate the executioner caspase-3 and apoptosis. Although initial studies found that caspase-12 deficient cells could resist ER stress-induced apoptosis *(34)*, more recent studies showed that *caspase-12^{-/-}* mice were not protected from cell death induced by

ER stress *(37)*. The requirement for caspase-12 for apoptosis in human cells is open to question, as the human caspase-12 gene contains several inactivating mutations *(38)*. In addition, ER stress-induced Ca^{2+} release can also promote Ca^{2+} uptake into the mitochondrial matrix, leading to depolarization of mitochondrial inner membrane and Cytochrome *c* release and formation of the apoptosome to activate caspase-3, DNA fragmentation, and cell death *(39)*.

2. Materials

2.1. Cell Culture, Stress-Inducing Compounds, and Lysis

1. Dulbecco's Modified Eagle's Medium (DMEM) supplemented with Pen:Strep and 10% Fetal Bovine Serum (FBS) (Invitrogen).
2. Thapsigargin (Calbiochem) is dissolved at 500 µM in DMSO.
3. Tunicamycin (Calbiochem) is dissolved at 10 mg/mL in DMSO.
4. Trizol (Invitrogen).
5. Modified Radioimmune-precipitation Assay (RIPA) buffer: 50 mM sodium fluoride, 0.5–1.0% NP-40, 10 mM sodium (mono) phosphate, 150 mM sodium chloride, 25 mM Tris (pH 8.0), 1 mM phenylmethylsulfonylfluoride (PMSF), 2 mM ethylenediaminetetraacetic acid (EDTA), and 0.2 mM sodium vanadate; prepared in ddH_2O. Immediately prior to use RIPA buffer is supplemented at 1:100 with 50 mM PMSF and protease inhibitor cocktail (P3840 Sigma).

2.2. Western Immunoblot Analysis and Immunoprecipitation of IRE1α and PERK

1. Protein Loading buffer (PLB), prepared by combining 3 mL of 20% SDS, 3-mL glycerol, 1.9 mL of 1 M Tris (pH 6.8), 1.5 mL of 2-mercaptoethanol, 200 µg bromophenol blue (15 mM final concentration), and ddH_2O to 10.0 mL.
2. PVDF membrane (#162-0177 BioRad).
3. Tris-Buffered Saline with Tween (TBST) containing 145 mM NaCl, 10 mM Tris (pH 8.0) and 0.1% Tween 20 prepared in ddH_2O.
4. Nonfat powdered milk (Carnation).
5. Cell lysis buffer for immunoprecipitation-Western blot analysis is 1% NP-40, 50 mM Tris–HCl (pH 7.5), 150 mM NaCl, 0.05% SDS, 0.5 mM sodium vanadate, 100 mM NaF, 50 mM beta-glycerophosphate, and 1 mM phenylmethylsulfonyl fluoride (PMSF) supplemented 1:100 with protease inhibitor cocktail (Sigma).

6. Protein A beads.

7. Primary antibodies: CREB2/ATF4 (sc-200, Santa Cruz), GADD153/CHOP (sc-793 or sc-575, Santa Cruz), eIF2α (#9722, Cell Signaling), phospo-eIF2α (#44-728G, Biosource International), polyclonal rabbit anti-PERK-PITK-289 antibody was provided by Dr. Yuguang Shi (Lilly Research Laboratories) or from Cell Signaling (#3179), GADD34 (sc-825 or sc-8327, Santa Cruz), GRP78/BiP (#610978, BD Transduction Laboratories; or sc-1050, Santa Cruz), caspase3 (sc-7148, Santa Cruz), PARP (#9542, Cell Signaling), NOXA (sc-22764, Santa Cruz), PUMA (#4976, Cell Signaling), murine anti-human IRE1α primary antibody was raised in our lab.

8. Secondary horseradish peroxidase (HRP)-conjugated antibodies: Goat anti-mouse IgG (H + L) (BioRad), and goat anti-rabbit IgG (H + L) (Promega).

9. Super Signal West Pico Chemiluminescent Substrate reagent (Pierce).

10. Restore Western blot stripping buffer (Pierce).

2.3. Reverse Transcription PCR, Conventional RT-PCR, and Quantitative PCR

1. Multiscribe Reverse Transcriptase kit (Bio-Rad).

2. SYBR green PCR master mix (Bio-Rad).

3. Murine conventional reverse transcription PCR (RT-PCR): *Xbp1* (total and spliced) forward 5′- CCTTGTGGTTG AGAACCAGG-3′ and reverse 5′-CTAGAGGCTTGG TGTATAC -3′ products for *Xbp1* un-spliced and spliced are 451 bp and 425 bp, respectively; *GRP78/BiP* forward 5′-CTGGGTACATTTGATCTGACTGG-3′ and reverse 5′-GCATCCTGGTGGCTTTCCAGCCATTC-3′ product is 397 bp.

4. Murine quantitative real-time PCR (qPCR): *Xbp1* (spliced) forward primer 5′-GAGTCCGCAGCAGGTG-3′. This primer was designed to span the 26 base intron, thus can only anneal to the spliced *Xbp1* transcript. The reverse primer sequence is 5′-GTGTCAGAGTCCATGGGA-3′, which is 70 bases downstream of the forward primer. *Xbp1* (spliced and un-spliced) forward primer 5′-AAGAACACGCTTGGGAA TGG-3′, and reverse 5′-ACTCCCCTTGGCCTCCAC-3′; *GRP78/BiP* forward primer 5′-CATGGTTCTCACTAA AATGAAAGG-3′ and reverse 5′-GCTGGTACAGTAACAA CTG-3′; *Atf4* forward 5′-ATGGCCGGCTATGGATGAT-3′ and reverse 5′-CGAAGTCAAACTCTTTCAGATCCATT-3′; *Chop* forward 5′-CTGCCTTTCACCTTGGAGAC-3′ and reverse 5′-CGT TTCCTGGGGGATGAGATA-3′; *β-actin* forward primer 5′-GATCTGGCACCACACCTTCT-3′, and reverse primer 5′-GGGGTGTTGAAGGTCTCAAA-3′.

We have found the use of GAPDH as an internal control for the induction of UPR-related mRNA transcripts to be less reliable than *β-actin* or 18s ribosomal mRNA for reasons that are not entirely clear.

2.4. Nuclear DNA Fragmentation Assay

1. DNA lysis buffer: 0.5 mL of 1 M Tris–HCL (pH 7.4), 0.1 mL of 5 M NaCl, 1.0 mL of 0.5 M EDTA (pH 8.0) and 2.5 mL 10% SDS; bring up to 50 mL with ddH_2O. Note: supplement with 100 μg/mL proteinase K immediately prior to use.

2. Phenol–chloroform (1:1).

3. 70% Ethanol.

4. RNAse A (Applied Biosystems).

---

3. Methods

Most studies examining mechanisms of ER stress-induced apoptosis are performed with the well-described pharmacological toxins tunicamycin or thapsigargin that cause severe ER stress. Tunicamycin interferes with N-linked glycosylation of polypeptides early in the secretory pathway, and thapsigargin is an effective inhibitor of Ca^{+2} ion pump proteins located in the membrane of the ER. Of the pathways implicated in ER stress-induced apoptosis, only the significance of the PERK/eIF2α pathway has been confirmed in several physiological models including pancreatic beta cell death in diabetes and macrophage cell death in atherosclerotic lesions *(6, 40–42)*. Following tunicamycin or thapsigargin treatment the ER stress-inducible proapoptotic transcription factor CHOP plays a primary role in ER stress-induced death. Due to the fact that CHOP null murine embryonic fibroblasts are only partially protected from ER stress-induced apoptosis *(15)*, it is likely that other stress-induced proapoptotic factors remain to be elucidated. Reliable methods have been established to characterize ER stress-induced apoptosis in vitro and in vivo. Here, we will discuss established experimental methods for the identification and characterization of ER stress-induced apoptosis mediated through PERK/eIF2α/CHOP signaling and by several BH3-only BCL2 family members.

3.1. Cell Culture and Sample Preparation

1. $1.5–2.0 \times 10^5$ murine embryonic fibroblasts (MEFs) are plated in 10-cm tissue culture dishes 16–24 h prior to stress induction.

2. ER stress and apoptosis can be induced by treating cultures with 5.0–10.0 μg/mL tunicamycin (Tm) or 5.0–10.0 μM thapsigargin (Tg). To measure the induction of ER stress and

effectively monitor the early and terminal stages of apoptosis, markers for the UPR and cell death should be observed over a range of time points. Whereas appreciable inductions of UPR markers such as eIF2α phosphorylation and CHOP and ATF4 occur within 4–6 h after the introduction of stressful stimuli, markers of apoptosis such as caspase activation and DNA fragmentation might not be observed until 8–24 h *(8, 43)*.

3. To examine cultures for the induction of UPR markers samples should be harvested 1, 4, and 8 h following the addition of Tm or Tg. Care should be taken to note whether or not cells are attached to the tissue culture plates at the end of treatment. For attached cells, the medium is aspirated from the dishes and each is washed with 5 mL of ice-cold 1× PBS.

4. For mRNA analysis of UPR markers, 1 mL of Trizol (Invitrogen) should be added in a fume hood to each dish and mRNA can be harvested according to the manufacturer's protocol with the exception that an equal volume of isopropanol should be used to precipitate the mRNA.

5. For immunoblot analysis of UPR markers, 4 mL of fresh 1× PBS is added and cells are scraped with a Teflon scraper or rubber policeman into the PBS and transferred to clearly labeled 15-mL Falcon tubes. Cells are pelleted at 300–400 × *g* in a swinging bucket centrifuge for 3–5 min, resuspended in 1-mL PBS, transferred into labeled 1.8-mL Eppendorf tubes and repelleted. The PBS is then aspirated from each pellet which is then resuspended in 30–60 μL of modified RIPA buffer prepared as described (*see* **Subheading 2.1, step 5**). The RIPA-suspended pellets are stored on ice for 20 min. with periodic vortexing to mechanically dissociate cell membranes, and then spun at 4°C at 15,000–20,000 × *g* for 15 min. Cleared lysates (containing soluble cellular proteins) are then assayed with the Bradford assay to determine the concentration of protein in each whole-cell lysate. Lysates must be stored at –80°C.

6. To examine stress-induced cell cultures for markers of apoptosis (PARP, caspase3, NOXA, or PUMA) attached and floating cells are pooled, by scraping into the medium. The media containing the cells are then transferred into clearly labeled 15-mL Falcon tubes and pelleted at 300–400 × *g* in a swinging bucket centrifuge for 3–5 min. Total mRNAs can be isolated for qPCR of NOXA, PUMA, or other BH3-only BCL2 family members using Trizol or whole cell lysates can be prepared for Western immunoblot analysis of PARP and caspases in modified RIPA buffer as described (*see* **Subheading 2.1, step 5**).

3.2. Western Immunoblot Analysis and Immunoprecipitation of IRE1α and PERK

3.2.1. Western Immunoblot Analysis

1. 25, 50, or 80 μg of whole cell lysate is diluted in 3× PLB and denatured by boiling for 4 min and iced until cool.

2. Denatured proteins are separated by SDS-PAGE on 7.5 or 10% polyacrylamide gels and transferred to PVDF membrane.

3. Transferred membranes are washed in ddH$_2$0 briefly and blocked for 1 h at room temperature in 5% nonfat powdered milk in TBST.

4. PVDF membranes can be cut to examine the relative levels of more than one protein if there is sufficient difference in the size of the proteins of interest. Membranes are incubated with primary antibodies diluted in fresh 5% milk-TBST overnight at 4°C on a rocking platform.

5. Membranes are washed three times for 10 min in TBST to remove excess primary antibodies and secondary antibodies are added at 1:3,000 (Rabbit-HRP) or 1:7,500 (Mouse-HRP) for 1 h at room temperature on a rocking platform.

6. Membranes are washed three times for 10 min in TBST to remove unbound secondary antibodies and visualized by incubating with Super Signal chemiluminescent reagent for 3–5 min and exposed to autoradiographic film.

3.2.2. Immunoprecipitation of IRE1α or PERK

1. 1 μL of IRE1α or PERK antibody is incubated with 20-μL protein A beads in NP-40 lysis buffer for 1 h on an end-over-end rotator at room temperature.

2. The antibody-bound beads are washed by adding 500 μL lysis buffer in a 1.5-mL microcentrifuge tube by rotating them at 4°C for 10 min, and then spun in a microcentrifuge for 30 s at 8,200 × *g*. The supernatant is aspirated and the beads are washed two more times in a similar manner, and then placed on ice.

3. Equal amounts (μg) of whole cell lysate samples prepared as described (*see* **Subheading 3.1**) are added into the tubes with washed antibody-bound beads and incubated for 3 h at room temperature or overnight at 4°C with constant rotation.

4. After incubation, the beads are washed three times with 1-mL NP-40 lysis buffer and once with 1 mL of 1× PBS.

5. The PBS is aspirated from the tube containing the beads and then 10 μL of PLB is added to the tube which is then denatured at 95°C for 5 min.

6. Denatured proteins are separated by SDS-PAGE on 7.5 or 10% polyacrylamide gels and transferred to PVDF membrane.

7. The blots are incubated with the same IRE1α or PERK primary antibody at a 1:1,000 dilution in TBST with 5% (w/v) nonfat powdered milk overnight at 4°C.

8. Following incubation with primary antibodies membranes are washed, probed with secondary antibodies, and visualized (*see* **Subheading 3.2.1**). Murine IRE1α protein is detected as one major band migrating with a molecular mass of ~120 kDa and murine PERK protein is detected migrating with a molecular mass of ~170 kDa. After Tm treatment, the size of IRE1α and PERK protein is slightly increased due to phosphorylation.

3.3. Conventional Reverse Transcription PCR and Quantitative PCR

Upon activation of the UPR, the ER membrane-resident primary UPR transducer IRE1α is activated and functions as an endoribonuclease to remove a 26 base intron from the human or murine X-box binding protein 1 (*Xbp1*) mRNA to encode a potent UPR *trans*-activator. Due to the difficulty of detecting endogenous IRE1α protein, quantitative analysis of spliced and total *Xbp1* mRNA is the most convenient and reliable method to measure the activation of the IRE1α-mediated UPR pathway.

It has been well established that induction of GRP78/BiP is a marker of ER stress and a central regulator of the activation of the UPR transducers IRE1α, PERK, and ATF6. An increased level of GRP78/BiP protein expression is a good marker for UPR activation following prolonged stress treatment. The observable increase in protein accumulation usually occurs more slowly than the accumulation of mRNA transcripts due in part to the long half-life of GRP78/BiP protein. Due to the durability of "pre-stress" GRP78/BiP protein and the late increase of protein expression that can be detected in whole-cell lysates, immunoblot analysis for GRP78/BiP is not a very sensitive measure to detect low levels of ER stress and UPR activation. Hence, the induction of ER stress and activation of the UPR can often remain undiscovered without examination of both GRP78/BiP protein and mRNA. Accurate and sensitive methods have been developed for quantification of spliced and total *Xbp1* and *GRP78/BiP* mRNA in mammalian cells or tissue by using conventional RT-PCR or qPCR.

1. Total RNA is isolated from cell cultures with Trizol reagent. Trizol can be used according to the manufacturer's protocol with the exception that mRNAs should be precipitated with a volume of isopropanol equal to the recovered aqueous phase; this will dramatically enhance the yield.

2. Synthesis of cDNA from murine total RNA is performed using iScript cDNA synthesis kit (Bio-Rad). The reaction mixture (20 μL) contains 500 ng total RNA, 4 μL of 5× iScript reaction mix, and 1 μL reverse transcriptase (RT).

3. The reverse transcription reaction is incubated at 25°C for 10 min, followed by incubation at 48°C for 30 min, and then RT is inactivated at 85°C for 5 min.

4. The reaction mixture is diluted tenfold by addition of 180 µL of nuclease-free water. The diluted cDNA mix from the reverse transcription reaction is subjected to semi-quantitative reverse transcription PCR or quantitative real-time PCR to determine levels of specific mRNAs.

5. For conventional semi-quantitative RT-PCR for *Xbp1* or *GRP/BiP*, 10 µL of diluted cDNA template (25 ng) is mixed with 200 µM of each dNTP, 300 nM forward and reverse primers, 5 µL of 10× PCR reaction buffer with 15 mM $MgCl_2$, 2.6 Units of Taq polymerase, and nuclease-free water to a total volume of 50 µL. PCR cycle begins with a 2-min incubation at 95°C, then 25 cycles of 30 s at 94°C, 30 s at 55°C, 45 s at 72°C, followed by a 7-min incubation at 72°C. The reaction can then be held at 4°C.

6. PCR products are separated by electrophoresis on a 2% agarose gel and visualized by ethidium bromide staining.

7. For the qPCR reaction a 2-µL aliquot of diluted cDNA template (12.5 ng) is mixed with 10 µL iQ SYBR Green Supermix (Bio-Rad), 150 nM forward and reverse real-time PCR primers for *Xbp1* or *GRP78/BiP GRP78/BiP* or *β-actin* (internal control), and nuclease-free water to a final volume of 20 µL.

8. The thermal cycle parameters for *Xbp1* or *GRP78/BiP* are an initial step of 95°C for 10 min, followed by 40–48 cycles of 95°C for 15 s and 59°C for 1 min; for the final step the reaction is terminated by holding at 4°C.

9. The data are analyzed with the iCycler iQ real-time PCR detection system (Bio-Rad) according to the manufacturer's instructions.

3.4. Nuclear DNA Fragmentation Assay

During apoptosis, cells undergo many distinct morphological and biochemical changes. One striking event that occurs late in apoptosis involves endonuclease cleavage of DNA between nucleosomes, thereby producing a mixture or "ladder" of different sized DNA fragments. DNA fragmentation has been well recognized as a marker of the final stage of apoptosis *(44)*. To detect ER stress-induced apoptosis in mammalian cultures:

1. Attached and floating cells are pooled by scraping and gently pelleted at 300–400 × *g* in a table-top swinging bucket centrifuge.

2. Cells are solubilized in DNA lysis buffer supplemented with 100 µg of proteinase K/mL to remove histone proteins, for 2 h at 50°C.

3. DNA is extracted twice with phenol–chloroform mixed 1:1, and twice with chloroform alone.

4. DNA is precipitated by adding 0.8 volumes (w/r/t the aqueous phase from 3) of isopropanol and centrifuging at high speed for 15 min.

5. The isopropanol/aqueous phase (containing the nucleic acid) is removed and the precipitated DNA is washed with 70% ethanol and allowed to dry briefly at room temperature.

6. The pellet is resuspended in 30–50 µL of TE, pH 8, supplemented with 0.25 mg/mL RNAse A and 20–30 µg of DNA (A^{260} determined) is resolved on a 1.5% agarose gel (*see* **Note 1**).

4. Note

1. A laddered appearance of the DNA is indicative of apoptosis; a smeared appearance suggests the observed cell death is the result of some other form of cell death such as necrosis.

Acknowledgments

Portions of this work were supported by NIH grants DK042394, HL052173, and HL057346. RJK is an Investigator of the Howard Hughes Medical Institute.

References

1. Kaufman, R. J. (2002). Orchestrating the unfolded protein response in health and disease. *J Clin Invest* **110**, 1389–1398

2. Mori, K. (2000). Tripartite management of unfolded proteins in the endoplasmic reticulum. *Cell* **101**, 451–454

3. Ron, D., and Walter, P. (2007). Signal integration in the endoplasmic reticulum unfolded protein response. *Nat Rev Mol Cell Biol* **8**, 519–529

4. Schroder, M., and Kaufman, R. J. (2005). The Mammalian unfolded protein response. *Annu Rev Biochem* **74**, 739–789

5. Harding, H. P., Zhang, Y., Bertolotti, A., Zeng, H., and Ron, D. (2000). Perk is essential for translational regulation and cell survival during the unfolded protein response. *Mol Cell* **5**, 897–904

6. Scheuner, D., Song, B., McEwen, E., Liu, C., Laybutt, R., Gillespie, P., et al. (2001). Translational control is required for the unfolded protein response and in vivo glucose homeostasis. *Mol Cell* **7**, 1165–1176

7. Harding, H. P., Zhang, Y., Zeng, H., Novoa, I., Lu, P. D., Calfon, M., et al. (2003). An integrated stress response regulates amino acid metabolism and resistance to oxidative stress. *Mol Cell* **11**, 619–633

8. Fribley, A. M., Evenchik, B., Zeng, Q., Park, B. K., Guan, J. Y., Zhang, H., et al. (2006). Proteasome inhibitor PS-341 induces apoptosis in cisplatin-resistant squamous cell carcinoma cells by induction of Noxa. *J Biol Chem* **281**, 31440–31447

9. Ron, D., and Habener, J. F. (1992). CHOP, a novel developmentally regulated nuclear

protein that dimerizes with transcription factors C/EBP and LAP and functions as a dominant- negative inhibitor of gene transcription. *Genes Dev* **6**, 439–453

10. Matsumoto, M., Minami, M., Takeda, K., Sakao, Y., and Akira, S. (1996). Ectopic expression of CHOP (GADD153). induces apoptosis in M1 myeloblastic leukemia cells. *FEBS Lett* **395**, 143–147

11. McCullough, K. D., Martindale, J. L., Klotz, L. O., Aw, T. Y., and Holbrook, N. J. (2001). Gadd153 sensitizes cells to endoplasmic reticulum stress by down-regulating Bcl2 and perturbing the cellular redox state. *Mol Cell Biol* **21**, 1249–1259

12. Ohoka, N., Yoshii, S., Hattori, T., Onozaki, K., and Hayashi, H. (2005). TRB3, a novel ER stress-inducible gene, is induced via ATF4-CHOP pathway and is involved in cell death. *EMBO J* **24**, 1243–1255

13. Sok, J., Wang, X. Z., Batchvarova, N., Kuroda, M., Harding, H., and Ron, D. (1999). CHOP-Dependent stress-inducible expression of a novel form of carbonic anhydrase VI. *Mol Cell Biol* **19**, 495–504

14. Yamaguchi, H., and Wang, H. G. (2004). CHOP is involved in endoplasmic reticulum stress-induced apoptosis by enhancing DR5 expression in human carcinoma cells. *J Biol Chem* **279**, 45495–45502

15. Zinszner, H., Kuroda, M., Wang, X., Batchvarova, N., Lightfoot, R. T., Remotti, H., et al. (1998). CHOP is implicated in programmed cell death in response to impaired function of the endoplasmic reticulum. *Genes Dev* **12**, 982–995

16. Marciniak, S. J., Yun, C. Y., Oyadomari, S., Novoa, I., Zhang, Y., Jungreis, R., et al. (2004). CHOP induces death by promoting protein synthesis and oxidation in the stressed endoplasmic reticulum. *Genes Dev* **18**, 3066–3077

17. Wu, J., Rutkowski, D. T., Dubois, M., Swathirajan, J., Saunders, T., Wang, J., et al. (2007). ATF6alpha optimizes long-term endoplasmic reticulum function to protect cells from chronic stress. *Dev Cell* **13**, 351–364

18. Tirasophon, W., Welihinda, A. A., and Kaufman, R. J. (1998). A stress response pathway from the endoplasmic reticulum to the nucleus requires a novel bifunctional protein kinase/endoribonuclease (Ire1p). in mammalian cells. *Genes Dev* **12**, 1812–1824

19. Wang, X. Z., Harding, H. P., Zhang, Y., Jolicoeur, E. M., Kuroda, M., and Ron, D. (1998). Cloning of mammalian Ire1 reveals diversity in the ER stress responses. *EMBO J* **17**, 5708–5717

20. Calfon, M., Zeng, H., Urano, F., Till, J. H., Hubbard, S. R., Harding, H. P., et al. (2002). IRE1 couples endoplasmic reticulum load to secretory capacity by processing the XBP-1 mRNA. *Nature* **415**, 92–96.

21. Lee, A. H., Iwakoshi, N. N., and Glimcher, L. H. (2003). XBP-1 regulates a subset of endoplasmic reticulum resident chaperone genes in the unfolded protein response. *Mol Cell Biol* **23**, 7448–7459

22. Shen, X., Ellis, R. E., Lee, K., Liu, C. Y., Yang, K., Solomon, A., et al. (2001). Complementary signaling pathways regulate the unfolded protein response and are required for *C. elegans* development. *Cell* **107**, 893–903.

23. Yoshida, H., Matsui, T., Yamamoto, A., Okada, T., and Mori, K. (2001). XBP1 mRNA is induced by ATF6 and spliced by IRE1 in response to ER stress to produce a highly active transcription factor. *Cell* **107**, 881–891.

24. Urano, F., Bertolotti, A., and Ron, D. (2000). IRE1 and efferent signaling from the endoplasmic reticulum. *J Cell Sci* **113**, 3697–3702.

25. Yoneda, T., Imaizumi, K., Oono, K., Yui, D., Gomi, F., Katayama, T., et al. (2001). Activation of caspase-12, an endoplastic reticulum (ER). resident caspase, through tumor necrosis factor receptor-associated factor 2- dependent mechanism in response to the ER stress. *J Biol Chem* **276**, 13935–13940.

26. Nishitoh, H., Saitoh, M., Mochida, Y., Takeda, K., Nakano, H., Rothe, M., et al. (1998). ASK1 is essential for JNK/SAPK activation by TRAF2. *Mol Cell* **2**, 389–395

27. Nishitoh, H., Matsuzawa, A., Tobiume, K., Saegusa, K., Takeda, K., Inoue, K., et al. (2002). ASK1 is essential for endoplasmic reticulum stress-induced neuronal cell death triggered by expanded polyglutamine repeats. *Genes Dev* **16**, 1345–1355

28. Davis, R. J. (2000). Signal transduction by the JNK group of MAP kinases. *Cell* **103**, 239–252

29. Hetz, C., Bernasconi, P., Fisher, J., Lee, A. H., Bassik, M. C., Antonsson, B., et al. (2006). Proapoptotic BAX and BAK modulate the unfolded protein response by a direct interaction with IRE1alpha. *Science* **312**, 572–576

30. Urano, F., Wang, X., Bertolotti, A., Zhang, Y., Chung, P., Harding, H. P., et al. (2000). Coupling of stress in the ER to activation of

JNK protein kinases by transmembrane protein kinase IRE1. *Science* **287**, 664–666

31. Gunn, K. E., Gifford, N. M., Mori, K., and Brewer, J. W. (2004). A role for the unfolded protein response in optimizing antibody secretion. *Mol Immunol* **41**, 919–927

32. Zong, W. X., Li, C., Hatzivassiliou, G., Lindsten, T., Yu, Q. C., Yuan, J., et al. (2003). Bax and Bak can localize to the endoplasmic reticulum to initiate apoptosis. *J Cell Biol* **162**, 59–69

33. Krajewski, S., Tanaka, S., Takayama, S., Schibler, M. J., Fenton, W., and Reed, J. C. (1993). Investigation of the subcellular distribution of the bcl-2 oncoprotein: residence in the nuclear envelope, endoplasmic reticulum, and outer mitochondrial membranes. *Cancer Res* **53**, 4701–4714

34. Nakagawa, T., and Yuan, J. (2000). Crosstalk between two cysteine protease families. Activation of caspase-12 by calpain in apoptosis. *J Cell Biol* **150**, 887–894

35. Rao, R. V., Hermel, E., Castro-Obregon, S., del Rio, G., Ellerby, L. M., Ellerby, H. M., et al. (2001). Coupling endoplasmic reticulum stress to the cell death program. Mechanism of caspase activation. *J Biol Chem* **276**, 33869–33874

36. Tan, Y., Dourdin, N., Wu, C., De Veyra, T., Elce, J. S., and Greer, P. A. (2006). Ubiquitous calpains promote caspase-12 and JNK activation during endoplasmic reticulum stress-induced apoptosis. *J Biol Chem* **281**, 16016–16024

37. Saleh, M., Mathison, J. C., Wolinski, M. K., Bensinger, S. J., Fitzgerald, P., Droin, N., et al. (2006). Enhanced bacterial clearance and sepsis resistance in caspase-12-deficient mice. *Nature* **440**, 1064–1068

38. Fischer, H., Koenig, U., Eckhart, L., and Tschachler, E. (2002). Human caspase 12 has acquired deleterious mutations. *Biochem Biophys Res Commun* **293**, 722–726

39. Crompton, M. (1999). The mitochondrial permeability transition pore and its role in cell death. *Biochem J* **341**(Pt 2)., 233–249

40. Feng, B., Yao, P. M., Li, Y., Devlin, C. M., Zhang, D., Harding, H. P., et al. (2003). The endoplasmic reticulum is the site of cholesterol-induced cytotoxicity in macrophages. *Nat Cell Biol* **5**, 781–792

41. Harding, H. P., Zeng, H., Zhang, Y., Jungries, R., Chung, P., Plesken, H., et al. (2001). Diabetes mellitus and exocrine pancreatic dysfunction in perk-/- mice reveals a role for translational control in secretory cell survival. *Mol Cell* **7**, 1153–1163

42. Zhou, J., Lhotak, S., Hilditch, B. A., and Austin, R. C. (2005). Activation of the unfolded protein response occurs at all stages of atherosclerotic lesion development in apolipoprotein E-deficient mice. *Circulation* **111**, 1814–1821

43. Fribley, A., Zeng, Q., and Wang, C. Y. (2004). Proteasome inhibitor PS-341 induces apoptosis through induction of endoplasmic reticulum stress-reactive oxygen species in head and neck squamous cell carcinoma cells. *Mol Cell Biol* **24**, 9695–9704

44. Schwartzman, R. A., and Cidlowski, J. A. (1993). Apoptosis: the biochemistry and molecular biology of programmed cell death. *Endocr Rev* **14**, 133–151

Chapter 15

Detection of Uncoupling Protein-2 (UCP2) As a Mitochondrial Modulator of Apoptosis

Zoltan Derdak, Tamako A. Garcia, and Gyorgy Baffy

Summary

There is an increasing evidence that uncoupling protein-2 (UCP2), a recently identified molecular sensor and suppressor of mitochondrial reactive oxygen species (ROS), plays an important role in regulating apoptosis in different cell systems. A great technical difficulty that many groups have encountered is the reliable detection of endogenously or exogenously expressed UCP2 protein. The goal of this chapter is to introduce the reader to techniques that we have successfully used over the years to detect UCP2 protein in various mouse and human specimens. These techniques include mitochondrial isolation and submitochondrial fractionation followed by Western blotting and UCP2 immunohistochemistry. We find that sample preparation is a key to success and it allows one to produce relevant and important data using commercially available antibodies.

Key words: UCP2, Mitochondria, Oxidative stress, Superoxide, Mitochondrial isolation, Submitochondrial fractionation, Digitonin, Alkali treatment, Immunohistochemistry, Apoptosis

1. Introduction

UCP2 is a member of the mitochondrial carrier protein superfamily located in the mitochondrial inner membrane *(1, 2)*. UCP2 was identified by "reversed cloning" based on its sequence similarity (59% amino-acid identity) to UCP1 or thermogenin, the archetypal member of uncoupling proteins *(3)*. The primary task of UCP1 is to facilitate nonshivering thermogenesis. UCP1 is highly abundant in brown adipose tissue where it disconnects substrate oxidation from ATP generation by dissipating the proton electrochemical gradient of the inner mitochondrial membrane (the driving force for ATP synthesis), transferring substrate

Peter Erhardt and Ambrus Toth (eds.), *Apoptosis*, Methods in Molecular Biology, vol. 559
DOI 10.1007/978-1-60327-017-5_15
© Humana Press, a part of Springer Science+Business Media, LLC 2004, 2009

energy to heat instead of ATP *(4)*. In contrast to UCP1, the evolutionary role and biological function of UCP2 is not well established and remains the subject of intense debate *(5, 6)*.

One suggested function of the ubiquitously expressed UCP2 is the negative regulation of mitochondrial reactive oxygen species (ROS) *(7)*. Mitochondrial electron transport is the primary source of intracellular ROS *(8)*. Electron spin-off from the respiratory complexes may result in reduction of molecular oxygen into superoxide, an inherent process that is more likely to occur when the proton electrochemical gradient is high *(8)*. Based on the proposed model, UCP2 initiates proton leak across the mitochondrial inner membrane, which facilitates conventional electron transport and decelerates superoxide production *(9)*. Indeed, modulation of intracellular ROS levels by UCP2 has been described in a variety of cells both in vitro and in vivo *(10–13)*.

ROS have a major role in the regulation of mitochondria-mediated apoptosis *(14)*. Not surprisingly, UCP2-mediated negative control of ROS affects apoptosis in many different biological systems. Protection from apoptosis by UCP2 has been described in endothelial cells *(15)*, testicular germ cells *(16)*, and cardiomyocytes *(17)*. UCP2 also protects transformed cells from apoptosis, as recently shown for HepG2 hepatocarcinoma cells *(18)*. Apparently, the level of mitochondrial uncoupling is critical for its antiapoptotic effect. Excessive amounts of plasmid-encoded UCP2 *(19)* or high doses of artificial uncoupling agents such as the pure protonophore carbonylcyanide-4-trifluoromethoxyphenylhydrazone or FCCP (unpublished observation) may result in severely compromised mitochondrial ATP production and necrotic cell death. An important determinant of the degree of uncoupling is the absolute amount of UCP2 protein, although UCP2 also needs proper activation for proper biological function *(20)*.

By virtue of its ability to curb mitochondrial ROS production, UCP2 is therefore of increasing interest as a modulator of apoptosis. However, detection and quantification of UCP2 protein is often challenging. Although UCP2 has a wide tissue distribution, it is typically present in low or very low amounts *(21, 22)*. Much of our knowledge on UCP2 is based on assessment of mRNA levels and anti-UCP2 antibodies have been verified (i.e., tested on tissues from UCP2-deficient mice) in relatively few studies *(21–25)*. As we recently learned from the work of Pecqueur et al. *(22)*, UCP2 is a protein much less abundant than we would expect based on mRNA levels. These authors confirmed the presence of UCP2 protein in mitochondria isolated from spleen, stomach, lung, intestine, and white adipose tissue, whereas they failed to detect it in mitochondria isolated from organs that usually contain high levels of UCP2 mRNA (skeletal muscle, heart, kidney, brain, and brown adipose tissue).

Apparently, techniques that utilize antibodies against the UCP2 protein are superior over mRNA-based methods in predicting the biological effects of UCP2. Our laboratory has gathered significant experience by using a commercially available antibody (goat polyclonal anti-UCP2 antibody (C-20), Santa Cruz) for Western blotting and immunohistochemistry to detect UCP2 in human and mouse tissue and cell specimens *(25, 26)*. In tissues of UCP2-ablated mice, this antibody detects no nonspecific bands in the range of 25–37 kDa, where UCP2 is located (~33 kDa). To assess UCP2 expression in tissues with particularly low abundance (e.g., liver), we often use mitochondrial isolation to enrich the signal. To this end, we homogenize the tissue or cells in isotonic mitochondrial buffer and use various centrifugation steps to sediment the mitochondria. Further enrichment is possible by performing submitochondrial fractionation; applying digitonin to isolated mitochondria will separate mitoplasts from the outer membrane and the intermembrane space. This assay can be combined with alkali treatment to separate membranous and nonmembranous compartments, which helps verify the correct localization of plasmid-encoded or exogenous UCP2. This information is in fact the cornerstone of any overexpression studies. We have successfully used human cell lines and mouse tissue preparations obtained by these isolation and fractionation techniques to identify UCP2 by Western blotting. Technical modifications to a standard immunoblot protocol can be found in **Subheading 4**. The C-20 antibody has also proved useful for assessing tissue and cell-specific distribution and intensity of UCP2 expression by immunohistochemistry. These proven protocols for the detection of UCP2 are discussed in detail below.

2. Materials

2.1. Mitochondrial Isolation from Soft Tissue or Cells

1. Isotonic mitochondrial buffer: 210 mM mannitol, 70 mM sucrose, 1 mM EDTA, 10 mM HEPES, pH 7.5, store at 4°C and supplement with protease inhibitors (Complete, Mini, EDTA-free protease inhibitor cocktail from Roche) just before usage.

2. Albumin from bovine serum (Sigma): prepare 50 mg/mL stock solution in ddH_2O and store it at –20°C.

3. Kontes Pellet Pestle Cordless Motor (Fisher) for isolating mitochondria from cells.

4. Glass homogenizer (Corning Life Sciences) for isolating mitochondria from soft animal tissue.

5. Cell disruption buffer: 50 mM Tris–HCl, pH 7.4, 100 mM NaCl, 1% NP-40, 10% glycerol, 1 mM EDTA, store it at 4°C, supplement it with 10 mM β-glycerophosphate, 2 mM Na_3VO_4, 1 mM NaF, and protease inhibitor cocktail (Roche) just before use.

2.2. Submitochondrial Fractionation

1. Freshly isolated, pelleted mitochondria in an Eppendorf tube (~100 µL/assay) kept on ice.

2. 1.2 mg/mL digitonin (Sigma) solution in ddH_2O; dissolve digitonin by heating to 95–98°C and then cool it to RT. Prepare it freshly and store it on ice. Digitonin is toxic so wear coat, gloves, mask, and goggles for the preparation.

3. 0.1 M sodium carbonate (Sigma) solution, pH 11.5. Prepare it freshly and store it on ice.

4. Cell disruption buffer (*see* **Subheading 2.1, step 5**).

2.3. UCP2 Immunohistochemistry

1. Native slides from paraffin-embedded formalin-fixed tissue (e.g., liver, colon).

2. NordicWare® Microwave Tender Cooker (Biogenex).

3. 10 mM sodium citrate (Sigma) solution, pH 6.0; store it at RT.

4. 3% H_2O_2 solution in distilled water; dilute 30% H_2O_2 (Fisher) carefully in ddH_2O; wear labcoat, gloves, and goggles.

5. ImmEdge Pen (Vector).

6. Mouse serum (Sigma).

7. Antibody diluent (Zymed).

8. Anti-UCP2 goat polyclonal antibody (C-20) (Santa Cruz).

9. Avidin/Biotin Kit (Zymed).

10. ImmunoPure mouse anti-goat IgG, biotin-conjugated antibody (Pierce).

11. Standard Vectastain Elite ABC Kit (Vector).

12. DAB Substrate Kit (Vector).

13. Gill 3 Hematoxylin (Richard-Allan Scientific).

14. Poly-Mount (Polysciences).

15. Xylene, 100%–95%–70% ethanol, 1× phosphate-buffered saline (PBS), Tris-buffered saline with Tween-20 (TBS-T).

3. Methods

3.1. Mitochondrial Isolation from Soft Tissue (e.g., liver) or Cells

The purpose is to separate intact mitochondria from other cellular components. Isolated mitochondria can be homogenized in cell disruption buffer and the lysate be used to detect endogenous UCP2 expression by Western blotting. This approach is intended to increase the signal-to-noise ratio of the immunoblot since UCP2 is an inner mitochondrial membrane protein. This method can also be used to verify the mitochondrial localization of exogenously expressed UCP2. Moreover, mitochondria obtained by this approach remain intact and can be used for functional assays as well (such as membrane potential measurement by JC-1 assay).

1. Always use fresh tissue (50–100 mg, within 1 h after sacrifice) or freshly harvested cells (1×10^7 cells at least, *see* **Note 1**). Never freeze the samples. Keep them on ice. From this point on, all steps should be carried out at 4°C, including centrifugation.

2. Wash the samples twice with 2 volumes of isotonic mitochondrial buffer. If using harvested cells, this step can be performed in 1.5-mL Eppendorf tubes, using 2×5 min centrifugation at $500 \times g$. Tissue samples can be carefully washed in a 6-well plate on ice, using a pipette.

3. Homogenize the samples with ~10 volumes of isotonic mitochondrial buffer containing 2 mg/mL albumin (*see* **Note 2**) by using Kontes Pellet Pestle Cordless Motor (for cells) or glass homogenizer (for soft tissue). The advantage of the Pellet Pestle is that cells can be homogenized in a 1.5-mL Eppendorf tube and there is no need to transfer the homogenate in the next step (*see* **Note 3**). Ensure total homogenization of the samples by moving the pestle up and down 5–10 times or by making 5–10 twists counterclockwise with the glass homogenizer. Keep the homogenate on ice.

4. If you work with tissue then transfer the homogenate with a pipette to a 1.5-mL Eppendorf tube after the tissue is perfectly disrupted. Centrifuge the samples at $600 \times g$ for 5 min.

5. Carefully transfer the supernatant into a new 1.5-mL Eppendorf tube. Centrifuge it at $11,000 \times g$ for 10 min.

6. Discard the supernatant and resuspend the pellet in 10 volumes of isotonic mitochondrial buffer.

7. Repeat **steps 5** and **6**.

8. For Western blotting (*see* **Note 4** for conditions), pelleted mitochondria can be resuspended in cell disruption buffer and the sample is ready to be assayed. Samples can be also stored at −80°C until further analysis. For submitochondrial fractionation, pelleted mitochondria should be subjected to digitonin/alkali treatment immediately after isolation (*see* **Subheading 3.2**).

3.2. Submitochondrial Fractionation

The purpose of this method is to separate mitoplasts (inner mitochondrial membrane-bound matrix) from the outer membrane and the intermembrane space by using digitonin. Mitoplast proteins can be analyzed by Western blotting and this technique further increases the signal-to-noise ratio, which is very useful when UCP2 expression is investigated. In addition, this assay can be complemented by alkali treatment during which membranes are separated from the soluble fraction of mitochondria. Fractions from digitonin and alkali treatment can be analyzed by Western blotting and detection of proteins specific to the inner membrane (such as subunit IV of cytochrome *c* oxidase) and intermembrane space (such as cytochrome *c*) can verify the purity of fractions. This is a great method to check the correct localization of plasmid-encoded UCP2 protein or other mitochondrial proteins (*see* **Fig. 1** and **Note 5**)

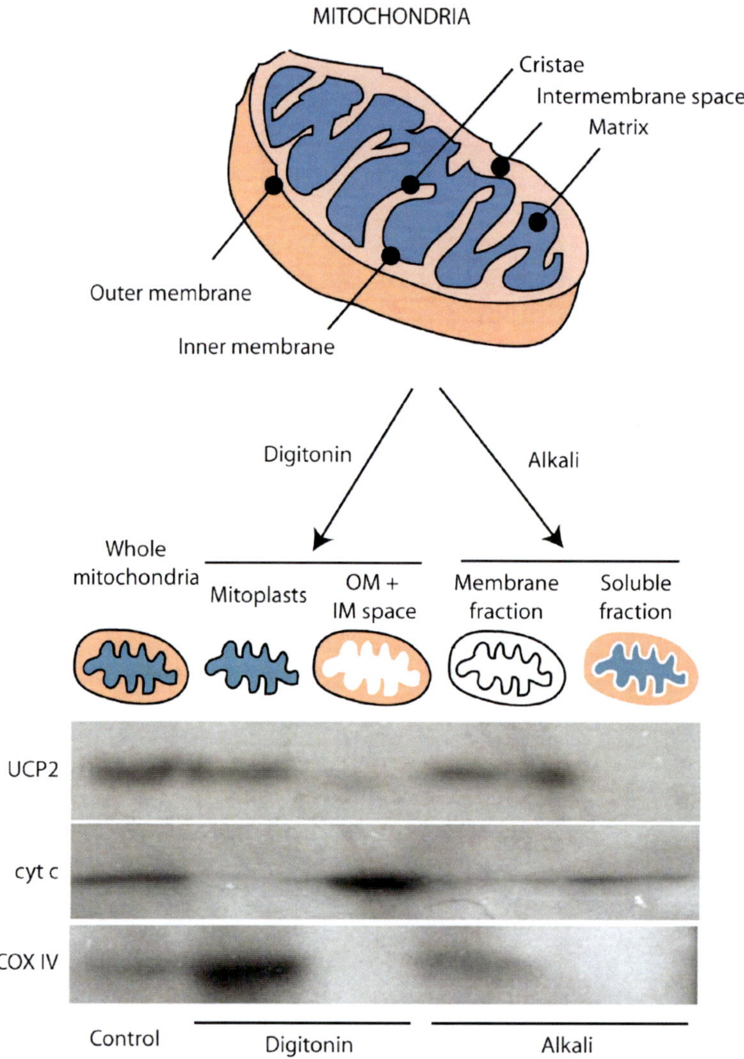

Fig. 1. Detection of exogenously expressed UCP2 in the mitochondrial inner membrane. Mitochondria of HCT116 human colon adenocarcinoma cells transfected with hUCP2-pcDNA3.1/Zeo(–) plasmid containing the full-length human *ucp2* cDNA (pUCP2) were isolated **(Subheading 3.1)** and subfractionated by parallel treatment with digitonin and alkali **(Subheading 3.2)**. Mitochondrial fractions were loaded on 12–15% gels for SDS-PAGE and Western blotting was performed by standard protocol (*see* **Note 4**). Purity of the preparation was verified by detecting cytochrome *c* and subunit IV of cytochrome *c* oxidase (COX IV). Whole mitochondria served as control. Immunoblotting indicates presence of UCP2 in the inner membrane fraction (identified by COX IV), but not in the intermembrane (IM) space (identified by cytochrome *c*). *OM*, mitochondrial outer membrane.

3.2.1. Digitonin Treatment

1. Dissolve 100 μL of purified mitochondria in 500 μl of digitonin solution (1.2 mg/mL) using a 1.5-mL Eppendorf tube.

2. Incubate the tube on ice for 25 min.

3. Centrifuge the suspension at $10,000 \times g$ for 10 min at 4°C to generate mitoplasts. The supernatant contains the intermembrane space fraction and outer membrane.

4. Save the supernatant in a separate prechilled tube and store it at –80°C until further analysis. Resuspend mitoplasts in cell disruption buffer and store at –80°C.

3.2.2. Alkali Treatment

1. Wash and resuspend 100 μL of purified mitochondria in 500 μL of freshly prepared 0.1 M sodium carbonate, pH 11.5.

2. Incubate the suspension at 0°C for 30 min.

3. Recover the mitochondrial membranes by centrifugation at $100,000 \times g$ for 30 min at 4°C.

4. This supernatant represents the soluble fraction of mitochondria. Transfer it to a new prechilled tube and store it at –80°C until further analysis.

5. Reconstitute the mitochondrial membranes in cell disruption buffer and store at –80°C.

3.3. UCP2 Immunohistochemistry

The purpose of this method is to detect UCP2 in paraffin-embedded, formalin-fixed tissue (*see* **Note 6**), using the ABC method. We used this method to detect UCP2 in liver and colon tissue of mice and in human colon and bile duct cancer specimens (*see* **Note 7** and **Fig. 2**).

1. Put the slides in a metal rack and bake them at 60°C overnight.

2. Deparaffinize the slides (let slides cool down before emerging) by the following regimen:

    (a) Xylene for 5 min, repeat

    (b) Xylene 20 dips

    (c) 100% ethanol 20 dips, repeat

    (d) 95% ethanol 20 dips

    (e) 70% ethanol 20 dips

    (f) Water rinse

    (g) Hold slides in water

3. Add 600 mL of distilled water to a 1,000-mL beaker.

4. Fill up a plastic staining container with citrate buffer.

5. Bring the water and the citrate buffer to the boil in a microwave oven.

6. Place the plastic staining container with hot citrate buffer into the NordicWare® Microwave Tender Cooker and fill up the cooker with boiling water.

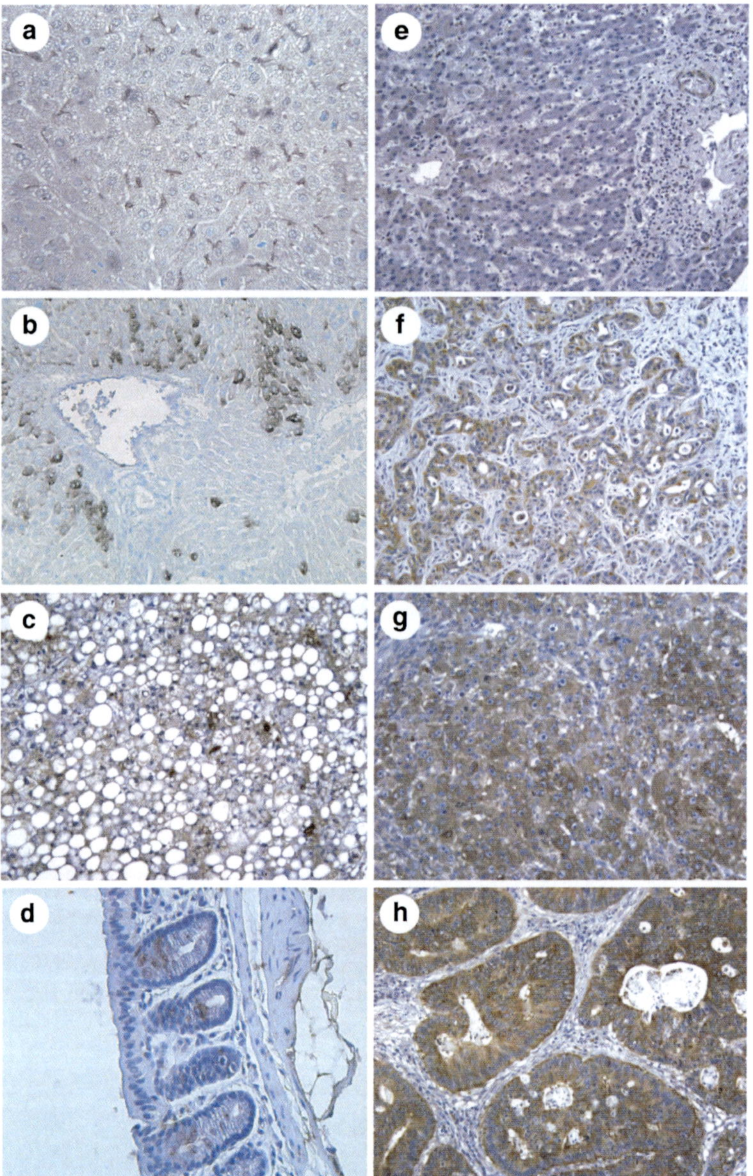

Fig. 2. Detection of endogenously expressed UCP2 in mouse and human tissues. Abundance and distribution of UCP2 protein was assessed by immunohistochemistry using goat polyclonal anti-UCP2 antibody (C-20; 1:100 dilution; Santa Cruz, CA) as primary antibody and peroxidase for visualization **(Subheading 3.3)**. Negative control slides were treated with goat IgG instead of primary antibodies (not shown). (**a**) Normal mouse liver with UCP2 residing in Kupffer cells. (**b**) Mouse liver remnant 2 days following partial hepatectomy with hepatocellular expression of UCP2 in the periportal areas. (**c**) Fatty liver from *ob/ob* mice exhibiting increased hepatocellular UCP2 expression. (**d**) Mouse colon crypts with increased UCP2 expression following 4 weeks of high fat diet (40% fat of total caloric intake) enriched in polyunsaturated (*n*–3) fatty acids. (**e**) Normal human liver with minimally detectable UCP2 expression in contrast to marked UCP2 abundance in human cholangiocarcinoma (**f**), human hepatocellular carcinoma (**g**), and human colon adenocarcinoma (**h**). *See also* **Notes 6–8**.

7. Place deparaffinized slides in a plastic rack and set rack into the plastic staining container.

8. Top off plastic staining container with citrate buffer and place lid to prevent evaporation and drying out of specimens.

9. Put plastic container into the NordicWare® Microwave Tender Cooker and close it tightly. Put pressure cooker in the microwave oven and microwave for 11 min at power level 10.

10. After the pressure cooker has pressurized, take out the container and let the slides cool down for 30 min.

11. Remove the slides and wash them with distilled water for three times for 5 min each.

12. Transfer the slides into a small plastic Coplin jar and block them with 3% H_2O_2 solution for 15 min.

13. Wash the slides with distilled water for three times for 5 min each.

14. Use the ImmEdge Pen to place a hydrophobic barrier around the section and put the slides back to distilled water to prevent specimens from drying.

15. Place the slides to humidified chamber.

16. Block the slides for 15 min with 5% mouse serum diluted in PBS. Do not wash.

17. Pour off some of the serum and add 100 µL of diluted anti-UCP2 goat polyclonal antibody (1:100, using Zymed antibody diluent). Incubate the slides overnight at 4°C in a humidified chamber (*see* **Note 8**).

18. Next day wash the slides with TBS-T for three times for 5 min each.

19. Using the Avidin/Biotin Kit treat the specimens with avidin for 15 min on a flat surface.

20. Wash the slides once with TBS-T for 5 min.

21. Incubate the specimens with biotin for 15 min on a flat surface.

22. Wash the slides once with TBS-T for 5 min.

23. Pipette 100 µL of diluted ImmunoPure Mouse Anti-Goat IgG, Biotin Conjugated antibody (1:500, using Zymed antibody diluent) onto the specimens.

24. Incubate the slides for 30 min at RT. During this time prepare the Standard Vectastain Elite ABC Kit: in the ABC Reagent mixing bottle mix one drop of reagent A and one drop of reagent B with 2.5-mL TBS-T. Incubate this mixture for 30 min at RT.

25. Wash the slides with TBS-T for three times for 5 min each.

26. Incubate the slides with the Standard Vectastain Elite ABC Kit for 30 min.

27. Wash the slides with TBS-T for three times for 5 min each.

28. Prepare DAB chromogen just prior to use, following the manufacturer's instructions (Vector).

29. Apply DAB chromogen for 5 min or until color develops. Place the slides in distilled water to stop the reaction, rinse them several times. Collect DAB and put it into the chemical waste.

30. Counterstain the slides with Hematoxylin for about 12 s.

31. Rinse the slides in water at least for 5 min and dehydrate them following the manufacturer's instructions (Richard-Allan Scientific).

32. Cover the specimens with coverslip using Poly-Mount mounting medium (1–2 drops).

33. Examine UCP2 staining under microscope.

4. Notes

1. The number of cells needed for the isolation largely depends on the down-stream application. We find that a fully confluent T-150 flask (for HCT116 cells) provides sufficient amount of protein for Western blotting. We typically load 30–50 µg mitochondrial protein to detect UCP2. We load 100–300 µg protein if we work with total cellular lysate.

2. The role of delipidated BSA in the isotonic mitochondrial buffer during the extraction is that it removes free fatty acids present in the tissue that can cause uncoupling of respiration in the mitochondria *(27)*.

3. The Kontes Pellet Pestle Cordless Motor is ideal to break up cells, but it is insufficient to disintegrate fresh tissue.

4. To detect UCP2 by Western blotting using mitochondrial protein fraction, we follow standard protocol with tailored modifications including these: we prepare 12% SDS-polyacrylamide gel and run the samples at low voltage (34 V) overnight, until proteins of ~33 kDa molecular weight are localized in the middle of the gel. For transfer we use a buffer containing 48 mM Tris–HCl, 40 mM glycine, 0.0375% SDS, and 15% methanol, and apply 2 mA/cm^2 electric current (~0.3 A for a large gel) for 1 h. To further increase the signal-to-noise ratio, it may be useful to completely dry out the membrane right after transfer. This step will ensure that the protein becomes completely membrane bound. For proper drying we typically soak the membrane in methanol and air-dry it for 20 min. For blocking we use 5% dry milk in TBS-T and block the membrane for 1 h at RT. Our protocol

is based on TBS-T and we use this blocking buffer to dilute the primary antibody (C-20, Santa Cruz) at a rate of 1:300. We incubate the membrane overnight at 4°C. The dilution rate for the secondary antibody is 1:10,000.

5. Submitochondrial fractionation can be very useful to investigate the intramitochondrial localization of an exogenously expressed protein. Here we use UCP2-overexpressing HCT 116 human colon adenocarcinoma cells to illustrate the technique. We performed digitonin and alkali treatment on isolated mitochondria, then loaded the four resulting fractions (mitoplasts, intermembrane space plus outer mitochondrial membrane fraction, pure membrane fraction, and soluble fraction) into one gel to detect UCP2 (*see* **Fig. 1**). We detected to control proteins to verify the purity of preparations and to demonstrate that the newly introduced UCP2 had an expression pattern similar to the cytochrome *c* oxidase IV, which is also an inner membrane protein. These proteins are subunit IV of cytochrome oxidase *c* or complex IV (detected by a mouse monoclonal antibody from Invitrogen, 1:2,000) and cytochrome *c* (detected by a mouse monoclonal antibody from BD Pharmingen, 1:1,000). Since cytochrome *c* has a low molecular weight (15 kDa), we prepared a 15% SDS polyacrylamide gel for its detection.

6. We strongly recommend 10% neutralized buffered formalin as a fixative for UCP2 immunohistochemistry. Avoid prolonged storage of samples in the fixative. We encountered serious problems such as high background with other fixatives (e.g., Histofix).

7. Based on our previous findings, UCP2 expression correlates with neoplastic changes in human colon cancer *(26)*. Human colon tumor tissue microarray can be used therefore as a possible positive control, besides many others, most commonly human or murine spleen with relatively abundant UCP2. Negative control slides can be treated as the actual samples, except goat IgG should replace the primary antibody.

8. Simple humidified chamber can be created by putting some wet paper towels into a plastic slide box.

Acknowledgments

Experimental studies providing the basis of this work were supported by National Institutes of Health grants DK-61890 and RR-17695 to G.B. We thank Rose M. Tavares at the Department of Pathology, Rhode Island Hospital, for her help with immunohistochemistry studies. The authors of this article declare no financial conflict of interest.

References

1. Krauss S., Zhang C. Y., Lowell B. B. (2005). The mitochondrial uncoupling-protein homologues. *Nat. Rev. Mol. Cell. Biol.* **6**:248–61.

2. Brand M. D., Esteves T. C. (2005). Physiological functions of the mitochondrial uncoupling proteins UCP2 and UCP3. *Cell Metab.* **2**:85–93.

3. Fleury C., Neverova M., Collins S., Raimbault S., Champigny O., Levi-Meyrueis C., et al. (1997). Uncoupling protein-2: a novel gene linked to obesity and hyperinsulinemia [see comments]. *Nat. Genet.* **15**:269–72.

4. Heaton G. M., Wagenvoord R. J., Kemp A., Jr., Nicholls D. G. (1978). Brown-adipose-tissue mitochondria: photoaffinity labelling of the regulatory site of energy dissipation. *Eur. J. Biochem.* **82**:515–21.

5. Nedergaard J., Cannon B. (2003). The "novel" uncoupling proteins UCP2 and UCP3: What do they really do? Pros and cons for suggested functions. *Exp. Physiol.* **88**:65–84.

6. Jezek P., Zackova M., Ruzicka M., Skobisova E., Jaburek M. (2004). Mitochondrial uncoupling proteins – facts and fantasies. *Physiol. Res.* **53**(Suppl 1):S199–211.

7. Brand M. D., Affourtit C., Esteves T. C., Green K., Lambert A. J., Miwa S., et al. (2004). Mitochondrial superoxide: production, biological effects, and activation of uncoupling proteins. *Free Radic. Biol. Med.* **37**:755–67.

8. Turrens J. F. (2003). Mitochondrial formation of reactive oxygen species. *J. Physiol.* **552**:335–44.

9. Korshunov S. S., Skulachev V. P., Starkov A. A. (1997). High protonic potential actuates a mechanism of production of reactive oxygen species in mitochondria. *FEBS Lett.* **416**:15–8.

10. Arsenijevic D., Onuma H., Pecqueur C., Raimbault S., Manning B. S., Miroux B., et al. (2000). Disruption of the uncoupling protein-2 gene in mice reveals a role in immunity and reactive oxygen species production. *Nat. Genet.* **26**:435–9.

11. Mattiasson G., Shamloo M., Gido G., Mathi K., Tomasevic G., Yi S., et al. (2003). Uncoupling protein-2 prevents neuronal death and diminishes brain dysfunction after stroke and brain trauma. *Nat. Med.* **9**:1062–8.

12. Teshima Y., Akao M., Jones S. P., Marban E. (2003). Uncoupling protein-2 overexpression inhibits mitochondrial death pathway in cardiomyocytes. *Circ. Res.* **93**:192–200.

13. Ryu J. W., Hong K. H., Maeng J. H., Kim J. B., Ko J., Park J. Y., et al. (2004). Overexpression of uncoupling protein 2 in THP1 monocytes inhibits beta2 integrin-mediated firm adhesion and transendothelial migration. *Arterioscler. Thromb. Vasc. Biol.* **24**:864–70.

14. Orrenius S., Gogvadze V., Zhivotovsky B. (2007). Mitochondrial oxidative stress: implications for cell death. *Annu. Rev. Pharmacol. Toxicol.* **47**:143–83.

15. Lee K. U., Lee I. K., Han J., Song D. K., Kim Y. M., Song H. S., et al. (2005). Effects of recombinant adenovirus-mediated uncoupling protein 2 overexpression on endothelial function and apoptosis. *Circ. Res.* **96**:1200–7.

16. Zhang K., Shang Y., Liao S., Zhang W., Nian H., Liu Y., et al. (2007). Uncoupling protein 2 protects testicular germ cells from hyperthermia-induced apoptosis. *Biochem. Biophys. Res. Commun.* **360**:327–32.

17. Hang T., Jiang S., Wang C., Xie D., Ren H., Zhuge H. (2007). Apoptosis and expression of uncoupling protein-2 in pressure overload-induced left ventricular hypertrophy. *Acta Cardiol.* **62**:461–5.

18. Collins P., Jones C., Choudhury S., Damelin L., Hodgson H. (2005). Increased expression of uncoupling protein 2 in HepG2 cells attenuates oxidative damage and apoptosis. *Liver Int.* **25**:880–7.

19. Mills E. M., Xu D., Fergusson M. M., Combs C. A., Xu Y., Finkel T. (2002). Regulation of cellular oncosis by uncoupling protein 2. *J. Biol. Chem.* **277**:27385–92.

20. Esteves T. C., Brand M. D. (2005). The reactions catalysed by the mitochondrial uncoupling proteins UCP2 and UCP3. *Biochim. Biophys. Acta* **1709**:35–44.

21. Alves-Guerra M. C., Rousset S., Pecqueur C., Mallat Z., Blanc J., Tedgui A., et al. (2003). Bone marrow transplantation reveals the in vivo expression of the mitochondrial uncoupling protein 2 in immune and non immune cells during inflammation. *J. Biol. Chem.* **278**:42307–12.

22. Pecqueur C., Alves-Guerra M. C., Gelly C., Levi-Meyrueis C., Couplan E., Collins S., et al. (2001). Uncoupling protein 2, in vivo distribution, induction upon oxidative stress, and evidence for translational regulation. *J. Biol. Chem.* **276**:8705–12.

23. Zhang C., Baffy G., Perret P., Krauss S., Peroni O., Grujic D., et al. (2001). Uncoupling protein-2 negatively regulates insulin secretion and is a major link between obesity, beta cell dysfunction, and type 2 diabetes. *Cell* **105**:745–55.

24. Krauss S., Zhang C. Y., Lowell B. B. (2002). A significant portion of mitochondrial proton leak in intact thymocytes depends on expression of UCP2. *Proc. Natl Acad. Sci. USA* **99**:118–22.

25. Horimoto M., Fulop P., Derdak Z., Wands J. R., Baffy G. (2004). Uncoupling protein-2 deficiency promotes oxidant stress and delays liver regeneration in mice. *Hepatology* **39**:386–92.

26. Horimoto M., Resnick M. B., Konkin T. A., Routhier J., Wands J. R., Baffy G. (2004). Expression of uncoupling protein-2 in human colon cancer. *Clin. Cancer Res.* **10**:6203–7.

27. Graham J. M. (1993). Isolation of mitochondria, mitochondrial membranes, lysosomes, peroxisomes, and Golgi membranes from rat liver. *Methods Mol. Biol.* **19**:29–40.

Chapter 16

Multiple Approach to Analyzing the Role of MicroRNAs in Apoptosis

Riccardo Spizzo and George A. Calin

Summary

MicroRNAs (miRNAs) are noncoding RNAs whose hallmarks are the very short sequences and the ability to repress the translation and/or transcription of target genes. miRNAs can have diverse functions, including regulation of cellular differentiation, proliferation, and embryogenesis. Over the past 5 years, an increasing number of studies have linked different miRNAs with programmed cell death or apoptosis. The principal aim of this chapter is to describe a method that (1) identifies miRNAs involved in apoptosis, using a validated array profiling approach, (2) assesses the direct involvement of candidate miRNAs in apoptosis, and (3) identifies the molecular mechanisms possibly involved in apoptotic response. To disclose the possible molecular targets of miRNAs, we propose the generation of a database created using a list of presumptive miRNA targets and the changes in the transcriptome after ectopic expression of the miRNAs. Our proposed method for doing this is suitable for both discovery of apoptotic pathways that regulate miRNAs and finding new miRNAs able to induce apoptosis.

Key words: microRNA, Apoptosis, Microarray, Luciferase assay, Target prediction software

1. Introduction

MicroRNAs (miRNAs) are noncoding RNAs first described in *Caenorhabditis elegans* in 1993 *(1, 2)*. They are characterized by a very short sequence (19–22 nt), and they regulate gene expression by blocking the translational process and/or causing degradation of the target messenger RNA (mRNA) *(3)*. Furthermore, a recent study showed that miRNAs induce upregulation of the translation of target mRNAs during cell cycle arrest *(4)*. miRNAs can have diverse functions, including regulation of cellular

Peter Erhardt and Ambrus Toth (eds.), *Apoptosis*, Methods in Molecular Biology, vol. 559
DOI 10.1007/978-1-60327-017-5_16
© Humana Press, a part of Springer Science+Business Media, LLC 2004, 2009

differentiation *(5, 6)*, proliferation *(7, 8)*, embryogenesis *(9)*, and apoptosis *(10–12)*.

Over the past 5 years, an increasing number of reports has linked different miRNAs with programmed cell death or apoptosis. Researchers have compiled two main bodies of evidence concerning the relationship between miRNAs and apoptosis. The first showed how proapoptotic (TP53) *(10, 13, 14)* and antiapoptotic (interleukin-6/signal transducer and activator of transcription 3) *(15)* pathways directly regulate miRNA expression at the transcriptional level. The second showed how the miRNAs can directly and autonomously induce *(16)* or repress *(17)* apoptosis in different cell models by targeting antiapoptotic or proapoptotic proteins, respectively. Thus, miRNAs can play the previously unknown mechanism of regulation of apoptosis. *(18)*. **Table 1** lists the principal miRNAs related to apoptosis.

The principal aim of this chapter is to describe a method that identifies miRNAs involved in apoptosis using a validated array profiling approach *(42, 43)*, assesses the direct involvement of candidate miRNAs in apoptosis, and identifies the underlying molecular mechanism possibly involved in miRNA-induced apoptotic response (**Table 2**). Investigators have successfully used miRNA array profiling to identify oncomirs *(44)* in different tumor types *(45, 46)* and to identify miRNAs regulated by proapoptotic proteins, such as TP53 *(13)*. Researchers have developed and optimized other techniques for miRNA identification and detection, including cloning strategies *(47)*, Northern blotting *(2)*, bead-based flow cytometry *(48)*, quantitative real-time polymerase chain reaction (PCR) *(49)*, in situ hybridization *(50)*, the Invader assay *(51)*, RNA-primed array-based Klenow enzyme assay *(52)*, and direct miRNA assay *(53)*. Real-time PCR analysis of mature miRNA is described in this chapter too.

To disclose the possible targets of miRNAs, we propose a dual approach: (1) an in silico method using software programs available online that identify presumptive miRNA targets and (2) an in vitro method to analyze changes in the transcriptome after ectopic expression of miRNAs in a cellular model. We believe that comparison of the lists generated by the two methods provide valuable clues for identification of the molecular mechanisms regulated by miRNAs.

Our proposed method is suitable for both discovery of apoptotic pathways that regulate miRNAs and finding new miRNAs able to induce apoptosis. Thus, in this chapter, we do not describe any specific models, but instead we focus only on the materials and methods necessary to reproduce our approach. We precisely describe only miRNA-specific techniques; we do not describe in depth other, more common techniques, such as Western blotting, bacterial plasmid cloning, real-time PCR for coding genes, or coding gene microarray.

Table 1
Principal miRNAs related to programmed cell death

miRNA	Effect on apoptosis	Cellular model	Molecular mechanism	Reference
mir-1, mir-133	Proapoptotic (mir-1), antiapoptotic (mir-133)	Rat cardiomyocyte	HSP60 and HSP90 (mir-1), CASP9 (mir-130)	(19)
mir-34a	Proapoptotic	Colon cancer cell line	E2F3	(10)
		Neuroblastoma cell line	NA	(20)
		Lung cancer cell line	NA	(13)
		Lung cancer cell line	E2F, TP53	(21)
		Colon cancer cell line		(22)
Cluster mir-17-18-19-20-92	Antiapoptotic	Mouse lymphoma model	NA	(17)
		Lung cancer cell line	NA	(23)
		Prostate cancer cell line	NA	(24)
mir-21	Antiapoptotic	Glioblastoma cell lines	NA	(25)
		Cholangiocyte	PTEN	(26)
		Myeloma cells	mir-21 transcription regulation by Stat-3	(15)
		Breast cancer cell line		(27)
		Human glioma cells		(28)
		Breast cancer cell line	Bcl2 regulation	(29)
		Vascular neointimal model	NA	(30)
		Fetal cerebral cortex-derived progenitors	TP53 linkage and PDCD4 regulation PTEN and Bcl2 Jagged-1, ELAVL2	(31)
mir-335	Proapoptotic	Fetal cerebral cortex-derived progenitors	Jagged-1, ELAVL2	(31)
mir-15a, mir-16	Proapoptotic	Leukemia cell line	Bcl2	(16)
let-7c, mir-10a, mir-144, mir-150, mir-155, mir-193	Regulation of TRAIL response	Breast cancer cell line	NA	(32)
Let-7a	Antiapoptotic	Cholangiocyte	Increased activation of Stat-3 by direct NF2 downregulation	(33)
mir-143	Proapoptotic	Alfa-mangostin–induced apoptosis in colon cancer cell lines	NA (caspase-independent)	(34)
mir-29b	Antiapoptotic	Cholangiocyte	Mcl-1	(35)
mir-8	Antiapoptotic	*Drosophila*	Atrophin	(36)

(continued)

Table 1
(continued)

miRNA	Effect on apoptosis	Cellular model	Molecular mechanism	Reference
Bantam miRNA	Antiapoptotic	*Drosophila*	Downregulation of proapoptotic protein hid Transactivation by Hippo-Yorkie pathway	*(37)* *(38)*
mir-278	Antiapoptotic	*Drosophila*	–	*(39)*
mir-14	Antiapoptotic	*Drosophila*	Elevated levels of apoptotic effector Drice	*(40)*
mir-LAT (herpes simplex virus 1)	Antiapoptotic	Neuroblastoma cell line	Downregulation of TGF-beta1 and SMAD3 expression	*(41)*

NA, not available; TRAIL, tumor necrosis factor-related apoptosis-inducing ligand; TGF, transforming growth factor

Table 2
Summary of proposed method for identification of miRNAs involved in apoptosis

Aim	Methods	
Identification of miRNAs involved in apoptotic response	1. Identification of miRNAs involved in apoptotic response	1.1. Sample collection
	2. MiRNA array profiling	2.1. RNA extraction 2.2. Sample preparation 2.3. Synthesis of biotin-labeled first-strand cDNA targets 2.4. Array hybridization 2.5. Posthybridization miRNA microarray 2.6. Array scanning
	3. Identification of significant deregulated miRNAs	3.1. Data normalization 3.2. Raw data and statistical analysis by comparison of treated versus control miRNA signatures
	4. Validation of microarray results	4.1. DNase digestion 4.2. Gene-specific reverse transcription for mature miRNA

(continued)

Table 2
(continued)

Aim	Methods	
Proof of direct involvement of miRNAs in apoptosis	5. Transfection of miRNAs[a] in vitro	5.1. Transfection of miRNAs
	6. Assay to test the apoptotic effect	6.1. Annexin V assay, caspase 3 assay and others
Finding molecular targets of miRNAs	7. Multiple approaches to identifying possible pathways targeted by miRNAs	7.1. In silico identification of miRNA targets using multiple algorithms
		7.2. In vitro identification of miRNA targets using coding-gene array analysis after miRNA[a] transfection (*see* **Subheading 3.7.2**)
		7.3. Cross-analysis of the lists in **Subheading 3.7.1** and **3.7.2**
	8. Validation of the possible targets	8.1. Western blot assay to verify downregulation of expression of the protein after transfection of miRNA (*see* **Subheading 5.1**)
		8.2. Luciferase reporter assay of a direct interaction between the miRNA and the 3′ UTR of the target mRNA

[a] If the profiling experiments (*see* **Subheading 3.2**) show downregulation of expression of the involved miRNA, the transfection experiments will be performed with an antagomir instead of a precursor miRNA

2. Materials

2.1. Identification of miRNAs Involved in Apoptotic Response (see Note 1)

Sample Collection
1. Cell culture medium.
2. Phosphate-buffered saline [PBS].
3. Solution of containing trypsin (0.25%) and ethylenediaminetetraacetic acid (1 mM).

2.2. miRNA Array Profiling

2.2.1. RNA Extraction

1. TRIzol reagent (Invitrogen): Store at room temperature. Additional care should be taken when using TRIzol because it is toxic when it comes in contact with skin or is swallowed. The use of TRIzol should be restricted to a chemical hood.
2. 100% chloroform (Sigma-Aldrich): Store at room temperature (*see* **Note 2**). Chloroform should be handled under a chemical hood.
3. 100% ethanol (Sigma-Aldrich).
4. RNase and DNase-free water (IDT) (*see* **Note 3**).

2.2.2. Synthesis of Biotin-Labeled First-Strand cDNA Targets

1. Reagents

   a) 0.5 pM/µL 3′ NNNNNNNN-$(dA)_{12}$ T (biotin) $(dA)_{12}$-biotin 5′ oligonucleotide primer.

   b) 5× first-strand buffer.

   c) 0.1 M dithiothreitol.

   d) Superscript II RNaseH⁻ reverse transcriptase (200 U/µL; Invitrogen).

   e) 10 mM dNTP mix (Invitrogen).

2. Equipment and Materials Supplied by User

   a) Pipette tips (sterile, RNase-free, and aerosol-resistant).

   b) Microcentrifuge tubes (sterile, RNase-free, 1.7 mL).

   c) Micropipettes (10, 20, 200, and 1,000 µL).

   d) Nanodrop ultraviolet spectrophotometer.

   e) Microcentrifuge (room temperature and 4°C).

   f) Water bath (70, 65, and 37°C).

   g) Sterile, nuclease-free conical tubes (15 and 50 mL).

   h) SpeedVac concentrator.

   i) Vortex.

   j) Pipet Aid and disposable pipettes.

3. Required Reagents/Kits Supplied by User

   a) 0.5% NEN blocking reagent (PerkinElmer).

   b) Streptavidin/Alexa Fluor 647 conjugate (staining solution: a 1:500 dilution in [TNB] (Molecular Probes, Invitrogen).

   c) Nuclease-free water (Ambion).

   d) 1× PBS, pH 7.4 (Invitrogen).

   e) 1 M Tris-HCl, (pH 7).

   f) 5 M NaCl.

   g) Tween-20 (Sigma).

   h) Formamide (Sigma).

   i) 50× Denhardt's solution (Sigma).

4. Other Equipment and Materials Supplied by User

   a) Tecan HS 4800 hybridization station.

   b) Axon GenePix 4000B.

   c) Computer configured for the Axon GenePix 4000B scanner.

   d) New Brunswick Innova 4080 shaking incubator.

   e) Sigma/Qiagen centrifuge (4–15°C) (Qiagen).

   f) Centrifuge plate rotor: 2 × 96 (Qiagen).

   g) Pipette tips (sterile, RNase-free, and aerosol-resistant).

h) Microcentrifuge tubes (sterile, RNase-free, 1.7 mL).

i) Micropipettes.

j) Powder-free gloves.

k) Microcentrifuge.

l) Microtiter plate lid, black (Corning).

m) Bioarray processors (GE Healthcare).

n) Bioarray rack (GE Healthcare).

o) Small reagent reservoir (GE Healthcare).

p) Large reagent reservoir (GE Healthcare).

q) Bioarray removal tool (GE Healthcare).

r) Bioarray position tool (GE Healthcare).

2.2.3. Array Hybridization

1. Required Reagents/Kits

   a) 0.5% NEN blocking reagent (PerkinElmer).

   b) Streptavidin/Alexa Fluor 647 conjugate (staining solution: a 1:500 dilution in TNB) (Molecular Probes).

   c) Nuclease-free water (Ambion).

   d) 1× PBS, pH 7.4 (Invitrogen).

   e) 1 M Tris-HCl, pH 7.6 (Sigma).

   f) 5 M NaCl (Sigma.

   g) Tween-20 (Sigma).

   h) Formamide (Sigma).

   i) 50× Denhardt's solution (Sigma).

2. Other Equipment and Materials

   a) Tecan HS 4800 hybridization station.

   b) Axon GenePix 4000B scanner.

   c) Computer configured for the Axon GenePix 4000B scanner.

   d) New Brunswick Innova 4080 shaking incubator.

   e) Sigma/Qiagen centrifuge (4–15°C) (Qiagen).

   f) Centrifuge plate rotor: 2 × 96 (Qiagen).

   g) Pipette tips (sterile, RNase-free, and aerosol-resistant).

   h) Microcentrifuge tubes (sterile, RNase-free, 1.7 mL).

   i) Micropipettes.

   j) Powder-free gloves.

   k) Microcentrifuge.

   l) Microtiter plate lid, black (Corning).

   m) Bioarray processors (GE Healthcare).

   n) Bioarray rack (GE Healthcare).

o) Small reagent reservoir (GE Healthcare).

p) Large reagent reservoir (GE Healthcare).

q) Bioarray removal tool (GE Healthcare).

r) Bioarray position tool (GE Healthcare).

2.2.4. Posthybridization miRNA Microarray

Stock Solutions for Posthybridization Array Processing

1. TNT buffer (20 L): 0.1 M Tris-HCl, pH 7.6, 0.15 M NaCl, 0.05% Tween-20. Rinse a 25-L carboy out with 150 mL of isopropanol. Rinse the carboy twice with 3 L of deionized water and completely drain the carboy. Add 2 L of 1 M Tris-HCl, 600 mL of 5 M NaCl, 120 mL of Tween-20, and 17.39 L of deionized water. Mix well by swirling. Filter TNT through a 0.2-μm filter. This solution can be stored for up to 2 weeks at room temperature.

2. TNT buffer (0.75×): Add 25 mL of deionized water to 75 mL of TNT buffer (described above) per 100 mL of buffer required.

3. TNB buffer (0.5 L): 0.1 M Tris-HCl, pH 7.6, 0.15 M NaCl, 0.5% NEN blocking reagent (PerkinElmer). Add 435 mL of nuclease-free water, 50 mL of 1 M Tris-HCl, pH 7.6, and 15 mL of 5 M NaCl. Slowly add 2.5 g of NEN blocking reagent in 0.5-g increments until all 2.5 g of the reagent are dissolved while warming in a water bath at 60°C. Filter TNB buffer through a 0.88-μm filter. Aliquot the TNB buffer into 50-mL tubes and store at –20°C. This solution can be stored for up to 12 weeks at –20°C. Thaw immediately before use.

2.2.5. Array Scanning

Equipment:Axon GenePix 4000B scanner.

2.3. Identification of Significant Deregulated miRNAs

2.3.1. Data Normalization, Raw Data, and Statistical Analysis by Comparison of Treated and Control miRNA Signatures

Statistical analysis of the data requires different software programs, some of which are freely available on the Internet (open source), whereas others require purchasing a license. The method described by Stefano Volinia and colleagues uses the R, significance analysis of microarray (SAM), and prediction analysis of microarrays software programs (all of which are freely available on the Internet), whereas the approach used by Manuela Ferracin and Massimo Negrini requires purchasing a license for the GeneSpring software program (Silicon Genetics, Redwood City, CA, USA).

2.4. Validation of Microarray Results

2.4.1. DNase Digestion

Required Reagents/Kits: TURBO DNase kit (Ambion).

2.4.2. Gene-Specific Reverse Transcription for Mature miRNA

Required Reagents/Kits: TaqMan MicroRNA Reverse Transcription Kit (Applied Biosystems).

2.5. Proof of Direct Involvement of miRNAs in Apoptosis

In Vitro Transfection: Required Reagents/Kits
1. Lipofectamine 2000 (Invitrogen).
2. miRNA mimic/antagomir (Ambion).

2.6. Assay to Determine the Apoptotic Effect of miRNA

Annexin V Assay kit (Becton Dickinson).

2.7. Multiple Approaches to Identifying Possible Pathways Targeted by miRNAs

The miRGen online database (http://www.diana.pcbi.upenn.edu/miRGen.html) is a useful resource for combining and intersecting different algorithms that predict miRNA targets.

2.7.1. In Silico Identification of miRNA Targets Using Multiple Algorithms

2.7.2. In Vitro Identification of miRNA Targets Using Coding-Gene Array Analysis After miRNA Transfection (see Subheading 2.5, item 1)

Different microarray platforms are currently employed; these platforms can be used interchangeably.

2.7.3. Cross-Analysis of the Lists of miRNA Targets Generated as Described in Subheadings 2.7.1 and 2.7.2

We commonly use the Excel and Word software programs (Microsoft, Redmond, WA, USA) to compare lists of miRNA targets.

2.8. Validation of Possible miRNA Targets

Luciferase reporter assay of the direct interaction between the miRNA and the 3′ untranslated region (UTR) of the target mRNA.

Required Reagents/Kits:
1. Two primers designed for the 3′ UTR of the predicted mRNA target, both containing an *Xba*I site at the 5′ end.
2. Genomic DNA to be used as a template for the PCR.
3. Taq polymerase and other reagents for the PCR (dNTPs, buffer, $MgCl_2$).
4. A kit used for gel extraction.
5. Ligase and competent bacteria.
6. pGL3 control vector (Promega), pRLTK vector (Promega), and *Xba*I restriction enzyme.
7. A kit used for luciferase detection (Promega).

3. Methods

3.1. Identification of miRNAs Involved in Apoptotic Response

Sample Collection: We recommend performing these experiments at least three times both for the treatment and the control (drug or transfection) and to hybridize the labeled RNAs on the miRNA chip (*see* **Note 4**).

Cells growing in monolayers can be directly lysed in tissue culture wells or dishes. Cells can be also trypsinized before RNA purification (*see* **Note 5**). In addition, the RNA purification can be postponed; if it is, the cells can be trypsinized and collected as a pellet. The pellet can be snap-frozen in liquid nitrogen and then stored at –80°C (*see* **Note 6**). If the samples must be sent to another laboratory or to a core facility before the RNA extraction, the frozen pellets can be sent on dry ice, or the cells can be collected using stabilization reagents and delivered at room temperature or 2–8°C (*see* **Note 7**).

3.2. miRNA Array Profiling

3.2.1. RNA Extraction

We commonly perform RNA extraction using the standard TRIzol reagent protocol (*see* **Note 8**). A total of 2.5–5.0 µg of RNA is sufficient for each replicate of each sample. Before the sample preparation for microarray hybridization, quantity and quality assessment of the RNA must be performed. Regarding the quality of RNA, as usual, higher is better (*see* **Note 9**).

3.2.2. Sample Preparation

The microarray platform for miRNA (microRNACHIP) described herein is based on the technology described by Liu et al. *(42)*. We do not describe the procedure used for microRNACHIP printing; Liu and colleagues previously described this method in detail *(54)*. The microRNACHIP currently used at the microarray facility at The Ohio State University Comprehensive Cancer Center is version 4.0 (microRNACHIPv4) and contains 847 probes targeting 474 human miRNAs and 373 mouse miRNAs. Furthermore, the chip contains probes targeting other noncoding RNA, such as ultraconserved sequences *(55, 56)* and human predicted miRNAs *(57–59)*. The miRNA oligo probes that have been spotted on the chip are 40 nt long and target both the mature and precursor forms of each miRNA. This technology is based on direct reverse transcription with a biotin-labeled random octamer primer to obtain labeled target cDNA. After miRNA sample preparation, the biotin-labeled target cDNAs are hybridized with the probes on the array slides in a semiautomated manner. Finally, the biotinylated target/probe complexes are stained with a streptavidin/Alexa Fluor 647 conjugate.

3.2.3. Synthesis of Biotin-Labeled First-Strand cDNA Targets

1. Prepare each total RNA sample for manual target preparation.

   5 µg of total RNA (the optimal concentration should be determined for each source of total RNA; *see* **Note 10**).

2 μL of 0.5 μg/μL primer.

Add X μL nuclease-free water to a 12-μL final volume.

12-μL total volume.

5 μg of total RNA in 10 μL of RNase-free water.

2 μL of an oligo primer (5′ biotin-AAA-AAA-AAA-AAA-[T-biotin]-AAA-AAA-AAA-AAA-NNN-NNN-NN 3′ [0.5 μg/μL] in which N stands for a random octamer).

2. Incubate for 10 min in a 70°C water bath and then immediately place the tube on ice.

3. Centrifuge for 5 s to collect the sample at the bottom of the tube and immediately place the tube on ice.

4. With the tube remaining on ice, add the following reagents to the 12-μL total RNA/control mRNA/primer mix.

    4 μL of 5× first-strand buffer.

    2 μL of 0.1 M dithiothreitol.

    1 μL of 10 mM dNTP mix.

    1 μL of SuperScript II RNase H⁻ reverse transcriptase (200 U/μL).

    20-μL final volume.

5. Incubate 90 min in a 37°C water bath.

6. Centrifuge for 5 s to collect the sample at the bottom of the tube.

7. RNA template degradation. After 90 min of incubation for the first-strand synthesis, add 3.5 μL of 0.5 M NaOH/50 mM ethylenediaminetetraacetic acid into a 20-μL reaction mix and incubate at 65°C for 15 min to denature the DNA/RNA hybrids and degrade the RNA template. Next, neutralize the reaction with 5 μL of 1 M Tris-HCI, pH 7.6 (Sigma). Each labeled target should be in a 28.5-μL volume. The sample preparation will thus end, and the samples must be stored at −80°C until use.

3.2.4. Array Hybridization

1. Prime all channels of the Tecan HS 4800 hybridization stations.

2. Load a preprinted miRNA array face-up to the Tecan HS 4800 hybridization station (*see* **Note 11**). Close the hybridization chambers of HS 4800.

3. Run the hybridization program for the chip hybridization with labeled cDNA targets.

    a. Prime the chip in the hybridization chamber at 23°C with 6× SSPE with 0.05% Tween-20 for 1 min.

    b. Inject 150 μL of a prehybridization mix of 6× SSPE/2× Denhardt's solution/30% formamide and prehybridize the chip at 25°C for 30 min.

    c. Inject the hybridization mix of labeled biotin-cDNA in 6× SSPE/2× Denhardt's solution/30% formamide and hybridize at 25°C for 18 h (*see* **Note 12**).

    d. Wash in 0.75× TNT buffer at 23°C for 5 min.

    e. Wash in 0.75× TNT buffer at 37°C for 10 min.

    f. Rinse with water in the HS 4800 at 23°C for 30 s.

    g. Unload the chips from the machine for posthybridization washing.

3.2.5. Posthybridization miRNA Microarray

1. Open the hybridization chamber of the HS 4800 and remove the slide chips as quickly as possible and place them in a slot of a bioarray rack placed in a large reagent reservoir containing 0.75× TNT prewarmed to 37°C. Move the slide into place using the bioarray position tool with the tooth side down. Wash the slides in 0.75× TNT prewarmed to 37°C with agitation in a New Brunswick Innova 4080 shaking incubator at 37°C for 40 min with agitation at 50 rpm.

2. Block the slide in TNB blocking buffer at room temperature for 30 min.

3. Stain the slide with a streptavidin/Alexa Fluor 647 conjugate in TNB buffer at 1:500 at room temperature for 30 min.

4. After staining, wash in 1× TNT at room temperature for 40 min with three time buffer changes.

5. Briefly rinse the chips with distilled water and spin-dry them at 1,000 rpm for 1 min.

3.2.6. Array Scanning

Scan the slides at a resolution of 10 μm using an Axon GenePix 4000B scanner at a power setting of 100 and PMT 800 (*see* **Note 13**).

1. Open the .gal file provided by the builder of the slide.

2. Drag the file onto the features just collected to fit all of the blocks.

3. Set each block and press the F5 key for automatic alignment of the grid on the features.

The image data may be extracted using the GenePix software and saved as a .gpr file for further data analysis.

3.3. Identification of Significant Deregulated miRNAs

Ideally, the same raw data should be analyzed by two distinct bioinformaticians using two independent methods of analyses. Two strategies have been used by us and developed in two different laboratories *(45, 60)*.

3.3.1. Data Normalization

The data analysis strategy developed by Stefano Volinia and colleagues *(61)* can be summarized as follows. Mean values of the replicate spots of each miRNA are background-subtracted, normalized, and subjected to further analysis. miRNAs are retained when present in at least 20% of analyzed samples, or in at least the smallest group of comparison in the data set, and when the miRNAs have a fold change in signal intensity of more than 1.5 from the median of all the samples. Prior to statistical analysis, a value of 22 (4.5 on the log2 scale) is assigned to miRNA probes with absent signal. This value is the mean minimum intensity level above the background detected in miRNA chip experiments. Three different data-normalization procedures are applied: global median, housekeeping, and cyclic loess.

A second data analysis strategy developed by Manuela Ferracin and Massimo Negrini *(43)* consists of raw data normalization and analysis using the GeneSpring software program. Expression data are median-centered using the GeneSpring normalization option or global median normalization option of the BioConductor open-source software program (http://www.bioconductor.org).

3.3.2. Raw Data and Statistical Analysis by Comparing Treated and Control miRNA Signatures	We have compared the SAM and Prediction Analysis of Microarrays (http://www-stat.stanford.edu/~tibs/PAM/index.html) tests performed for the different classes of comparison (usually, "cancer" vs. "normal" or "cancer set 1" vs. "cancer set 2") over the respective expression tables. Cyclic loess exhibited the best improvement in reducing variability of hybridization and yielded the highest number of significant miRNAs across the different tests in agreement with a recent evaluation of different normalization methods for CodeLink microarrays *(62)*. Cyclic loess uses the MA plot and loess smoothing to estimate intensity-dependent differences in each pair of slides in the data set and then removes these differences by centering the loess line at zero. This procedure is iterated until intensity-dependent differences between slides are removed from all of the arrays. We implemented cyclic loess using the BioConductor affy/normalization loess with miRNA nomenclature according to the miRNA database at the Wellcome Trust Sanger Institute (Hinxton, Cambridge, United Kingdom). Differentially expressed miRNAs between classes are identified using the *t*-test procedure, which is included in significance analysis of microarrays (SAM). The SAM program calculates a score for each gene based on the change in its expression relative to the standard deviation for all measurements. The miRNA signatures are identified by applying the nearest shrunken centroids to the entire dataset. This method identifies a subgroup of genes that best characterizes two sample groups (i.e., tumor vs. normal). The prediction error is calculated using tenfold cross-validation. Using this specific approach, we were able to identify a common miRNA expression signature of human solid tumors that defines cancer gene targets *(45)*.

The analysis by Manuela Ferracin and Massimo Negrini is based on the use the GeneSpring analysis of variance and support vector machine tools for cross-validation and test-set prediction. These tools identify and confirm the miRNAs able to best separate the groups of comparison using two different methods of raw-data normalization. All data are submitted to the ArrayExpress database (http://www.ebi.ac.uk/arrayexpress/) using MIAMExpress software program. Using this specific strategy, we were able to find a common miRNA signature for breast cancer *(60)* as well as, for the first time, an miRNA signature of hypoxia *(63)*.

The raw-data analysis output using this method consists of (1) dendrograms showing the clustering of multiple samples according to the transfection status and type of transfectant (e.g., miRNAs, scrambled oligos, empty vectors) and (2) gene lists containing genes differentially expressed at high statistical significance ($P < 0.05$ or, better, $P < 0.01$). The level of differential expression of these genes that can be considered biologically significant is a key issue. If fold differences in gene expression of 10–20 are biologically important, then we support the view that much smaller differences (less than twofold) may also be significant. The explanation for this resides in the multiplicity of targets of specific miRNAs and the large number of altered miRNAs, meaning that two or more target protein-coding genes in the same pathway likely are consequently disturbed. An opposite view is that in some cases, the described changes in miRNA expression between normal and tumor cells may be biologically irrelevant; the main interactions of one miRNA with various targets may not accumulate but antagonize each other (e.g., repression of both proapoptotic and antiapoptotic genes) resulting in unmodified phenotype *(64)*.

3.4. Validation of Microarray Results

Microarray data can be confirmed using Northern blotting and quantitative real-time reverse transcriptase (RT)-PCR analysis based on the assay for active miRNA *(49)* or on quantitative RT-PCR analysis of precursor miRNA (*see* **Note 14**) *(65)*. Below, we describe a modified protocol for analysis of mature miRNA molecules that is based on the Applied Biosystems protocol (*see* **Note 15**).

3.4.1. DNase Digestion

Before performing the miRNA assay by RealTime RT-PCR assay, we recommend treatment of DNase in each sample to be tested (*see* **Note 16**).

1. Prepare each total RNA sample for DNase treatment in 0.5-mL labeled tubes.

   2 µg of total RNA.

   2 µL of 10× Turbo DNase buffer.

   1 µL of Turbo DNase.

   X µL of diethylpyrocarbonate (DEPC)-treated water to 20 µL.

   20-µL total volume.

2. Incubate for 30 min at 37°C (*see* **Note 17**).

3. Quickly spin each tube to collect all of the material at the bottom.

4. Add 2 μL of DNase inhibitor to each tube (*see* **Note 18**).

5. Incubate for 2 min at room temperature, gently vortexing every 30 s.

6. Place all of the samples in a bench centrifuge and centrifuge them at maximum speed for 2 min.

7. Remove the tubes from the centrifuge and quickly collect the supernatant from each tube into a new tube without touching the white bottom pellet and place the samples on ice (*see* **Note 19**).

8. Optional: measure the RNA concentration in each sample after the DNase digestion (*see* **Note 20**).

3.4.2. Gene-Specific Reverse Transcription

We usually perform a reverse transcription for at least two miRNAs up to six plus one housekeeping gene (RU6) in each tube using 100 ng of DNase-treated RNA (*see* **Note 21**).

1. Prepare the following mixture for each sample (*see* **Note 22**).

   1.5 μL of buffer (Applied Biosystems).

   0.15 μL of dNTPs (Applied Biosystems).

   0.2 μL of RNase inhibitor (Applied Biosystems).

   1 μL of reverse transcriptase (Applied Biosystems).

   1.15 μL of DEPC-treated water.

   4-μL total volume.

2. Add 2 μL (200 ng) of DNase-treated total RNA to each tube.

3. Add 3 μL of gene-specific RT primer for each miRNA to each tube to a maximum volume of 9 μL (two miRNAs plus the U6 housekeeping gene).

4. Spin the tubes and place them in a thermocycler machine using the following program: 16°C for 30 min, 45°C for 30 min, and 85°C for 5 min to denature the enzyme.

3.4.3. Real-Time PCR Reaction for miRNAs

1. After the thermocycler program, briefly spin the samples to collect all of the cDNAs at the bottom.

2. Dilute each sample by adding 40 μL of DEPC-treated water.

3. Prepare the following master mix for each sample and for each miRNA to be tested (each sample will be tested in triplicate for each miRNA).

   10 μL of Taqman Universal Master Mix no AmpErase UNG (Applied Biosystems).

   1 μL of miRNA primer-probe (Applied Biosystems).

   4 μL of water.

   15-μL total volume.

4. First dispense the master mix into a 96-well plate and then add 5 μL of cDNA (final volume; *see* **Note 23**).

5. Seal the plate using optically clear film and briefly spin the plate. Check for any air bubbles at the bottom of the wells and remove them if necessary by flicking the wells.

6. Place the plate in the real-time instrument and start the program (95°C for 10 min, 40 cycles at 95°C for 15 s, and 60°C for 1 min; collect the data during the 60°C step).

7. Analyze the data using DeltaCt/DeltaCt analysis.

3.5. Proof of Direct Involvement of miR-NAs in Apoptosis

At the end of all the previous steps, the final output is a list of miRNA genes differentially expressed. Biologists and clinicians involved in the study have to determine the meaning of these data. Therefore, to understand the significance of the data, we propose performing at least one of two steps. The first step is the analysis of the biological consequences of exogenous expression of miRNA upregulated during apoptosis (or, vice versa, shutting down overexpressed miRNAs), whereas the second step is the identification of mRNA targets of differentially expressed miRNAs.

In Vitro Transfection of miRNAs: To confirm the direct involvement of miRNAs in the apoptotic process, downregulation of a downregulated miRNA (*see* **3.3.2**) or reexpression of an upregulated miRNA (*see* **3.3.2**) should be able to produce apoptosis.

Different methods can be used to express or downregulate the expression of miRNAs in the cells. As our first approach, we commonly use small RNA molecules that mimic endogenous precursor miRNAs (precursor miRNA; Ambion) or knock down the target miRNA (antagomir; Ambion).

1. The day before the transfection, plate a number of cells sufficient to reach a level of confluence of about 50–60% the day after (*see* **Note 24**).

2. Use the small interfering RNA transfection protocol for Lipofectamine 2000 (Invitrogen). We commonly use a miRNA mimic/antagomir concentration of 100 μM in the final medium volume without antibiotics.

3. Perform the transfection using Opti-MEM medium (Invitrogen); after 4 h, replace the medium with complete fresh medium (*see* **Note 25**).

4. Confirm the up/downregulation of miRNA of interest by Real Time PCR as a control of good transfection (*see* **3.4** for methods).

3.6. Assay to Determine the Apoptotic Effect of miRNA

Different assays can be used to determine the apoptotic effect after in vitro transfection of miRNA precursor/antagomir (*see* **Note 26**).

Annexin V Assay: of the available assays used to test the apoptotic effect of miRNAs, we have commonly used the Annexin V assay with flow cytometry. We have usually performed this assay at different time points and using six multiwell plates. Many other assays are available such as caspase 3. At least two independent assays should be used and confirm the apoptotic effects.

3.7. Multiple Approaches for Identification of Possible Pathways Targeted by miRNAs

Two different approaches can be used to discover the molecular pathways regulated by miRNAs. The first is an in silico approach that applies algorithms able to predict miRNA targets. The second is an empiric approach based on a study of the protein-coding transcriptome changes (microarray analysis) related to reexpression or downregulation of expression of the miRNA.

3.7.1. In Silico Identification of miRNA Targets Using Multiple Algorithms

In animals, miRNAs do not bind perfectly to their targets, dramatically increasing the number of their potential targets. Thus far, researchers have identified only few miRNA/mRNA interactions important to cancer pathogenesis (e.g., *let-7* and RAS oncogenes, *miR-15* and *miR-16* and the BCL2 oncogene, *miR-17-5p* and *miR-20a* and the tumor suppressor E2F1 transcription factor, *miR-127* and the oncogene BCL6, and *miR-21* and PTEN) *(64, 66)*. After the discovery of miRNA mechanism of action, investigators developed several different target prediction programs *(67)*. Unfortunately, the results of miRNA::target prediction generated using different software programs are quite different, making identification of true-positive results difficult. The four most used predictive software programs for miRNA targets are DianaMicroT *(68)*, TargetScan *(69)*, miRanda *(70)*, and PicTar *(71)*. A clear advance was the recent development of the miRGen online database (http://www.diana.pcbi.upenn.edu/miRGen.html); its target interface provides access to unions and intersections of the four most widely used target prediction programs and to experimentally supported targets from TarBase database *(72)*. The first approach to identify miRNA::target interactions can be generation of a set of predicted targets (if available) using the intersection of all of the algorithms in the miRGen database.

3.7.2. In Vitro Identification of miRNA Targets Using Coding-Gene Array Analysis After miRNA Transfection (see Subheading 3.5)

The second approach to discover the molecular pathways regulated by miRNAs is based on evaluation of the protein-coding transcriptome modifications caused by expression or downregulation of expression of the miRNA using microarray analysis. Different microarray analysis platforms exist, and the technology is commonly used nowadays.

The Gene Array experiment is performed after transfection of miRNA mimic or antagomir as described in **Subheading 3.5**. Proof of an effect of the miRNA (e.g., increased apoptosis) is mandatory in such experiments. Different time points can be used to distinguish early and late effects after miRNA precursor/

antagomir transfection. If only a few time points can be used, we suggest using early time points to better understand the specific mechanisms related to the activation or inactivation of the miRNA. This experiment has to be performed using biological replicates of the same treatments. The output of such an analysis is a list of differentially expressed genes, such as that described in **Subheading 3.3**. If a proteomic facility is available, protein extracts can be analyzed for the same time points with various proteomic approaches, starting from 2D gels to masspectrometry.

3.7.3. Cross-Analysis of the Lists of Genes Generated as Described in Subheadings 3.7.1 and 3.7.2

The finding of common genes in the two lists generated as described above in **Subheadings 3.7.1** and **3.7.2** results in the hypothesis that these genes are regulated by miRNAs at the mRNA level *(3)*. Another way to compare the two lists is to compare the ontologies of all the genes obtained in **Subheadings 3.7.1** and **3.7.2**, looking for those already related to apoptosis; this can provide some hints about which apoptotic pathways are regulated by miRNAs. A third way to correlate the transcriptome modifications (**Subheading 3.7.2**) and putative targets (**Subheading 3.7.1**) is to identify known transcription factors that regulate the genes that are differentially expressed in the list in **Subheading 3.7.2** and look for those that are present in the list in **Subheading 3.7.1**. This type of comparison is based on the hypothesis that miRNAs regulate one or more transcription factors whose modification causes the transcriptome changes in the list in **Subheading 3.7.2**. This analysis can be performed using a computational approach with the TRANSFAC database (http://www.gene-regulation. com; BIOBASE Corporation, Beverly, MA, USA) *(73)*.

3.8. Validation of Possible Targets of miRNAs

After generation of a list of possible miRNA target genes, empirical validation of miRNA::target interactions is necessary. We believe that two different sets of evidence must be collected: the decrease in the expression of the protein of the hypothetical target gene after transfection of the miRNA precursor and the proof of the direct interaction between the miRNA and mRNA in an artificial model.

3.8.1. Verification of Downregulation of the Expression of the Protein of Predicted Target Gene After Transfection of miRNA Using Western Blot Assay

Western blot assay can be simply performed with protein lysates after treatment of cells with an miRNA mimic or antagomir. Because the effect of the miRNA can be the translational process, one must keep in mind the half-life of the protein when performing such experiments. Thus, different time points are required to determine the effect of the miRNA on the potential target. If there is hypothesis of a direct effect of the miRNA on the mRNA expression level of the target gene, real-time PCR, semiquantitative PCR, or Northern blotting can be performed, as well. As reported previously, the effect of an miRNA on a target gene can be cell-specific, and different cell lines may not show any modifications of the level of expression of the target protein. Such findings have yet to be explained, but this suggests first using the same cell model as that described in **Subheadings 3.1** and **3.2** and eventually testing such interactions using other models.

3.8.2. Luciferase Reporter Assay of a Direct Interaction Between an miRNA and the 3′ UTR of the Target mRNA

After validation of an interaction between the miRNA and target gene using Western blotting, proving that this interaction is direct and that no other intermediaries (i.e., transcription factors) are required is necessary. To prove this, one must build an artificial model in which the 3′ UTR of the potential target is cloned in the 3′ tail of a reporter gene. In this model, if the interaction exists, the direct complementarity between the miRNA and the 3′ UTR sequence of the gene has been proven. Different reporter systems can be used, such as luciferase reporter or GFP (green fluorescent protein). We commonly use the luciferase assay.

The luciferase assay is a well-established, easily performed technique that we *(16)* and others *(74)* have used successfully to define the direct interaction between miRNA and mRNA. It consists of simply measuring the luciferase activity in cell lysates after transfection of a reporter plasmid containing the predicted region of interaction between an mRNA and an miRNA (putative target miRNA; *see* **Note 27**) *(16)*. Different cell lines can be used to assess such interactions, for example, HeLa, 293, or the same cell line in which the miRNA of interest has been associated with apoptosis. The last offers the advantage of using a model in which the miRNA being tested has already showed an effect (i.e., differential expression after apoptotic stimulus; presumably, the ability to induce apoptosis itself and, hopefully, downregulation of expression of the protein of the predicted target gene).

Briefly, in the luciferase assay the 3′ UTR segment of the target gene is amplified by PCR using as template human genomic DNA and inserted into the pGL3 control vector (Promega) using the *Xba*I site immediately downstream from the stop codon of luciferase. Also, we usually generate inserts with deletions/mutations located in the site of perfect complementarity (defined as seed, located in the 5′ end of the miRNA). The presence of wild-type and mutant inserts are confirmed using sequencing (*see* **Note 28**). The cells are plated on 24-well plates the day before the transfection at a confluence of 70–80% in medium without antibiotics. The next day, the cells are transfected using Lipofectamine 2000 (Invitrogen) according to the manufacturer's protocol with 0.4 μg of the firefly luciferase reporter vector (pGL3) and 0.08 μg of the control vector containing Renilla luciferase, pRL-TK (Promega), in each well. The required amount of miRNA mimic/antagomir is added to the mixture to reach a concentration of 100 nM in the final transfection volume. A standard transfection protocol is shown below.

1. We perform each transfection in a **24-well** plate, usually performed in six replicates.

One well	MiRNA (stock 50 μM) (μL)	Scrambled (stock 50 μM) (μL)	pGL3 vector (stock 0.4 μg/μL) (μL)	pRLTK (stock 0.1 μg/μL) (μL)	Opti-MEM (μL)
MiRNA + pGL3 contr	1	–	1	0.8	50
MiRNA + pGL3 3′ UTR wt	1	–	1	0.8	50
MiRNA + pGL3 3′ UTR mut	1	–	1	0.8	50
Scrambled + pGL3 contr	–	1	1	0.8	50
Scrambled + pGL3 3′ UTR wt	–	1	1	0.8	50
Scrambled + pGL3 3′ UTR mut	–	1	1	0.8	50

contr, control; wt, wild-type; mut, mutation

2. At least 1 h before the transfection, we replace the medium with fresh medium (0.4 mL).

Tube A (one tube for each pGL3 vector)

3. First add the miRNAs, then the plasmids, and finally the

Lipofectamine 2000	2 μL
Opti-MEM	48 μL

(*see* **Note 29**)

Opti-MEM.

4. Incubate for 10 min at room temperature.

Tube B

5. Add 50 μL of tube B to each tube A (*see* **Note 30**).

6. Incubate for 20 min.

7. Add 100 μL of the mixture of tube A and tube B to each replicate well in the 24-well plate.

8. After 4 h, replace the transfection medium with fresh medium without antibiotics.

9. Twenty-four hours after the transfection, the activity of firefly and *Renilla luciferase* is measured consecutively using dual luciferase assays (Promega).

10. Thaw all reagents (e.g., active lysis buffer, substrates for Renilla and firefly luciferase) at least 1 h before measurement of cell lysates. The reagents must be used at room temperature.

11. Before starting the protocol, prepare the active lysis buffer to the final concentration of 1× (dilute five times in water).

12. Aspirate the media from each well, wash each well with 0.5 mL of PBS (keep it at room temperature), aspirate the PBS, and add 100 µL of 1× active lysis buffer to each well.

13. Cover the plate with aluminum foil and place it on a shaker for at least 20 min.

14. Collect 20 µL protein lysates from each well and dispense it in a 96-well opaque plate if using a 96-well luminometer (*see* **Note 31**).

We usually repeat this protocol for each miRNA::target interaction at least three times. We collect all of the data for the three experiments and pull the results. We perform the final statistical analysis using these data.

4. Notes

1. All of the solutions and plastic supplies used during cell collection must be tested using cell culture analysis, and all of the tubes and tips used to collect the samples must be RNase and DNase-certified or autoclaved before use.

2. A few milliliters of RNase/DNase-free water can be overlaid onto chloroform to decrease evaporation.

3. Instead of buying ready to use RNase/DNase-free water, one can easily treat 18.2 MΩ-cm water overnight at room temperature with DEPC and autoclave it the next day.

4. The other choice is to treat cell lines once and, after RNA purification, split the same sample in three replicates.

5. We suggest not proceeding with more than 1×10^7 cells.

6. The first procedure is recommended, as the freezing process is quicker, and modification of the miRNoma (the full spectrum of miRNAs expressed in a cell) is less affected.

7. Different companies sell these reagents, such as Qiagen and Ambion. We do not have any experience with these stabilization reagents and are not sure how they may affect the miRNoma. Thus, we recommend using the former procedure.

8. In the last step, "RNA wash with 75% ethanol," mixing of the sample should be avoided, because floating the RNA pellet can decrease the number of small RNA molecules, such as miRNAs. When using an RNA extraction kit, carefully check the kit's protocol, as some protocols have a 200-bp RNA length cut-off value; both mature and precursor molecules

would be lost during washing of the column using such protocols.

9. Using RNA column purification instead of TRIzol increases the chance of obtaining RNA samples with good quality. The cut-off size is set to 18 bp, which is very close to the miRNA size (19–24 nt), at least for Qiagen columns. This specificity is based on the ratio of water to ethanol in the samples before the transfer of the samples to the columns. For these reasons, this RNA purification technique may affect the miRNoma. We have never compared the TRIzol protocol with column purification on the microarray platform that we describe here, and we do not know whether some differences between these RNA purifications exist. Thus, we strongly recommend the TRIzol purification at this point.

10. The lowest concentration allowed is 0.5 μg/μL.

11. Each slide has a code scratched on it identifying the batch. The code is located on the opposite face of the printing side.

12. Increase the volume of the sample preparation to 100 μL with prehybridization solution to guarantee more uniform hybridization.

13. The slide has to be placed face down and with the scratched code toward the operator. The direction of the slide is very important for the following superimposition of the .gal file.

14. The total RNA batch used for confirmation must be the same used for microarray data.

15. The difference that we adopted is to perform multiple gene-specific reverse transcriptions for more than one miRNA, as it was intended in the original protocol instead.

16. We believe that this step is required to prevent false-positive signals from genomic DNA in the real-time step.

17. We usually perform such incubations in a water bath set at 37°C.

18. Before adding the inhibitor, one must vortex the stock solution very well to homogenously resuspend it.

19. We usually stop the centrifuge after 2 min and remove four samples to collect the supernatant while the others are still spinning. After collecting the first four samples, we stop the centrifuge and remove four more samples. We usually collect 18 μL of supernatant to reduce the risk of touching the white pellet. Each sample has a concentration of about 100 ng/μL.

20. The samples commonly contain almost the same amount of genomic DNA.

21. We prefer to use an amount of total RNA larger than the amount that is usually recommended (i.e., 1–10 ng) because miRNAs can be expressed at low levels; thus, verifying differences in their expression levels can be very difficult.

22. Do not vortex the reagents; just flick them.

23. Use the same cDNA for all of the different miRNAs to be tested.

24. The number of cells depends on the assay performed. For example, for an assay requiring more than 1 day in culture after treatment, one may consider the possibility of reducing the number of cells. If this is the case, one must keep in mind that the half-lives of these short RNA molecules range from 48 to 72 h, so repeating the transfection 48 h after the first treatment may be necessary.

25. Different protocols can be used. Usually, the efficiency of such small RNA molecules is very high and can be tested by adding a small RNA stained with a red fluorescent dye (Invitrogen).

26. Refer to previous chapters of this book to find detailed protocols describing apoptotic assays.

27. The cloned region can be a small region of the 3′ UTR or the complete sequence of the 3′ UTR. The latter option is probably better because the conformation of the target sequence is much closer to that of the endogenous one. Before trying to clone the entire 3′ UTR, check for any *Xba*I sites inside it, because *Xba*I will be used to clone the target sequence into the reporter vector (pGL3).

28. The exact predicted sequence of interaction can be retrieved using the software program used to predict the miRNA::gene interactions.

29. After addition of the Lipofectamine, the solution usually becomes cloudy.

30. Dispense the solution drop by drop from tube B to tube A and then flick tube A.

31. Spinning all of the lysates to remove the debris is not necessary because they do not affect the reading of luciferase. Make sure that the 96-well plates are opaque.

References

1. Wightman, B., Ha, I., and Ruvkun, G. (1993). Posttranscriptional regulation of the heterochronic gene lin-14 by lin-4 mediates temporal pattern formation in *C. elegans*. *Cell* **75**, 855–862.

2. Lee, R. C., Feinbaum, R. L., and Ambros, V. (1993). The C. elegans heterochronic gene lin-4 encodes small RNAs with antisense complementarity to lin-14. *Cell* **75**, 843–854.

3. Valencia-Sanchez, M. A., Liu, J., Hannon, G. J., and Parker, R. (2006). Control of translation and mRNA degradation by miRNAs and siRNAs. *Genes Dev* **20**, 515–524.

4. Vasudevan, S., Tong, Y., and Steitz, J. A. (2007). Switching from repression to activation: microRNAs can up-regulate translation. *Science* **318**, 1931–1934.

5. Georgantas, R. W., III, Hildreth, R., Morisot, S., Alder, J., Liu, C. G., Heimfeld, S., Calin, G. A., Croce, C. M., and Civin, C. I. (2007). CD34+ hematopoietic stem-progenitor cell microRNA expression and function: a circuit diagram of differentiation control. *Proc Natl Acad Sci USA* **104**, 2750–2755.

6. Garzon, R., Pichiorri, F., Palumbo, T., Iuliano, R., Cimmino, A., Aqeilan, R., Volinia, S., Bhatt, D., Alder, H., Marcucci, G., Calin, G. A., Liu, C. G., Bloomfield, C. D., Andreeff, M., and Croce, C. M. (2006). MicroRNA fingerprints during human megakaryocytopoiesis. *Proc Natl Acad Sci USA* **103**, 5078–5083.

7. Nakamoto, M., Jin, P., O'Donnell, W. T., and Warren, S. T. (2005). Physiological identification of human transcripts translationally regulated by a specific microRNA. *Hum Mol Genet* **14**, 3813–3821.

8. Cheng, A. M., Byrom, M. W., Shelton, J., and Ford, L. P. (2005). Antisense inhibition of human miRNAs and indications for an involvement of miRNA in cell growth and apoptosis. *Nucleic Acids Res* **33**, 1290–1297.

9. Leaman, D., Chen, P. Y., Fak, J., Yalcin, A., Pearce, M., Unnerstall, U., Marks, D. S., Sander, C., Tuschl, T., and Gaul, U. (2005). Antisense-mediated depletion reveals essential and specific functions of microRNAs in Drosophila development. *Cell* **121**, 1097–1108.

10. Chang, T. C., Wentzel, E. A., Kent, O. A., Ramachandran, K., Mullendore, M., Lee, K. H., Feldmann, G., Yamakuchi, M., Ferlito, M., Lowenstein, C. J., Arking, D. E., Beer, M. A., Maitra, A., and Mendell, J. T. (2007). Transactivation of miR-34a by p53 broadly influences gene expression and promotes apoptosis. *Mol Cell* **26**, 745–752.

11. He, L., He, X., Lim, L. P., de Stanchina, E., Xuan, Z., Liang, Y., Xue, W., Zender, L., Magnus, J., Ridzon, D., Jackson, A. L., Linsley, P. S., Chen, C., Lowe, S. W., Cleary, M. A., and Hannon, G. J. (2007). A microRNA component of the p53 tumour suppressor network. *Nature* **447**, 1130–1134.

12. Raver-Shapira, N., Marciano, E., Meiri, E., Spector, Y., Rosenfeld, N., Moskovits, N., Bentwich, Z., and Oren, M. (2007). Transcriptional Activation of miR-34a Contributes to p53-Mediated Apoptosis. *Mol Cell* **26**, 731–743.

13. Raver-Shapira, N., Marciano, E., Meiri, E., Spector, Y., Rosenfeld, N., Moskovits, N., Bentwich, Z., and Oren, M. (2007). Transcriptional activation of miR-34a contributes to p53-mediated apoptosis. *Mol Cell* **26**, 731–743.

14. Welch, C., Chen, Y., and Stallings, R. L. (2007). MicroRNA-34a functions as a potential tumor suppressor by inducing apoptosis in neuroblastoma cells. *Oncogene* **26**, 5017–5022.

15. Loffler, D., Brocke-Heidrich, K., Pfeifer, G., Stocsits, C., Hackermuller, J., Kretzschmar, A. K., Burger, R., Gramatzki, M., Blumert, C., Bauer, K., Cvijic, H., Ullmann, A. K., Stadler, P. F., and Horn, F. (2007). Interleukin-6 dependent survival of multiple myeloma cells involves the Stat3-mediated induction of microRNA-21 through a highly conserved enhancer. *Blood* **110**, 1330–1333.

16. Cimmino, A., Calin, G. A., Fabbri, M., Iorio, M. V., Ferracin, M., Shimizu, M., Wojcik, S. E., Aqeilan, R. I., Zupo, S., Dono, M., Rassenti, L., Alder, H., Volinia, S., Liu, C. G., Kipps, T. J., Negrini, M., and Croce, C. M. (2005). miR-15 and miR-16 induce apoptosis by targeting BCL2. *Proc Natl Acad Sci USA* **102**, 13944–13949.

17. He, L., Thomson, J. M., Hemann, M. T., Hernando-Monge, E., Mu, D., Goodson, S., Powers, S., Cordon-Cardo, C., Lowe, S. W., Hannon, G. J., and Hammond, S. M. (2005). A microRNA polycistron as a potential human oncogene. *Nature* **435**, 828–833.

18. Gupta, A., Gartner, J. J., Sethupathy, P., Hatzigeorgiou, A. G., and Fraser, N. W. (2006). Anti-apoptotic function of a microRNA encoded by the HSV-1 latency-associated transcript. *Nature* **442**, 82–85.

19. Xu, C., Lu, Y., Pan, Z., Chu, W., Luo, X., Lin, H., Xiao, J., Shan, H., Wang, Z., and Yang, B. (2007). The muscle-specific microRNAs miR-1 and miR-133 produce opposing effects on apoptosis by targeting HSP60, HSP70 and caspase-9 in cardiomyocytes. *J Cell Sci* **120**, 3045–3052.

20. Welch, C., Chen, Y., and Stallings, R. L. (2007). MicroRNA-34a functions as a potential tumor suppressor by inducing apoptosis in neuroblastoma cells. *Oncogene* **26**, 5017–5022.

21. Tarasov, V., Jung, P., Verdoodt, B., Lodygin, D., Epanchintsev, A., Menssen, A., Meister, G., and Hermeking, H. (2007). Differential regulation of microRNAs by p53 revealed by massively parallel sequencing: miR-34a is a p53 target that induces apoptosis and G1-arrest. *Cell cycle (Georgetown, Tex)* **6**, 1586–1593.

22. Tazawa, H., Tsuchiya, N., Izumiya, M., and Nakagama, H. (2007). Tumor-suppressive

miR-34a induces senescence-like growth arrest through modulation of the E2F pathway in human colon cancer cells. *Proc Natl Acad Sci USA* **104**, 15472–15477.

23. Matsubara, H., Takeuchi, T., Nishikawa, E., Yanagisawa, K., Hayashita, Y., Ebi, H., Yamada, H., Suzuki, M., Nagino, M., Nimura, Y., Osada, H., and Takahashi, T. (2007). Apoptosis induction by antisense oligonucleotides against miR-17–5p and miR-20a in lung cancers overexpressing miR-17–92. *Oncogene* **26**, 6099–6105.

24. Sylvestre, Y., De Guire, V., Querido, E., Mukhopadhyay, U. K., Bourdeau, V., Major, F., Ferbeyre, G., and Chartrand, P. (2007). An E2F/miR-20a autoregulatory feedback loop. *J Biol Chem* **282**, 2135–2143.

25. Chan, J. A., Krichevsky, A. M., and Kosik, K. S. (2005). MicroRNA-21 is an antiapoptotic factor in human glioblastoma cells. *Cancer Res* **65**, 6029–6033.

26. Meng, F., Henson, R., Lang, M., Wehbe, H., Maheshwari, S., Mendell, J. T., Jiang, J., Schmittgen, T. D., and Patel, T. (2006). Involvement of human micro-RNA in growth and response to chemotherapy in human cholangiocarcinoma cell lines. *Gastroenterology* **130**, 2113–2129.

27. Si, M. L., Zhu, S., Wu, H., Lu, Z., Wu, F., and Mo, Y. Y. (2007). miR-21-mediated tumor growth. *Oncogene* **26**, 2799–2803.

28. Corsten, M. F., Miranda, R., Kasmieh, R., Krichevsky, A. M., Weissleder, R., and Shah, K. (2007). MicroRNA-21 knockdown disrupts glioma growth in vivo and displays synergistic cytotoxicity with neural precursor cell delivered S-TRAIL in human gliomas. *Cancer Res* **67**, 8994–9000.

29. Frankel, L. B., Christoffersen, N. R., Jacobsen, A., Lindow, M., Krogh, A., and Lund, A. H. (2007). Programmed cell death 4 (PDCD4) is an important functional target of the microRNA miR-21 in breast cancer cells. *J Biol Chem* **283**, 1026–1033.

30. Ji, R., Cheng, Y., Yue, J., Yang, J., Liu, X., Chen, H., Dean, D. B., and Zhang, C. (2007). MicroRNA expression signature and antisense-mediated depletion reveal an essential role of MicroRNA in vascular neointimal lesion formation. *Circ Res* **100**, 1579–1588.

31. Sathyan, P., Golden, H. B., and Miranda, R. C. (2007). Competing interactions between micro-RNAs determine neural progenitor survival and proliferation after ethanol exposure: evidence from an ex vivo model of the fetal cerebral cortical neuroepithelium. *J Neurosci* **27**, 8546–8557.

32. Ovcharenko, D., Kelnar, K., Johnson, C., Leng, N., and Brown, D. (2007). Genome-scale microRNA and small interfering RNA screens identify small RNA modulators of TRAIL-induced apoptosis pathway. *Cancer Res* **67**, 10782–10788.

33. Meng, F., Henson, R., Wehbe-Janek, H., Smith, H., Ueno, Y., and Patel, T. (2007). The MicroRNA let-7a modulates interleukin-6-dependent STAT-3 survival signaling in malignant human cholangiocytes. *J Biol Chem* **282**, 8256–8264.

34. Nakagawa, Y., Iinuma, M., Naoe, T., Nozawa, Y., and Akao, Y. (2007). Characterized mechanism of alpha-mangostin-induced cell death: caspase-independent apoptosis with release of endonuclease-G from mitochondria and increased miR-143 expression in human colorectal cancer DLD-1 cells. *Bioorg Med Chem* **15**, 5620–5628.

35. Mott, J. L., Kobayashi, S., Bronk, S. F., and Gores, G. J. (2007). mir-29 regulates Mcl-1 protein expression and apoptosis. *Oncogene* **26**, 6133–6140.

36. Karres, J. S., Hilgers, V., Carrera, I., Treisman, J., and Cohen, S. M. (2007). The conserved microRNA miR-8 tunes atrophin levels to prevent neurodegeneration in Drosophila. *Cell* **131**, 136–145.

37. Brennecke, J., Hipfner, D. R., Stark, A., Russell, R. B., and Cohen, S. M. (2003). Bantam encodes a developmentally regulated microRNA that controls cell proliferation and regulates the proapoptotic gene hid in Drosophila. *Cell* **113**, 25–36.

38. Nolo, R., Morrison, C. M., Tao, C., Zhang, X., and Halder, G. (2006). The bantam microRNA is a target of the hippo tumor-suppressor pathway. *Curr Biol* **16**, 1895–1904.

39. Nairz, K., Rottig, C., Rintelen, F., Zdobnov, E., Moser, M., and Hafen, E. (2006). Overgrowth caused by misexpression of a microRNA with dispensable wild-type function. *Dev Biol* **291**, 314–324.

40. Xu, P., Vernooy, S. Y., Guo, M., and Hay, B. A. (2003). The Drosophila microRNA Mir-14 suppresses cell death and is required for normal fat metabolism. *Curr Biol* **13**, 790–795.

41. Gupta, A., Gartner, J. J., Sethupathy, P., Hatzigeorgiou, A. G., and Fraser, N. W. (2006). Anti-apoptotic function of a microRNA encoded by the HSV-1 latency-associated transcript. *Nature* **442**, 82–85.

42. Liu, C. G., Calin, G. A., Meloon, B., Gamliel, N., Sevignani, C., Ferracin, M., Dumitru, C. D., Shimizu, M., Zupo, S., Dono, M., Alder, H., Bullrich, F., Negrini, M., and Croce, C. M.

(2004). An oligonucleotide microchip for genome-wide microRNA profiling in human and mouse tissues. *Proc Natl Acad Sci USA* **101**, 9740–9744.

43. Calin, G. A., Ferracin, M., Cimmino, A., Di Leva, G., Shimizu, M., Wojcik, S. E., Iorio, M. V., Visone, R., Sever, N. I., Fabbri, M., Iuliano, R., Palumbo, T., Pichiorri, F., Roldo, C., Garzon, R., Sevignani, C., Rassenti, L., Alder, H., Volinia, S., Liu, C. G., Kipps, T. J., Negrini, M., and Croce, C. M. (2005). A MicroRNA signature associated with prognosis and progression in chronic lymphocytic leukemia. *N Engl J Med* **353**, 1793–1801.

44. Esquela-Kerscher, A., and Slack, F. J. (2006). Oncomirs - microRNAs with a role in cancer. *Nat Rev Cancer* **6**, 259–269.

45. Volinia, S., Calin, G. A., Liu, C. G., Ambs, S., Cimmino, A., Petrocca, F., Visone, R., Iorio, M., Roldo, C., Ferracin, M., Prueitt, R. L., Yanaihara, N., Lanza, G., Scarpa, A., Vecchione, A., Negrini, M., Harris, C. C., and Croce, C. M. (2006). A microRNA expression signature of human solid tumors defines cancer gene targets. *Proc Natl Acad Sci USA* **103**, 2257–2261.

46. Yanaihara, N., Caplen, N., Bowman, E., Seike, M., Kumamoto, K., Yi, M., Stephens, R. M., Okamoto, A., Yokota, J., Tanaka, T., Calin, G. A., Liu, C. G., Croce, C. M., and Harris, C. C. (2006). Unique microRNA molecular profiles in lung cancer diagnosis and prognosis. *Cancer Cell* **9**, 189–198.

47. Lagos-Quintana, M., Rauhut, R., Lendeckel, W., and Tuschl, T. (2001). Identification of novel genes coding for small expressed RNAs. *Science* **294**, 853–858.

48. Lu, J., Getz, G., Miska, E. A., Alvarez-Saavedra, E., Lamb, J., Peck, D., Sweet-Cordero, A., Ebert, B. L., Mak, R. H., Ferrando, A. A., Downing, J. R., Jacks, T., Horvitz, H. R., and Golub, T. R. (2005). MicroRNA expression profiles classify human cancers. *Nature* **435**, 834–838.

49. Chen, C., Ridzon, D. A., Broomer, A. J., Zhou, Z., Lee, D. H., Nguyen, J. T., Barbisin, M., Xu, N. L., Mahuvakar, V. R., Andersen, M. R., Lao, K. Q., Livak, K. J., and Guegler, K. J. (2005). Real-time quantification of microRNAs by stem-loop RT-PCR. *Nucleic Acids Res* **33**, e179.

50. Kloosterman, W. P., Wienholds, E., de Bruijn, E., Kauppinen, S., and Plasterk, R. H. (2006). In situ detection of miRNAs in animal embryos using LNA-modified oligonucleotide probes. *Nat Methods* **3**, 27–29.

51. Allawi, H. T., Dahlberg, J. E., Olson, S., Lund, E., Olson, M., Ma, W. P., Takova, T., Neri, B. P., and Lyamichev, V. I. (2004). Quantitation of microRNAs using a modified Invader assay. *RNA* **10**, 1153–1161.

52. Nelson, P. T., Baldwin, D. A., Kloosterman, W. P., Kauppinen, S., Plasterk, R. H., and Mourelatos, Z. (2006). RAKE and LNA-ISH reveal microRNA expression and localization in archival human brain. *RNA* **12**, 187–191.

53. Neely, L. A., Patel, S., Garver, J., Gallo, M., Hackett, M., McLaughlin, S., Nadel, M., Harris, J., Gullans, S., and Rooke, J. (2006). A single-molecule method for the quantitation of microRNA gene expression. *Nat Methods* **3**, 41–46.

54. Liu, C. G., Spizzo, R., Calin, G. A., and Croce, C. M. (2008). Expression profiling of microRNA using oligo DNA arrays. *Methods* **44**, 22–30.

55. Bejerano, G., Pheasant, M., Makunin, I., Stephen, S., Kent, W. J., Mattick, J. S., and Haussler, D. (2004). Ultraconserved elements in the human genome. *Science* **304**, 1321–1325.

56. Calin, G. A., Liu, C. G., Ferracin, M., Hyslop, T., Spizzo, R., Sevignani, C., Fabbri, M., Cimmino, A., Lee, E. J., Wojcik, S. E., Shimizu, M., Tili, E., Rossi, S., Taccioli, C., Pichiorri, F., Liu, X., Zupo, S., Herlea, V., Gramantieri, L., Lanza, G., Alder, H., Rassenti, L., Volinia, S., Schmittgen, T. D., Kipps, T. J., Negrini, M., and Croce, C. M. (2007). Ultraconserved regions encoding ncRNAs are altered in human leukemias and carcinomas. *Cancer Cell* **12**, 215–229.

57. Altuvia, Y., Landgraf, P., Lithwick, G., Elefant, N., Pfeffer, S., Aravin, A., Brownstein, M. J., Tuschl, T., and Margalit, H. (2005). Clustering and conservation patterns of human microRNAs. *Nucleic Acids Res* **33**, 2697–2706.

58. Xie, X., Lu, J., Kulbokas, E. J., Golub, T. R., Mootha, V., Lindblad-Toh, K., Lander, E. S., and Kellis, M. (2005). Systematic discovery of regulatory motifs in human promoters and 3' UTRs by comparison of several mammals. *Nature* **434**, 338–345.

59. Berezikov, E., Guryev, V., van de Belt, J., Wienholds, E., Plasterk, R. H., and Cuppen, E. (2005). Phylogenetic shadowing and computational identification of human microRNA genes. *Cell* **120**, 21–24.

60. Iorio, M. V., Ferracin, M., Liu, C. G., Veronese, A., Spizzo, R., Sabbioni, S., Magri, E., Pedriali, M., Fabbri, M., Campiglio, M.,

Menard, S., Palazzo, J. P., Rosenberg, A., Musiani, P., Volinia, S., Nenci, I., Calin, G. A., Querzoli, P., Negrini, M., and Croce, C. M. (2005). MicroRNA gene expression deregulation in human breast cancer. *Cancer Res* **65**, 7065–7070.

61. Bloomston, M., Frankel, W. L., Petrocca, F., Volinia, S., Alder, H., Hagan, J. P., Liu, C. G., Bhatt, D., Taccioli, C., and Croce, C. M. (2007). MicroRNA expression patterns to differentiate pancreatic adenocarcinoma from normal pancreas and chronic pancreatitis. *JAMA* **297**, 1901–1908.

62. Wu, W., Dave, N., Tseng, G. C., Richards, T., Xing, E. P., and Kaminski, N. (2005). Comparison of normalization methods for CodeLink Bioarray data. *BMC Bioinformatics* **6**, 309.

63. Kulshreshtha, R., Ferracin, M., Wojcik, S. E., Garzon, R., Alder, H., Agosto-Perez, F. J., Davuluri, R., Liu, C. G., Croce, C. M., Negrini, M., Calin, G. A., and Ivan, M. (2007). A microRNA signature of hypoxia. *Mol Cell Biol* **27**, 1859–1867.

64. Calin, G. A., and Croce, C. M. (2006). MicroRNA signatures in human cancers. *Nat Rev Cancer* **6**, 857–866.

65. Schmittgen, T. D., Jiang, J., Liu, Q., and Yang, L. (2004). A high-throughput method to monitor the expression of microRNA precursors. *Nucleic Acids Res* **32**, e43.

66. Johnson, S. M., Grosshans, H., Shingara, J., Byrom, M., Jarvis, R., Cheng, A., Labourier, E., Reinert, K. L., Brown, D., and Slack, F. J. (2005). RAS is regulated by the let-7 microRNA family. *Cell* **120**, 635–647.

67. Rajewsky, N. (2006). microRNA target predictions in animals. *Nat Genet* **38** Suppl, S8–S13

68. Kiriakidou, M., Nelson, P. T., Kouranov, A., Fitziev, P., Bouyioukos, C., Mourelatos, Z., and Hatzigeorgiou, A. (2004). A combined computational-experimental approach predicts human microRNA targets. *Genes Dev* **18**, 1165–1178.

69. Lewis, B. P., Burge, C. B., and Bartel, D. P. (2005). Conserved seed pairing, often flanked by adenosines, indicates that thousands of human genes are microRNA targets. *Cell* **120**, 15–20.

70. Enright, A. J., John, B., Gaul, U., Tuschl, T., Sander, C., and Marks, D. S. (2003). MicroRNA targets in Drosophila. *Genome Biol* **5**, R1.

71. Krek, A., Grun, D., Poy, M. N., Wolf, R., Rosenberg, L., Epstein, E. J., MacMenamin, P., da Piedade, I., Gunsalus, K. C., Stoffel, M., and Rajewsky, N. (2005). Combinatorial microRNA target predictions. *Nat Genet* **37**, 495–500.

72. Megraw, M., Sethupathy, P., Corda, B., and Hatzigeorgiou, A. G. (2007). miRGen: a database for the study of animal microRNA genomic organization and function. *Nucleic Acids Res* **35**, D149–D155.

73. Kel, A., Voss, N., Jauregui, R., Kel-Margoulis, O., and Wingender, E. (2006). Beyond microarrays: Finding key transcription factors controlling signal transduction pathways. *BMC Bioinformatics* **7** Suppl 2, S13.

74. Lewis, B. P., Shih, I. H., Jones-Rhoades, M. W., Bartel, D. P., and Burge, C. B. (2003). Prediction of mammalian microRNA targets. *Cell* **115**, 787–798.

Chapter 17

Assessment of Apoptotic Cell Phagocytosis by Macrophages

Kathleen A. McPhillips and Lars-Peter Erwig

Summary

Cells undergo apoptosis during development, tissue homeostasis, and disease, and are rapidly cleared by both professional and nonprofessional phagocytes. In the whole animal, this process is remarkably efficient and usually goes unnoticed. It is estimated that 2×10^{11} cells are cleared each day and it has been suggested that detection of apoptotic cells in tissues should lead one to at least question the presence of a local clearance defect. For the last two decades, in vitro phagocytosis assays have played a critical role in identifying the receptors and mechanisms involved in the recognition and ingestion of apoptotic cells. The methodology of phagocytosis assays can be broken down into four separate components: apoptosis induction in target cells, preparation of phagocytes, the interaction assay, and the quantitative assessment of apoptotic cell engulfment. Here, we attempt to provide a detailed description of all the individual components of this complex procedure. To date, this has not been done in its entirety but is vital for the accurate assessment of stimuli that influence the clearance process.

Key words: Apoptosis, Phagocytosis, Macrophage, Clearance, Neutrophil

1. Introduction

In vitro phagocytosis assays continue to play an essential role in the dissection of the receptors and signaling pathways involved in apoptotic cell uptake *(1–3)*. The standard assay has been changed little in the last two decades and, unlike many other techniques, it has not been replaced by alternative methods. Assessment of apoptotic cell uptake by flow cytometry has been developed and used by many laboratories, but this method is unable to distinguish between bound and ingested cells. This is often an important

Peter Erhardt and Ambrus Toth (eds.), *Apoptosis*, Methods in Molecular Biology, vol. 559
DOI 10.1007/978-1-60327-017-5_17
© Humana Press, a part of Springer Science + Business Media, LLC 2004, 2009

distinction when determining the mechanism of both the binding and engulfment of apoptotic cells.

If apoptotic cells are not phagocytosed, they lose their energy supply and their membrane integrity at rates that depend on cell type, stimulus, and intracellular signaling pathways. On the one hand, neutrophils generally maintain the characteristic nuclear alterations of apoptosis and membrane integrity for up to 24 h after apoptosis induction *(4)*. On the other hand, T cells undergoing apoptosis release cellular blebs 2 h after apoptosis induction. Also, T cells undergoing activation-induced cell death appear to show caspase-dependent nuclear changes simultaneously with non-caspase-dependent membrane disruption *(5)*. Examples of such disparity are so common that it is necessary to describe apoptosis induction for each cell type separately. Furthermore, primary macrophages of different origins and macrophage cell lines require cell type specific preparation for achieving optimal phagocytosis *(4, 6)*. Finally, the validity of the interaction assay between the phagocyte and the apoptotic cell depends on target cell and phagocyte specific conditions and appropriate quantitative assessment of apoptotic cell binding and engulfment. Here we will for the first time provide a detailed description of this complex assay in its entirety.

2. Materials

2.1. Preparation of Apoptotic Cells

2.1.1. Preparation of Apoptotic Neutrophils

1. Sodium citrate 3.8% in LPS free water.
2. Dextran T500 (Pharmacia) 6% in LPS free 0.9% saline (Pharmacia).
3. Percoll 90% in 0.9% NaCl.
4. Dextrose (5%) and 0.2% sodium chloride injection USP (5 mL aliquots).
5. Saline (0.9%) 100 mL.
6. **Table 1** lists the reagents used to make KRP-D.

Table 1
The components of KRP-D

Reagent	FW	g/100 mL	Aliquot (mL)
$MgSO_4$	120.37	0.318	6
KCl	74.55	1.15	4
$CaCl_2$	147	0.61	3
NaH_2PO_4 (mono)	137.99	0.92	6
Na_2HPO_4 (dibasic anhydrous)	141.96	5.68	4

2.1.2. Preparation of
Apoptotic Jurkat T Cells

1. Roswell Park Memorial Institute (RPMI Media Tech Inc.) supplemented with 10% heat-inactivated fetal bovine serum, 2 mM L-glutamine, 100 µg/mL streptomycin, and 100 U/mL penicillin (Jurkat T cell media).

2.2. Preparation of Phagocytes

1. X-Vivo medium (BioWhittaker Inc. Walkersville, MD) supplemented with 10% individual or pooled human serum, 2 mM L-glutamine, 100 µg/mL streptomycin, and 100 U/mL penicillin (human monocyte derived macrophages (HMDM) media).

2. Dulbecco's Modified Eagle's Medium (DMEM, Gibco/BRL) supplemented with 10% heat-inactivated fetal bovine serum (FBS, Hyclone), 2 mM L-glutamine, 100 µg/mL streptomycin, and 100 U/mL penicillin (J774 and resident peritoneal macrophage media).

3. J774 macrophages and Jurkat T cells are from American Type Culture Collections (ATCC).

4. 50 mL cold sterile HBSS for harvest of peritoneal macrophages (Hank's Balanced Salt Solution; Cellgro).

2.3. Assessment of Uptake

1. Diff-Quik® (VWR Scientific).
2. Inverted microscope with 40× objective.

---

3. Methods

3.1. Preparation of Apoptotic Cells

3.1.1. Preparation of
Apoptotic Neutrophils

1. Add 40 mL of whole blood to 4.4 mL of 3.8% citrate into 50 mL conical tube.

2. Spin citrated blood at $1,250 \times g$ for 20 min with the brake off (only for this spin, brake can be on low for subsequent spins).

3. Carefully remove the upper platelet rich plasma (PRP) layer into a new tube leaving 2.5 mL above the red cell layer (the white cells are sitting on top of the red cells). Keep the PRP.

4. To the red cell layer add 5 mL of dextran (prewarmed) and 0.9% saline to 50 mL.

5. Cap the tubes and mix by inverting five times. Loosen the caps and remove any large bubbles at the top of the tube.

6. Let the dextran sediment for 30–45 min with the caps loosened.

7. During the dextran sedimentation spin PRP at $3,000 \times g$ for 15 min to pellet the platelets.

8. Pipette off the platelet poor plasma (PPP) being careful not to pick up any platelets (leave 5–10 mL of PPP so as to be sure not to disrupt the platelet pellet) into a new tube and discard platelets.

9. Thaw the KRPD stocks in the heat block.

10. To make the KRPD, fill a tissue culture flask (T75) with 100 mL of 0.9% saline. Add all of the reagents listed above, but be sure to add the dextrose tube first and the dibasic last. Cap tightly and mix by shaking.

11. Prepare the gradients (4 gradients for 8 tubes of blood and 6 gradients for 10 tubes of blood) add the following into a 15 mL round bottom tube using a 1 mL pipette and allowing the Percoll and PPP to flow out by gravity (do not force the Percoll out).

12. For the 42% add: 0.8 mL cold Percoll + 40 μL of 1.1 mL PPP + 60 μL

13. For the 51% add: 1.0 mL cold Percoll + 20 μL of 0.9 mL PPP + 80 μL

14. Mix the gradients by flicking with your finger. Do not vortex.

15. After the dextran sedimentation use a 25 mL pipette to remove the upper layer. Leave some of the upper layer just above the red blood cell layer.

16. Transfer the upper layers from the 8 or 10 tubes equally into four 50 mL conical tubes if there are 8 tubes of blood or six 50 mL conical tubes if there are 10 tubes of blood.

17. Spin the cells at $1,000 \times g$ for 6 min and decant the upper layer. The cell pellet is soft – be careful not to decant the pellet.

18. To the cell pellet, add 2 mL of PPP and mix using a transfer pipette.

19. Transfer cell suspension into a 15 mL pointed bottom tube using a pipette.

20. Using a Pasteur pipette (fine end), draw up the 42% gradient mixture and cap with your finger.

21. With the Pasteur pipette still capped, place the tip of the pipette to the bottom of the cells in the tube. Allow the gradient mixture to slowly underlay the cell suspension by releasing your finger slightly. Once the gradient mixture has been released, recap the pipette with your finger (to help prevent mixing) and smoothly but quickly remove it from the tube.

22. Using a Pasteur pipette, draw up the 51% as you did for the 42%. Place the tip of the pipette to the bottom of the tube and allow the 51% gradient mixture to underlay the 42% by the same method as used for the 42%. Do this slowly – the interface is very fragile!

23. Balance the gradient tubes and spin the gradients at $1,070 \times g$ for 10 min.

24. After the spin, harvest the mono layer with a transfer pipette into a 50 mL conical tube (**Fig. 1**).

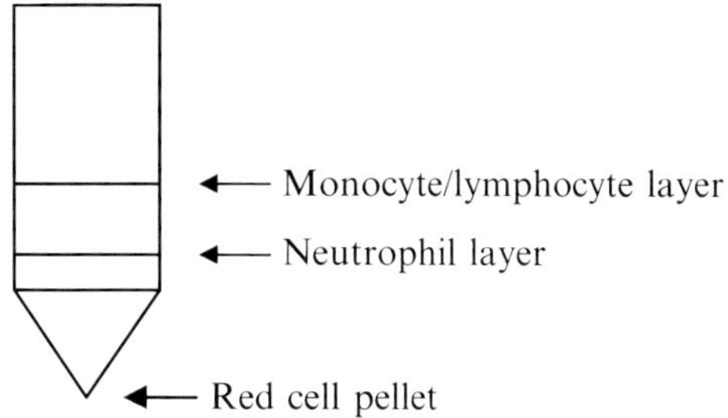

Fig. 1. Localization of monocyte/lymphocyte as well as neutrophil layers after gradient centrifugation.

25. Harvest the neutrophils into a separate 50 mL conical tube. If the RBCs do not pellet and are directly under the neutrophil layer, you can do either or both of the following: Respin the gradient for 5 more minutes at $1100 \times g$. If the RBCs still do not pellet, remove the RBCs from below the neutrophils using a Pasteur pipette. Remove as many RBCs as possible without losing too many neutrophils. There will always be a few RBCs in this layer.

26. Using a transfer pipette, gently resuspend the pellet with 1–2 mL of PPP. Make up to 50 mL with the remaining PPP and spin for 6 min at $1,030 \times g$.

27. Using a transfer pipette, gently resuspend the cell pellet with 1–2 mL of KRPD. Bring up to 50 mL with KRPD and spin for 6 min at $980 \times g$.

28. Using a transfer pipette, gently resuspend the pellet with 1–2 mL of KRPD and bring up to 30–40 mL with KRPD.

29. Count the neutrophils by adding 10 μL of PMN suspension to 990 μL of 2% glutaraldehyde (1:100 dilution). Spin for 6 min at $1,030 \times g$.

30. Apoptosis is induced by one of the following methods:
    - Resuspend the neutrophils at 2×10^6/mL in RPMI containing 1% BSA. Incubate the cells overnight in 5% CO_2 at 37°C.
    - Resuspend the neutrophils at 2×10^6/mL in DMEM and put in a tissue culture flask (T25 or T75 depending on the number of neutrophils). Expose the flask to ultraviolet irradiation at 254 nm for 10 min followed by culture for 2.5 h in 5% CO_2 at 37°C.

31. Apoptosis is quantified by light microscopic evaluation. Cytospin 150 μL of the neutrophil suspension for 2 min at $600 \times g$. After the slide is dry, stain it using DiffQuik (Wright-Geimsa stain). Count 100 cells and analyze for apoptosis using nuclear morphology. The cells should

typically be 70–90% apoptotic to use for the phagocytosis assay. Check the neutrophils every hour until the amount of apoptosis is between 70 and 90%.

3.1.2. Preparation of Jurkat T Cells

1. Human leukemia Jurkat T cells are cultured in humidified 5% CO_2 at 37°C.

2. Transfer the cells to a 50 mL conical tube.

3. Remove 10 μL for counting.

4. Spin the remaining cells at 1,000 × *g* for 10 min.

5. Resuspend in warm Jurkat media at 1 × 10^6/mL and place in a culture flask.

6. The nonadherent cells are made apoptotic by placing the tissue culture flask on a source of ultraviolet irradiation at 254 nm for 10 min followed by culture for 2.5 h in 5% CO_2 at 37°C. Apoptosis is quantified as described above for PMN. Check every 30 min until the cells are 70–90% apoptotic.

3.2. Preparation of Phagocytes

3.2.1. Preparation of Human Monocyte Derived Macrophages

1. Collect mononuclear cells from monocyte/lymphocyte layer (*see* **Subheading 3.1.1**) and dilute in 50 mL of sterile HBSS. Spin at room temperature for 5 min at 1,000 × *g*. Discard supernatant.

2. Resuspend the pellet in 25 mL of HBSS. Spin for 5 min at 1,000 × *g*. Discard supernatant.

3. Repeat **step 2**. Break up the pellet using 25 mL of HBSS. Count the cells 1:10 dilution (monocytes and lymphocytes). Spin for 5 min at 1,000 × *g*. Discard supernatant.

4. Resuspend the pellet in warm X-Vivo with no serum at 10 × 10^6cells/mL. Plate monocytes at a minimum of 5 × 10^6 cells and a maximum of 10 × 10^6 cells per well in a 24-well plate (*see* **Note 1**).

5. Place the plates in 10% CO_2 incubator for at least 1 h (to allow monocyte adherence). After 1 h pipet up and down gently three times and remove supernatant containing lymphocytes.

6. Add 1 mL of warmed X-Vivo with 10% autologous human serum.

7. Human serum is prepared from PRP obtained during the neutrophil/monocyte preparation.

8. Add 3 sterile glass beads and 1 mL of 2 M calcium chloride to the PRP tube and place in a 37°C waterbath.

9. After 2 h in the 37°C waterbath remove the noncoagulated serum. This serum is then added to X-Vivo (450 mL of X-Vivo and 50 mL of serum) to create HMDM medium.

10. Change the media every 3 days. Use for assays on days 6–8.

3.2.2. Preparation of Resident Peritoneal Macrophages

1. Isolate resident peritoneal macrophages from mice between the ages of 8 and 16 weeks.

2. Euthanize mice.

3. Sterilize mouse abdomen with ethanol, pinch skin lengthwise, make a small suprapubic incision in the skin, and rip the skin laterally using both hands to expose the peritoneum.

4. Insert 18 gauge needle attached to 10 mL syringe (HBSS) at a shallow angle to the body with the bevel up and above the fat in the belly.

5. Slowly inject the HBSS without rupturing the intestines or other organs.

6. Turn the needle bevel side down and lift the legs of the mouse about 4 cm off the ground.

7. Shake the mouse several times very carefully making sure not to damage the peritoneum.

8. Withdraw the HBSS/macrophage solution into the syringe while keeping the legs in the air.

9. Try to recover at least 7 mL from the peritoneum to get a good yield.

10. Pull the needle off the syringe and push the fluid directly (not along the wall of the tube) into 5 mL of HBSS in a 50 mL conical tube.

11. Keep the tube on ice and spin at 4°C for 10 min at $1,000 \times g$.

12. Remove the supernatant and resuspend the pellet in warm media.

13. Plate cells at 2.5×10^5 cells/well in a 24-well plate and use 24 h after plating (*see* **Note 2**).

3.2.3. Preparation of Mouse J774 Macrophage Cell Line

1. Culture murine J774 macrophages J774 media in humidified 10% CO_2 at 37°C. See notes for splitting suggestions.

2. Remove J774 cells from tissue culture flasks by banging the base of the flask on a hard surface or your hand approximately five times.

3. Take a 10 μL aliquot for counting.

4. Spin the remaining cells at $1,000 \times g$ for 10 min.

5. Resuspend the pellet in warm J774 media.

6. Plate 5×10^5 J774 macrophages in 1 mL of media into each well of a 24-well plate for 48 h. Keep the plate in humidified 10% CO_2 at 37°C. *See* **Note 3** for plating suggestions.

3.3. The Interaction Assay

1. Plate the phagocytes in a 24-well tissue culture plastic plate for the interaction assay as follows:

   • HMDMs: $5–10 \times 10^6$ cells/well for 6–8 days

- Resident peritoneal macrophages: 2.5×10^5 cells/well overnight
- J774 macrophages: 5×10^5 cells/well for 2 days

2. Plate the cells in duplicate or triplicate for each condition.

3. Prepare the apoptotic targets as described in **Subheading 3.1**.

4. Wash the phagocytes with warm media once to remove dead cells or debris.

5. Add 1×10^6 apoptotic cells (resuspended in phagocyte media) to each well. There should be 1 mL in each well (*see* **Note 4**).

6. Coculture the cells in 10% CO_2 at 37°C for 90 min (*see* **Note 5**).

7. Wash three times with PBS at room temperature.

8. Aspirate any residual PBS and shake off the excess over a sink.

9. Stain the cells using the Diff-Quik® system.

10. Wash off the excess stain using PBS.

11. Prior to counting add 1 mL PBS into each well and keep on while counting.

12. If the stain fades over time – restain using the last two reagents of the Diff-Quik® system.

3.4. Assessment of Uptake

1. After the phagocytosis assay assess:
   - The total number of phagocytes in the field.
   - The total number of ingested apoptotic bodies.
   - The total number of bound apoptotic cells.

2. As a general rule, if the phagocyte membrane completely surrounds the apoptotic body as demonstrated by the black arrowheads in **Fig. 2**, the body is counted as ingested.

3. If the apoptotic body is bound to the phagocyte, but the phagocyte has not completely surrounded the apoptotic cell, then the target is counted as bound as demonstrated by the double arrowheads (**Fig. 2**).

4. Count 200 macrophages in each well blinded. To get a sampling of all of the macrophages start in the upper left area and work your way around the well counting macrophages, bound, and ingested apoptotic bodies.

5. Avoid counting directly in the middle and on the "edges" of the well.

6. Calculate the phagocytic index using the following formula: no. of apoptotic bodies/200 total macrophages × 100.

7. Calculate the binding index using the following formula: no. of bound bodies/200 total macrophages × 100.

8. Because of variation between human donors and cell line passages, it is often best to express the phagocytic index as

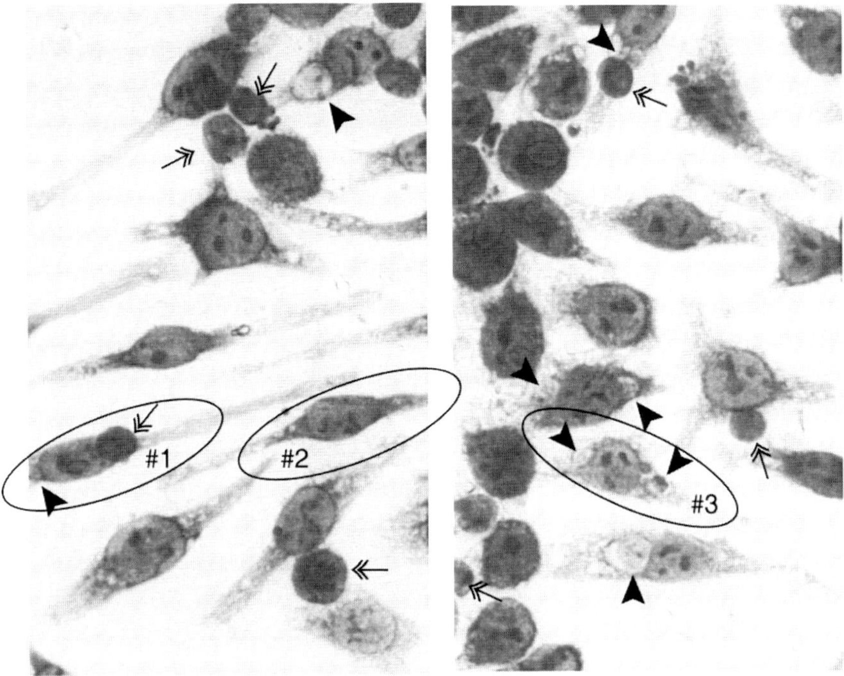

Fig. 2. J774 macrophages engulfing apoptotic Jurkat T cells. J774 macrophages were cultured for 2 days, washed with J774 media, and the apoptotic Jurkat T cells were added for 90 min. The cells were washed three times with PBS and stained using Diff-Quik. Pictures courtesy of Dr. Karla Kenyon.

a percent of control. The trends are often the same for the replicates, but the actual numbers often vary greatly between experiments (*see* **Note 6**).

9. Example cells are circled in **Fig. 2**. Cell 1 is counted as 1 macrophage, 1 ingested body, and 1 bound body. Cell 2 is counted as 1 macrophage (no ingested or bound bodies). Cell 3 is counted as 1 macrophage and 3 ingested bodies (0 bound bodies) (*see* **Note 7**).

4. Notes

1. When plating HMDMs for assays in larger wells increase the number of monocytes/lymphocytes to 10×10^6 minimum (20×10^6 maximum) for 12-well plates and 20×10^6 minimum for 6-well plates. Because each donor is unique, the yield of neutrophils and monocytes are also different among donors. Expect 1–5% of the monocyte/lymphocyte layer to adhere and become mature macrophages. A typical phagocytic index for this cell type ranges from 20 to 30.

2. Resident peritoneal macrophages are very sensitive to density. If not dense enough, they detach from the plate. If too dense, they will not engulf. A typical phagocytic index for this cell type ranges from 20 to 40.

3. J774 macrophages should be split 1:20 every 3 days. Monitor the growth closely because too much density will cause the cells to become activated. If you are using a 6 or 12-well plate, plate 2.5×10^5 or 1×10^5 cells, respectively, for 48 h. Expect a doubling of the cells every 24–48 h. A typical phagocytic index for this cell type ranges from 15 to 30.

4. Do pilot experiments for each type of macrophage and apoptotic target to determine the optimal density and ratio macrophages/apoptotic cell ratio for maximum phagocytosis.

5. Different cell types require different times of interaction for maximal phagocytosis. Epithelial cells, for example, require at least 4 h of interaction with an apoptotic cell before engulfment occurs.

6. There are two important factors when assessing uptake: being consistent and unbiased (always blinded).

7. It takes anywhere from 2 weeks to 6 months to become proficient at counting phagocytosis assays.

References

1. Haslett C, Guthrie LA, Kopaniak MM, Johnston RB Jr, Henson PM. (1985). Modulation of multiple neutrophil functions by preparative methods or trace concentrations of bacterial lipopolysaccharide. *Am J Pathol.* **119**:101–10.

2. Fadok VA, Laszlo DJ, Noble PW, Weinstein L, Riches DW, Henson PM. (1993). Particle digestibility is required for induction of the phosphatidylserine recognition mechanism used by murine macrophages to phagocytose apoptotic cells. *J Immunol.* **151**:4274–85.

3. Erwig LP, Henson PM. Clearance of apoptotic cells by phagocytes. (2008). *Cell Death Differ.* **15**:243–50

4. Erwig LP, McPhilips KA, Wynes MW, Ivetic A, Ridley AJ, Henson PM. (2006). Differential regulation of phagosome maturation in macrophages and dendritic cells mediated by Rho GTPases and ezrin-radixin-moesin (ERM) proteins. *Proc Natl Acad Sci USA* **103**:12825–30.

5. Hildeman DA, Mitchell T, Teague TK, Henson P, Day BJ, Kappler J, and Marrack PC. (1999). Reactive oxygen species regulate activation-induced T cell apoptosis. *Immunity* **10**:735–44.

6. McPhillips K, Janssen WJ, Ghosh M, Byrne A, Gardai S, Remigio L, Bratton DL, Kang JL, Henson P. (2007). TNF-alpha inhibits macrophage clearance of apoptotic cells via cytosolic phospholipase A2 and oxidant-dependent mechanisms. *J Immunol.* **178**:8117–26.

Part V

Analysis of Apoptosis in Different Organs

Chapter 18

Detection of Apoptosis in Mammalian Development

Lin Lin, Carlos Penaloza, Yixia Ye, Richard A. Lockshin, and Zahra Zakeri

Summary

Mammalian development is dependent on an intricate orchestration of cell proliferation and death. Deregulation in the levels, localization, and type of cell death can lead to disease and even death of the developing embryo. The mechanisms involved in such deregulation are many; alterations and or manipulations of these can aid in the detection, prevention and possible treatments of any effects this de-regulation may have. Here we describe how cell death can be detected during mammalian development, using diverse staining and microscopy methods, while taking advantage of the advancements in cell death mechanisms, derived from biochemical and teratological studies in the field.

Key words: Apoptosis, Macrophages, Embryology, Phagocytosis, Phosphatidylserine, Immunohistochemistry, Electron microscopy

1. Introduction

Development involves a delicate orchestration of cell division, movement, differentiation, and death. Cell death in mammalian embryogenesis occurs as early as inner cell-mass differentiation, and it continues to be part of the developing embryo during the formation and functional completion of the different organs *(1–3)*. The importance of cell death was first recognized by developmental biologists and teratologists *(4, 5)*. More recently, with the recognition of its role in many other areas such as aging, infectious diseases, and cancer, the field has received a burst of energy and, therefore, many investigators are searching for the signals that regulate cell death. The goal of this quest is, first, to

Peter Erhardt and Ambrus Toth (eds.), *Apoptosis*, Methods in Molecular Biology, vol. 559
DOI 10.1007/978-1-60327-017-5_18
© Humana Press, a part of Springer Science+Business Media, LLC 2004, 2009

identify the dead or dying cells that will be subject to study; and, second, to understand how the death is regulated.

The ability to identify and characterize the type of cell death has increased interest in where, when, and for what reasons cells die. Cell death has been divided into different classes based on the morphology and biochemical function of the cell during its demise, primarily as defined by synthesis or activation or modification of specific proteins. The types of cell death are categorized as apoptosis, autophagy, necrosis, mitotic catastrophe, and cell death partially characterized by one or more of the different types, i.e. apoptotic-like, necrotic-like *(6–9)*. Apoptosis and autophagy have been categorized as programmed cell death, referring to their strict genetic control. In programmed cell death, the cell participates in its own demise. This type of cell death has been further divided into type I and type II cell death. In type I (apoptosis), a characteristic coalescence and margination of chromatin, related to degradation of the DNA ultimately to a nucleosomal ladder, is an early and prominent feature *(1, 4, 10, 11)*.

A hallmark of apoptosis is also the activation of caspases, particularly the effector caspases 3 and 7 *(12)*, and in most instances alteration in mitochondrial function and the release of cytochrome *c (13)*. In type II (autophagy), although the nuclear collapse eventually occurs, it typically is late and modest, well preceded by many cytoplasmic changes. On the one hand, in cells with large amounts of cytoplasm, these changes likely include lysosomal degradation of massive amounts of cytoplasmic constituents *(14–16)* with formation of autophagic vacuoles and activation of lysosomal enzymes *(17, 18)*. On the other hand, necrosis and mitotic catastrophe are generally considered passive responses to massive cellular insult, and are not programmed. During necrosis, the cell typically loses energy resources or membrane integrity, swells, and osmotically lyses, losing or destroying contents in a chaotic manner. Nevertheless, some forms of necrosis are also under genetic control in some organisms and situations *(19, 20)*. Mitotic catastrophe is defined as death that occurs during mitosis, resulting from "a combination of deficient cell cycle-check points (in particular the DNA structure checkpoints and the spindle assembly checkpoint) and cellular damage" *(21)*. Although there are specific markers unique to each type of cell death, there are many gray areas and points of overlap, reflecting the different cell types that undergo cell death.

Although in the developing embryo many types of cell death (such as apoptosis, autophagy, and necrosis) are found, it is simpler to identify the more common apoptotic cell death. This type of cell death is also the most prominent type found in mammalian embryonic development, and for this reason we will focus our discussion on identifying it. The approaches described can also be used to identify cell death in adult tissues, as well as tools to monitor the events that take place. These approaches include vital staining, morphological examination by electron microscopy, and examination of DNA fragmentation, engulfment of cells by phagocytic cells (indicating activation of cell-surface markers or alteration in the state of the cell membrane proteins such as *phosphatidylserine*), activation of lysosomal enzymes (for recognition and destruction of the cells), and alteration in the pattern of gene expression.

Cell death in the mammalian embryo is often present in only very restricted regions and one can see only a few cells dying, though their death is very important. For these reasons, it is most efficient to examine cell death in situ. In the following section, we first describe methodology that can be used in whole embryo for vital staining or serial tissue sections to identify cell death by apoptosis combined with identification of specific gene products. These methods examine DNA fragmentation in situ, activation of the lysosomal enzyme acid phosphatase, and detection of phagocytic cells by the use of F4/80 surface antibody, which recognizes mature macrophages and monocytes *(22)*. Several gene products are upregulated during cell death. We have shown that the cell-cycle-dependent protein kinase 5 (CDK5) is strongly activated in dying cells *(22, 23)*. The expression of CDK5 is localized to cells undergoing apoptosis *(24)*. We will also look at activation of caspase and examination of the exposure of phosphatidylserine on the cell surface in transparent early embryos. We will also describe how to look for dead cells by morphology and electron microscopy.

2. Materials

2.1. Buffers

1. 1× Phosphate-buffered saline (PBS): 145.4 mM NaCl, 2.68 mM KCl, 10 mM Na_2HPO_4, 1.8 mM KH_2PO_4, pH 7.4.

2. 1× PBST: 0.1% Tween-20 (polyoxyethylenesorbitan monolaurate, Sigma Chemical Co.) in 1× PBS.

2.2. Embryo Fixation and Slide Preparation

1. 4% paraformaldehyde: for 50 mL of fixing solution, add 2 g of paraformaldehyde to 25 mL of DEP-H_2O (1 mL of diethyl pyrocarbonate [Sigma] is added to 1 L of H_2O, and the solution is mixed, allowed to stand for 1 h, and autoclaved for 1 h). Add 5 mL of 10 M NaOH and 5 mL of 10× PBS. Stir the mixture at 55–65°C (no more than 65°C) for 2 h. Filter the solution through a 0.45 µm filter (Gelman Sciences). Then adjust to pH 8.0 and use DEP-H_2O to adjust the final volume to 50 mL. This solution can be stored at 4°C no more than 1 week.

2. Poly-L-lysine-coated slides: place clean slides in autoclaved metal racks. Dip them in 0.2 M HCl, DEP-H_2O, and acetone for 30 s each. Leave slides under a hood to dry for 2 h. Dip slides in a solution of 50 mg poly-L-lysine (Sigma) in 1 L of 0.01 M Tris-HCl, pH 8.0 for 5 min, and dry overnight in a hood.

Coated slides are good to use within 4 weeks. An alternative is Vectabond®-coated slides: Place clean slides in autoclaved metal racks; dip them in 7 mL Vectabond reagent (Vector Laboratories) in 350 mL acetone for 5 min; wash by dipping them several times into dH$_2$O or DEP-H$_2$O and air-dry overnight. Slides are good for at least 4 weeks.

2.3. Lysosomal Activity

This assay is a slight modification of Sigma kit 386A or 387A, used to measure acid phosphatase in leukocytes.

1. Fast garnet stain: 0.6 mL NaNO$_3$ and 0.6 mL Fast Garnet stain are mixed and allowed to stand for 5 min. Then, 22.8 mL of prewarmed dH$_2$O, 3 mL of acetate solution, and 3 mL of naphthol AS-BI phosphate are added.

2. Citrate-acetone-formaldehyde solution: Make the mixture solution of citrate (pH 3.6), acetone, and 37% formaldehyde at volume ratio of 13:33:4. (The citrate solution is provided as part of the kit.)

3. 0.3% hydrogen peroxide solution: Dilute 30% v/v hydrogen peroxide (Sigma) with methanol. Make fresh! Caution! 30% H$_2$O$_2$ is extremely corrosive!

2.4. Detection of Apoptotic Markers by Immunohistochemistry

1. 1× PBST -gelatin solution: 0.1% gelatin (Fisher Scientific) in 1× PBST.

2. 5% Milk blocker: 5% dry milk in 1× PBST-gelatin solution.

3. Primary antibodies:

    a. *Phagocytic Cells.* F4 Primary antibody: Dilute F4/80 (Serotec) 1:10 in 1× PBST-gelatin solution.

    b. *Caspase-3 Activation*: anti active caspase-3 primary antibody: dilute anti caspase-3 (BD Pharmigen) 1:500 in 1× PBST solution.

    c. *Cdk5 Expression:* 1 µL Cdk5 antibody stock (Santa Cruz Biotechnology, Inc., 100 µg/mL) +1 mL 1× PBST (1:1,000 dilution).

4. Secondary antibody solutions:

    a. *Phagocytic Cells:* A 1:50 dilution of peroxidase labeled F*(ab)*$_2$ fragment of goat anti-rat IgG (H + L) (Jackson lmmunoResearch) in 1× PBST gelatin.

    b. *Caspase-3 Activation:* Cy3-conjugated IgG rabbit anti-biotin (Jackson lmmunoResearch): 1:250 dilution in 1× PBS.

    c. *Cdk5 Expression:* Anti-rabbit biotinylated secondary antibody: 1 drop of biotinylated antibody (Vectastain ABC Kit, Vector Lab Inc.) in 10 mL of 1× PBST.

2.5. DNA Fragmentation

2.5.1. Nonfluorescence Measurement of DNA Fragmentation

This technique uses the ApopTag kit S7101 from MP Biomedicals.

1. Working strength TdT solution: mix well 2 drops (76 µL) of reaction buffer (S7100-2) with 1 drop (38 µL) TdT enzyme (S7100-3) by vortexing, and keep on ice for no more than 6 h.

2. Working strength stop/wash buffer: add 1 mL of stop/wash buffer (S7100-4) into 34 mL distilled water. This reagent may be stored at 4°C for up to 1 year.

3. Methylene blue: 2.5% in 1× PBS, pH 7.4.

2.5.2. Fluorescence Measurement of DNA Fragmentation

This technique uses the ApopTag kit S7111 from MP Biomedicals.

1. Working strength TdT enzyme: mix 38 μL of reaction buffer (S7111-2) with 16 μL of TdT enzyme (terminal deoxynucleotidyl transferase) (S7111-3) by vortexing and keep on ice for no more than 6 h.

2. Working strength stop/wash buffer: add 1 mL of stop/wash buffer (S7111-4) into 34 mL distilled water. This reagent may be stored at 4°C for up to 1 year.

3. Working strength anti-digoxygenin-fluorescein: mix well 56 μL blocking solution (S7111-5) with 49 μL anti-digoxigenin-fluorescein (S7111-6) and keep on ice for no more than 3 h.

2.5.3. Double Labeling

1. Blocker: 1 drop goat serum (Vectastain ABC Kit, Vector Lab Inc.) in 10 mL of 1× PBST.

2. Diluted primary antibody: for Cdk5, it is 1 μg/mL. This is prepared as 40 μL BSA (bovine serum albumin) +1 μL Cdk5 antibody stock (Santa Cruz Biotechnology, Inc., 100 μg/mL) +1 mL of 1× PBST.

3. Biotinylated secondary antibody: 1 drop of biotinylated antibody (Vectastain ABC Kit, Vector Lab Inc.) in 10 mL of 1× PBST.

4. Cy3-conjugated IgG mouse anti-biotin (Jackson ImmunoResearch): 1:250 dilution in 1× PBS.

2.6. Electron Microscopy

1. 2.5% glutaraldehyde in 1× PBS, pH 7.4.

2. 1% osmium tetroxide in 1× PBS, pH 7.4.

2.7. Vital Stain

1. Nile blue sulfate (Sigma) 10% in 1× PBS, pH 7.4.

2.8. Annexin V

1. Fluorescein-labeled annexin V (Immunotech).

---

3. Methods

3.1. Embryo Fixation and Slide Preparation (see Note 1)

1. Mate male and female mice overnight and check the females for the presence of a vaginal plug. The time of plug detection is designated gestational day 0.5. Embryos are removed from pregnant females at different days of gestation depending on the specific period of interest.

2. Fix embryos or embryonic tissues in 4% paraformaldehyde overnight. After overnight fixation, embryos or tissues are

placed in 20% sucrose in 1× PBS overnight (filtered and stored at 4°C), blotted on a filter paper quickly to eliminate excess liquid around the tissue sample, and embedded in aluminum cups half-filled with OCT embedding compound. The samples are then quickly frozen by immersing the cup in isopentane placed in liquid N_2.

3. Cut frozen sections (5 μm) in a cryostat at –20°C, place on poly-L-lysine- or Vectabond-coated slides and store at –70°C prior to use. For all the steps outlined below, the slides are allowed to come to room temperature before use *(24, 25)*.

3.2. Measurement of Lysosomal Activity

To measure lysosomal activity, one can analyze acid phosphatase activity by the use of an acid phosphatase kit (Sigma, 386A or 387A).

1. Sections are passed through serial-graded solutions of 20%, 10%, and 5% sucrose in 1× PBS for 5 min each, postfixed with citrate–acetone–formaldehyde solution for 30 s, and washed in dH_2O for 1 min.

2. Treat sections with naphthol AS-BI phosphate and fast garnet stain for 1 h at 37°C.

3. Wash slides in running tap H_2O for 2 min, air-dry for 15 min and counterstain with methylene blue (1:100 dilution for 1 to 2 min depending on the tissue), rinse in dH_2O twice for 1 min each, and mount with Crystal Mount®. The acid phosphatase activity is detected as a distinct red focal precipitate. Timing is important, as sections can easily be under or overstained *(26)* (*see* **Fig. 1a**).

3.3. Detection of Apoptotic Markers (Expression of Active Proteins) by Immuno-histochemistry

1. For detection of the presence or activity of proteins associated with apoptosis, frozen sections are brought to room temperature and washed in 1× PBST for 10 min.

2. Inactivate endogenous peroxidase by immersing the sections in 0.3% hydrogen peroxide solution for 30 min and washing in 1× PBST twice for 5 min each, followed by blocking the non-specific binding of the primary antibody with normal serum (18 μL in 1 mL of 1× PBST) 30 min at 37°C.

3. After washing the slides two times with 1× PBST-gelatin for 5 min, apply the primary antibody and incubate the slides overnight at 4°C.

4. To detect the primary antibody, wash the slides twice with 1× PBST for 5 min each, apply the secondary antibody solution, and incubate slides for 2 h at room temperature.

5. The secondary antibody is removed by washing the slides three times in 1× PBST for 5 min each.

6. Slides are then counterstained with methylene blue (Sigma) for 30 s–3 min depending on the tissue, washed in dH_2O

three times for 1 min each, and mounted with Crystal Mount or Permount®.

3.3.1. Detection of Phagocytic Cells

One of the characteristics of apoptotic cell death is engulfment by macrophages. We have taken advantage of this fact and asked if antibodies specific for mature macrophages can recognize the phagocytic cells in these regions. Although mature macrophages do not circulate until later during development, we can detect phagocytic cells in the area of the limb cell death as early as our first detection of dying cells, i.e., ED (embryonic day) 10.5 and more intensely at ED 13.5 *(22)*. For detection of phagocytic cells, immunoreactivity is visualized by DAB (diaminobenzidine, provided as a prepared solution by Research Genetics, Inc.) staining for 2 min. The phagocytic cells are apparent by a dark brown staining of their cell membranes.

3.3.2. Detection of Caspase-3 Activation

Caspase-3 activation has become one of the best used techniques for detecting apoptotic cell death, as activation of caspase-3 is most likely an irreversible step. This activation occurs as a result of cleavage of pro-form of the protein. Antibodies specific for active caspase-3 have been raised and are used for analysis of cell death. Once fixed, caspase-3 activation is viewed as a fluorescent green punctate pattern in the cytoplasm *(24, 27)* (*see* **Fig. 1b**).

3.3.3. Detection of Cdk5 Expression

We have previously established a clear correlation between apoptotic morphology and activation of Cdk5. Thus, detection of Cdk5 activity could serve as a marker of apoptotic cell death. Using the previously mentioned antibodies, one can detect the levels of Cdk5 activity in cells and correlate the activity to apoptotic cell death. Immunoreactivity is visualized by DAB staining for 2 min. Cdk5 is apparent by a dark brown staining of the cells *(24, 25)* (*see* **Fig. 1c**).

3.4. DNA Fragmentation Assay

A hallmark of apoptotic cell death is the fragmentation of DNA when chromatin is cut between nucleosomes, resulting in distinct ladders when the purified DNA is electrophoresed. Although the ladders can be easily seen when cultured cells are forced to undergo apoptosis and at least one-third are dying, identifying ladders in a situation where only a few cells are dying is more challenging. However, one can use the end labeling terminal transferase deoxyuridine nick end labeling (TUNEL) to identify the dying cells. This technique is more applicable to the developing embryo, since one can see the fragmented DNA in cells without disrupting the morphology, so one can measure the level of cell death with considerable sensitivity and spatial specificity. Fragmented DNA can be detected using both fluorescent and colorimetric methods. We present the methodology for both *(24)*.

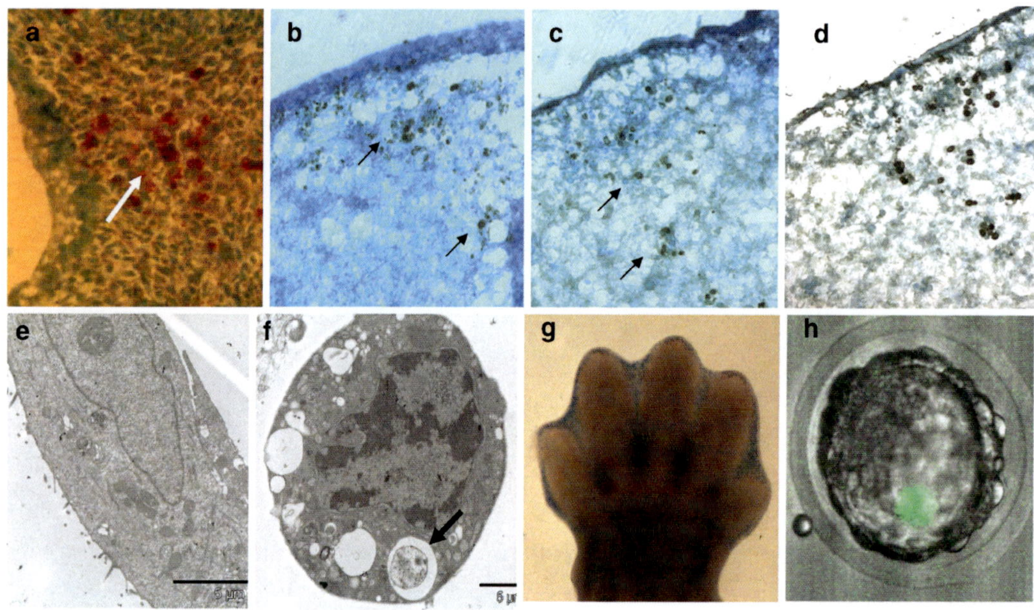

Fig. 1. (**a–h**) Identification of cell death in embryos. Embryos were fixed and processed for sectioning. (**a**) Detection of lysosomal activity (acid phosphatase). *Red* focal precipitates indicate the activation of acid phosphatase. (**b**) Detection of caspase-3 activation, precipitate indicates activation. (**c**) Cdk5 expression indicated by precipitate. (**d**) DNA fragmentation by TUNEL, *brown* indicates fragmentation. (**e**) Normal cell by electron microscopy (EM). (**f**) Dead cells with condensed and fragmented nucleus, and autophagosome (*arrow*) by EM. (**g**) Detection of dead/dying cells, by vital staining Nile *blue* in embryonic mouse limb. (**h**) Detection of phosphatidyl serine expression in early mouse embryo, *green* fluorescence indicates expression.

3.4.1. Nonfluorescent Method (Colorimetric TUNEL)

ApopTag Plus In Situ Apoptosis Detection Kit (Peroxidase) is used.

1. Tissue sections are treated and the endogenous peroxidase is quenched as above.

2. After rinsing with 1× PBS for 5 min, the excess liquid is gently tapped off and equilibration buffer (~100 µL/cm^2) is applied.

3. The section is covered with a plastic cover slip and incubated for 10 min (can go up to 30 min) in a humid chamber at room temperature.

4. The equilibration buffer is removed by gently tapping it off and the plastic cover slip is blotted dry to be reused.

5. The working strength of TdT solution is applied (~100 µL/cm^2) and the preparation is covered with the cover slip and incubated for at least 60 min in a humid chamber at 37°C.

6. Slides are placed in prewarmed stop/wash buffer (prewarm the stop/wash buffer to 37°C for 30 min) and incubated for 30 min at 37°C. The slides should be agitated by dipping in and out of buffer once every 10 min.

7. The slides are rinsed three times more in 1× PBS for 5 min/rinse.

8. To detect the attached digoxygenin-labeled UTP, apply anti-digoxygenin-peroxidase to the section (~100 µL/cm²), recover with plastic cover slips and incubate in a humid chamber for 30 min at room temperature.

9. Wash three times in 1× PBS for 5 min/rinse.

10. To detect the peroxidase, place the slides in a Coplin jar containing DAB for up to 20 min at room temperature.

11. Wash the DAB from the specimens by washing three times with tap water, 1 min/wash, and with dH₂O for 5 min.

12. Counterstain with methylene blue for 5 min.

13. Rinse slides with dH₂O for 30 s without agitation.

14. Dehydrate through ascending ethanol concentration (30, 50, 75, 95, and 100%) twice each for 5 min.

15. Do not let slides dry. Immediately mount using Permount® and a glass cover slip; be careful to avoid bubbles. The labeled cells have a brown staining that covers the nuclei and sometimes parts of the cytoplasm *(24)* (*see* **Fig. 1d**).

3.4.2. Fluorescence DNA Fragmentation Detection (Fluorescence TUNEL)

ApopTag Plus In Situ Apoptosis Detection Kit (fluorescein) is employed here.

1. Rinse slides twice in 1× PBS at 5 min rinse. Gently tap off excess liquid and carefully blot around sections.

2. Apply equilibration buffer (~100 µL/cm²) directly to specimen, cover with plastic cover slips, and incubate for 5 min in a humid chamber.

3. Tap off equilibration buffer. Rinse and dry plastic cover slips for later reuse.

4. Blot around sections and apply working strength TdT enzyme (~100 µL/cm²). Cover sections again with plastic cover slips and incubate for at least 60 min in a humid chamber at 37°C.

5. Prewarm Stop/Wash Buffer to 37°C for 30 min before next step. Remove plastic cover slips. Place slides in a Coplin jar containing prewarmed stop/wash buffer and incubate for 30 min at 37°C. Agitate slides by dipping in and out of buffer once every 10 min.

6. Tap off liquid. Rinse three times in 1× PBS at 3 min rinse.

7. Apply Oncor propidium iodide/Antifade (in kit).

8. Mount under glass coverslip. If storage is required, apply clear nail polish to edges of cover slip. Store at –20°C in the dark. At this point, you can stop and examine the slides for DNA fragmentation, using a fluorescence microscope for fluorescein, or continue as in **step 5** for double labeling for gene expression. The dead cells with fragmented DNA appear as green fluorescent dots *(25)*.

3.4.3. Detection of DNA Fragmentation and Gene Expression: Double-Labeling Fluorescence Assay (see Notes 2 and 3)

1. Follow the same procedure as in fluorescence DNA fragmentation up to the mounting step. Without applying propidium iodide and Permount, put slides in a Coplin jar containing 0.3% hydrogen peroxide for 20 min.

2. Wash slides two times in 1× PBST at 10 min wash.

3. Apply blocker to cover slides and incubate for 1 h.

4. Tap off excess blocker. Cover section with 100 µL/cm² diluted primary antibody (the final concentration of primary antibody depends on different target proteins.) Incubate at 4°C overnight in humidity chamber.

5. Wash three times in 1× PBST at 10 min wash. Agitate occasionally.

6. Apply biotinylated secondary antibody to cover. Incubate at 4°C overnight in a humid chamber.

7. Wash three times in 1× PBST at 10 min/wash.

8. Incubate with Cy3-conjugated IgG mouse anti-biotin for 30 min.

9. Wash three times in 1× PBST at 10 min wash.

10. Mount slides with 90% glycerol and cover slips.

The specific primary and secondary antibodies to be used for the detection of *CDK5* gene are described in the previous section. The steps outlined here are the general steps and can be applied to other genes of interest. The 3′ DNA ends are labeled with fluorescein isothiocyanate and will fluoresce green. In the example, we have used the expression of CDK5 labeled with Cy3 and fluorescing red. One can use confocal microscopy or other overlapping imaging systems to correlate the expression of the given gene and fragmented DNA *(25)*.

3.5. Detection of Cell Death by Electron Microscopy

Electron microscopic examination for detection of cell death is very useful in that not only one can detect the dying cells one can also learn much about the nature of cell death, i.e. whether it looks apoptotic or necrotic or if the dying cell has accumulated autophagosomes and may be dying by autophagy. Typically one can find a single dying cell (in normal physiological processes) or a cluster of dying cells (in pathological or experimental situations) that would be distinguished as follows: apoptotic cells are shrunken, rounded, and usually detached from their neighboring cells or substratum. Their cytoplasm is condensed and more electron-dense than normal, but the organelles appear to be intact. Although we now know that many apoptotic events are initiated by mitochondrial change, the membranes of mitochondria appear to be intact. The nuclei are rounded or fragmented, and the DNA is condensed, heterochromatic, and frequently clustered against the nuclear membrane (marginated). The cell itself ultimately fragments into several pieces, many of which contain nuclear fragments, and the fragments are readily taken up by phagocytes or neighboring cells. A necrotic cell is quite different: it bears evidence of osmotic lysis, with washed-out cytoplasm, amorphous precipitates of protein, and swollen, ruptured organelles such as mitochondria and nuclei. Depending on circumstances, there may be evidence of inflammation in the vicinity. In a cell dying by autophagy, the cytoplasm is filled with lysosomes, autophagosomes, and autophagic vacuoles, the latter of which contain mitochondria and other organelles. While the nucleus may be condensed, chromatin is typically not margin-ated. The cell shrinks as cytoplasmic material is digested, and it may or may not fragment.

While numerous images of cells may be collected for each of these typical forms of cell death, intermediates are possible. For instance, a cell may begin to die by apoptosis or autophagy, both of which processes require energy and control of ion movement; but during this process the cell may ultimately deplete its energy resources, leading to failure of its ion pumps, and ultimately lyse. Similarly, autophagy ultimately consumes the bulk of the cyto-plasm, including mitochondria, which may lead to a necrotic ter-mination. In some cells dying by autophagy, the terminal phases resemble a finally-invoked apoptosis *(14, 16)* (*see* **Fig. 1e, f**).

1. For electron microscopy, embryos are immediately immersed in 2.5% glutaraldehyde solution for a few days (at least 24 h). If the embryos are older than ED 11.5, they should be cut in sections and the placed in the fixative.

2. The samples should be postfixed in 1% osmium tetroxide solution for 1–3 h, then dehydrated in graded ethanol from 50% to 100% for 10 min each, and finally embedded in Spurr resin *(28)*.

3. Semi-thin sections can be cut by ultramicrotome and then stained with toluidine blue to localize the orientation and sit to cut for the thin sections. Select areas are cut in thin sections, col-lected on copper grids (Ernest F. Fullam, Inc.), and stained with uranyl acetate (5% in 70% ethanol) for 15 min and then in lead citrate for 10 min and observed under an electron microscope.

3.6. Detection of Cell Death by Vital Staining

In vivo staining of the live embryo by vital dyes takes advantage of the fact that there is an increase in permeability of cells while undergoing cell death, allowing vital dyes to selectively bind to internal structures (mostly acidic components of cells) when their membranes become leaky. Of the several vital dyes in use, Nile blue sulfate marks dead and dying cells in dark blue. The embryos are used before any fixation and without any sectioning, and therefore with intact morphology *(22)*.

Acridine orange is another dye effectively used for the detec-tion of cell death. Acridine orange is slightly permeable to cells, more so to dying cells. Living cells under high magnification can be seen to fluoresce pale green at 460 nm excitation and 650 nm emission peaks for RNA, while dead cells fluoresce more brightly, fluorescence dependent on the amount of RNA in the cell. This method is very effective for small embryos and nuclear shape usually can easily be seen. In general, for mammalian embryos Nile blue sulfate is recommended, and acridine orange has been used for *Drosophila* embryos.

1. Carefully remove the extra embryonic membranes so that the embryo remains intact!

2. Place the embryos in Nile blue sulfate stain. The staining time is 30–40 min at 37°C and the embryos should be examined under the dissecting microscope after this period. Note the blue stained dying cells. If the embryos are exposed to staining longer you will get increased and nonspecific staining *(22)* (*see* **Fig. 1g**).

3.7. Determination of Apoptosis by Phosphatidylserine

A hallmark for apoptotic cell death is the appearance of phosphatidylserine on the outer surface of the cell membrane in apoptotic cells and bodies. Phosphatidylserine, normally localized on the inner leaflet of the plasma membrane, translocates to the outer leaflet during apoptosis. We can use this translocation to determine whether or not a cell is apoptotic. This works for cells in culture and for very early transparent embryos *(29)*. Since the phosphatidylserine is exteriorized rather than synthesized or otherwise altered, cells must be fresh or, if fixed, they must remain impermeable to the reagent. These cells are stained with fluorescein-labeled annexin V (full concentration) for 5 min. Apoptotic cells are then visualized by confocal microscopy.

3.8. Conclusions

By and large, the classical methods of assessing apoptosis are valid for mammalian embryos. However, if one wishes to study the normal sequence of apoptosis rather than pathologically induced apoptosis, the number of apoptotic cells is very low and very restricted in space and time, but these deaths are physiologically important. Thus in situ methods are far more useful than those used for cultured cells, such as fluorescence-activated cell sorting or biochemical measurements of enzyme activity. Since many of these techniques have been worked out for systems other than mammalian embryos, all experiments should include preliminary evaluation of concentration ranges and positive and negative controls.

4. Notes

1. Tissue on slides may be fixed or unfixed, and frozen or paraffin sections. For unfixed tissue cryosections, the following steps should be followed: Fix the sections on slides in a Coplin jar

containing 10% neutral buffered formalin for 10 min at room temperature; blot off excess liquid and wash in 1× PBS twice for 5 min each change; postfix in ethanol/acetic acid (2:1) for 5 min at –20°C; drain, but do not let slides dry; rinse in 1× PBS twice for 5 min rinses. For paraffin-embedded tissue, the samples must be deparaffinized as follows. Prewarm xylene to 60°C for at least 30 min. The remaining steps are performed at room temperature. Incubate slides in prewarmed xylene twice for 10 min each change; rehydrate the tissue through descending ethanol concentration (100%, 95%, 70%, 50%, 30%) twice each for 5 min; wash slides in 1× PBS once for 5 min. Circle the area of tissue sections or cells with a peroxidase-antiperoxidase pen (PAP) (Research Products International) for later location.

2. The method of double labeling uses a two-step approach to confirm that the cell undergoing DNA fragmentation also expresses specific messages. In the first step, the free 3′-OH ends of fragmented DNA are labeled by the indirect fluorescence TUNEL technique described earlier. In the second step, specific protein expression is localized by means of a specific antibody. The technique described here combines the use of separate fluorescent markers to identify both DNA fragmentation and gene expression. Alternatively, DNA fragmentation may be labeled and protein expression may be studied separately using this methodology. To do the combined staining, it is necessary to perform each procedure separately as well. This controls that each step has worked on its own and provides information regarding the level of background noise that has to be subtracted from the combined procedure.

3. It is very important to have both positive and negative control slides for each part of this assay. Negative controls should be run as duplicates of each section. For this purpose, sham staining can be performed by substituting distilled water for the TdT enzyme in the DNA end-labeling study, and no primary antibody (PBS) or preimmune serum for the antibodies. As a positive control for the method, one can use short exposure of the control slide to DNase I, which causes fragmentation of DNA, and then look for DNA end labeling.

References

1. Coucouvanis, E. C., Sherwood, S. W., Carswell-Crumpton, C., Spack, E. G., Jones, P. P. (1993). Evidence that the mechanism of prenatal germ cell death in the mouse is apoptosis, *Exp Cell Res* **209**, 238–247.

2. Raff, M. C. (1992). Social controls on cell survival and cell death, *Nature* **356**, 397–400.

3. Walker, N. I., Harmon, B. V., Gobé, G. C., Kerr, J. F. (1988). Patterns of cell death, *Methods Achiev Exp Pathol* **13**, 18–54.

4. Glücksmann, A. (1951). Cell death in normal development, *Biol Rev Camb Phil Soc* **26**, 59–86.

5. Saunders, J. W., Jr. (1966). Death in embryonic systems, *Science* **154**, 604–612.

6. Okada, H., and Mak, T. W. (2004). Pathways of apoptotic and non-apoptotic death in tumour cells, *Nat Rev Cancer* **4**, 592–603.

7. Lockshin, R. A., and Zakeri, Z. (2007). Cell death in health and disease, *J Cell Mol Med* **11**, 1214–1224.

8. Penaloza, C., Lin, L., Lockshin, R., Zakeri, Z. (2006). Cell death in development: shaping the embryo, *Histochem Cell Biol* **126**, 149–158.

9. Zakeri, Z., and Lockshin, R. A. (2002). Cell death during development, *J Immunol Methods* **265**, 3–20.

10. Zakeri, Z. B. W., Tenniswood, M., Lockshin, R. A. (1995). Cell death: programmed, apoptosis, necrosis, or other? *Cell Death Differ* **2**, 87–96.

11. Christophe Chanoine, S. H. (2003). Xenopus muscle development: From primary to secondary myogenesis, *Dev Dynam* **226**, 12–23.

12. Kumar, S. (2007). Caspase function in programmed cell death, *Cell Death Differ* **14**, 32–43.

13. Garrido, C. G. L., Brunet, M., Puig, P. E., Didelot, C., Kroemer, G. (2006). Mechanisms of cytochrome c release from mitochondria, *Cell Death Differ* **13**, 1423–1433.

14. Halaby, R. Z. Z., and Lockshin, R. A. (1994). Metabolic events during programmed cell death in insect labial glands, *Biochem Cell Biol* **72**, 597–601.

15. Zakeri, Z. (1993). In vitro mammalian limb differentiation as an experimental model, *Prog Clin Biol Res.* **283A**, 361–370.

16. Zakeri, Z. F., Quaglino, D., Latham, T., Lockshin, R. A. (1993). Delayed internucleosomal DNA fragmentation in programmed cell death, *FASEB J* **7**, 470–478.

17. Mizushima, N. (2007). Autophagy: process and function, *Genes Dev* **21**, 2861–2873.

18. O'Connell, A. R., and Stenson-Cox, C. (2007). A more serine way to die: Defining the characteristics of serine protease-mediated cell death cascades, *Biochim Biophys Acta* **1773**, 1491–1499.

19. Kerr, J. (1969). An electron-microscope study of liver cell necrosis due to heliotrine, *J Pathol* **97**, 557–562.

20. Lockshin, R. A., and Zakeri, Z. (2001). Programmed cell death and apoptosis: origins of the theory, *Nat Rev Mol Cell Biol* **2**, 545–550.

21. Castedo, M., Perfettini, J. L., Roumier, T., Andreau, K., Medema, R., Kroemer, G. (2004). Cell death by mitotic catastrophe: a molecular definition, *Oncogene* **23**, 2825–2837.

22. Zakeri, Z. F., and Ahuja, H. S. (1994). Apoptotic cell death in the limb and its relationship to pattern formation, *Biochem Cell Biol* **72**, 603–613.

23. Zhang, Q., Ahuja, H. S., Zakeri, Z. F., Wolgemuth, D. J. (1997). Cyclin-dependent kinase 5 is associated with apoptotic cell death during development and tissue remodeling, *Dev Biol* **183**, 222–233.

24. Lin, L., Ye, Y., Zakeri, Z. (2006). p53, Apaf-1, caspase-3, and -9 are dispensable for Cdk5 activation during cell death, *Cell Death Differ* **31**, 141–150.

25. Zhu, Y. L. L., Kim, S., Quaglino, D., Lockshin, R. A., Zakeri, Z. (2002). Cyclin dependent kinase 5 and its interacting proteins in cell death induced in vivo by cyclophosphamide in developing mouse embryos, *Cell Death Differ* **9**, 421–430.

26. Zakeri, Z. Q. D., Latham, T., Woo, K., Lockshin, R. A. (1996). Programmed cell death in the tobacco hornworm, *Manduca sexta: alteration in protein synthesis, Microsc Res Tech.* **34**, 192–201.

27. Carracedo, J., Ramirez, R., Soriano, S., Martin-Malo, A., Rodriguez, M., Aljama, P. (2002). Caspase-3-dependent pathway mediates apoptosis of human mononuclear cells induced by cellulosic haemodialysis membranes, *Nephrol Dial Transplant* **17**, 1971–1977.

28. Coutinho, A. R., Mendes, C. M., Caetano, H. V., Nascimento, A. B., Oliveira, V. P., Hernandez-Blazquez, F. J., Sinhorini, I. L., Visintin, J. A., Assumpcao, M. E. (2007). Morphological changes in mouse embryos cryopreserved by different tecniques, *Microsc Res Tech* **70**, 296–301.

29. Zakeri, Z., Lockshin, R. A., Criado-Rodríguez, L. M., Martínez, A. C. (2005). A generalized caspase inhibitor disrupts early mammalian development, *Int J Dev Biol* **49**, 43–51.

Chapter 19

Detection of Apoptosis in the Central Nervous System

Youngsoo Lee and Peter J. McKinnon

Summary

Apoptosis occurs in the nervous system during normal development, but can also be induced by disease or after exogenous insults such as DNA damage. Depending on the magnitude and timing of the stimulus, apoptosis can be sporadic or widespread. Because of the highly ordered structure of the nervous system, immunohistochemical detection approaches provide a wealth of information about the spatiotemporal nature of apoptosis in this tissue. Therefore, immunohistochemistry offers valuable insights into the neuropathology of disease processes and the identification of specific cell populations that are susceptible to apoptosis. In this chapter, we outline standard approaches for the immunohistochemical analysis of apoptosis in the nervous system, with an emphasis on methodology useful for studies involving DNA damage-induced apoptosis.

Key words: Immunohistochemistry, DNA damage, p53, TUNEL, Caspase-3, γH2AX, 53BP1

1. Introduction

Studies of apoptosis in the nervous system are particularly informative during neural development because of the susceptibility of proliferative and differentiating neural cells to apoptosis. The development of the nervous system occurs in a reiterated manner involving widespread areas of proliferation adjacent to regions of differentiation and migration *(1–4)*. Neuroanatomically, this laminar arrangement is tremendously informative when immunohisto-chemistry is combined with mammalian model systems such as the mouse to analyze apoptosis. These approaches have proven valuable in deciphering tissue and cell-type specificity in diseases and after various insults. For example, studies of genotoxicity have

Peter Erhardt and Ambrus Toth (eds.), *Apoptosis*, Methods in Molecular Biology, vol. 559
DOI 10.1007/978-1-60327-017-5_19
© Humana Press, a part of Springer Science + Business Media, LLC 2004, 2009

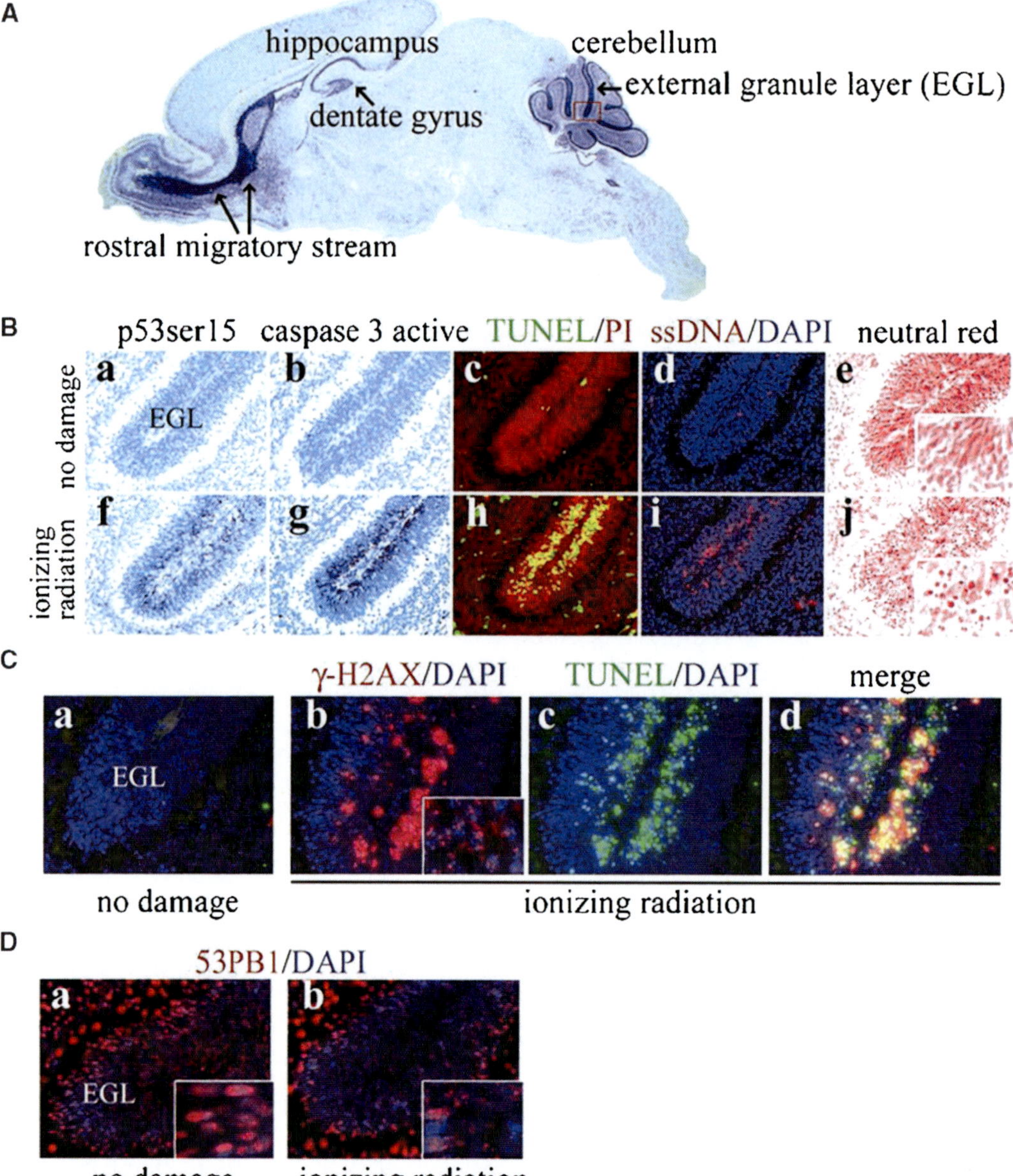

Fig. 1. Immunohistochemical analysis of apoptosis after DNA damage in the postnatal mouse brain. (**A**) Susceptible areas to DNA damage-induced apoptosis during postnatal brain development include the rostral migratory system, the hippocampus including the dentate gyrus, and the cerebellum particularly the external granular layer (EGL). A sagittal section of a postnatal day 7 mouse brain with Nissl staining is shown. The red box indicates an example of the cerebellar EGL shown in B–D. (**B**) Immunohistochemical detection of apoptosis process in the EGL of postnatal day 5 brain induced by ionizing radiation. The *lower panel* shows activation of p53 (**f**) and caspase 3 (**g**), TUNEL positive signal (**h**), single strand DNA immunopositive signal (*i*), and pyknosis visualized by Neutral red staining (**j**) induced by DNA damage. *Inset panels* in (**e**) and (**j**) are magnified views to illustrate the morphology of pyknotic cells. Apoptotic cells are not evident without any insult to the developing brain (**a–e**). p53 and caspase 3 immunoreactivity were visualized with the VIP substrate kit and the brains were counterstained with Methyl green (**a–b**, **f–g**). TUNEL was done using ApopTag Fluorescein in situ Apoptosis Detection Kit with DAPI counterstaining (**c**, **h**). ssDNA antibody was detected with Cy3 conjugated secondary antibody with PI counterstaining (**d**, **i**). (**C**) Phosphorylated H2AX (γH2AX) is a measure of DNA double strand breaks, and are visualized as nuclear punctate staining (*inset panel* in b), called foci. In the absence of DNA damage no γH2AX signal is found (**a**). γ-H2AX staining is detected after DNA damage using Cy3 secondary antibodies (**b**) and apoptosis is detected by TUNEL using fluorescein (**c**); panel (**d**) is a merge of panels (**b**) and (**c**). (**D**) Foci formation of 53BP1 after ionizing radiation induced DNA breaks. Similar to γH2AX, 53BP1 is another early responder to DNA double strand breaks. While 53BP1 is distributed evenly in the nucleus without any DNA damage, 53BP1 foci form in the nucleus after ionizing radiation (**b**) and were visualized using Cy3-coupled secondary antibody.

utilized apoptotic analyses as a major means of determining the consequences of this type of stress in the nervous system *(5–7)*. Organisms have developed efficient DNA repair mechanisms to maintain the integrity of genomic information. However, although repair of DNA lesions is an option, the tremendous proliferative capacity of germinal zones in the nervous system makes apoptosis and subsequent cellular replacement a more common strategy to deal with genomically compromised cells *(8–10)*.

Similar to the embryo, postnatal brain development also contains proliferating regions such as the dentate gyrus of the hippocampus or the cerebellar external granule layer (EGL), which are also vulnerable to DNA damage-induced apoptosis *(8)* (*see* **Fig. 1A**). Immature neurons are quite susceptible to DNA damage-induced apoptosis, while fully mature neurons are relatively resistant to this particular insult, and they do not typically undergo apoptosis *(8)*. The signaling pathway triggered by DNA damage in the nervous system generally results in p53 stabilization and activation to initiate transcription-dependent apoptosis *(5, 8–10)*. The response of p53 to DNA damage can be readily detected using p53 and phospho-p53 (serine 15) immunohistochemistry (*see* **Fig. 1B, a, f**). A major outcome of p53 signaling is the induction of PUMA (p53 upregulated mediator of apoptosis), a proapoptotic Bcl-2 family member, which is critical for DNA damage-mediated apoptosis in the nervous system *(11)*. Caspases become activated to effect apoptosis, and in the case of caspase-3, visualization by immunohistochemistry is relatively straightforward *(5)* (*see* **Fig. 1B, b, g**). Finally, the end-stage of programmed cell death can be detected by enzymatic labeling such as TUNEL, single strand DNA (ssDNA) immuno-reactivity, or simple Neutral Red staining to identify pyknotic cells (*see* **Fig. 1B, c–e, h–j**). With regard to DNA damage, immuno-histochemistry is also valuable for detecting DNA double strand breaks (DSBs), and available reagents such as antibodies against phosphorylated H2AX (γH2AX) and 53BP1, which form punctate nuclear focal staining in response to DSBs, are particularly useful in this regard *(8, 12, 13)* (*see* **Fig. 1C, D**; these foci generally persist until DNA damage is resolved). TUNEL and γH2AX foci can often overlap in proliferating areas of the developing brain (*see* **Fig. 1C**).

In the following, we describe methods suitable for the immuno-histochemical analysis of apoptosis, with some additional details regarding analysis of DNA damage-induced apoptosis.

2. Materials

1. Neutral Red staining: 1% Neutral red (Sigma-Aldrich) in 0.1 M acetic acid (pH 4.8). Shake or stir Neutral red solution overnight to completely dissolve Neutral red power, and filter the solution. Store Neutral red solution in an amber bottle at room

temperature. It is stable for several months. Maintain an acidic pH since solution color changes from red to yellow between pH 6.8 and 8.0. Care: Neutral red powder is possibly mutagenic.

2. Methyl Green staining: 0.5% Methyl green (Sigma-Aldrich) in 0.1 M sodium acetate solution (pH 4.2 adjusted with glacial acetic acid). Dissolve Methyl green overnight and then filter. It is stable for several months in an amber bottle at room temperature.

3. Paraformaldehyde: 4% paraformaldehyde (Sigma-Aldrich) in PBS (pH 7.4). After adding paraformaldehyde power, heat the solution to 65–70°C while stirring the solution continuously. Once paraformaldehyde has dissolved, add several drops of 5 M NaOH until the solution becomes clear. Cool the solution and filter. Before using the paraformaldehyde solution, adjust pH back to pH 7.4. This entire process should be done in a fume hood. Buffered paraformaldehyde solution is stable for about a month at 4°C. Alternatively make aliquots and freeze them down for long-term storage. Paraformaldehyde is toxic and needs appropriate disposal by neutralization with a formalin neutralizer such as Vytac (Richard-Allan Scientific).

4. Antigen-Retrieval solution

    a. Solution A: 0.1 M citric acid.

    b. Solution B: 0.1 M sodium citrate.

    c. Working Solution: 9 mL of Solution A/41 mL of Solution B in 450 mL of distilled water. Make the working solutions just before use and adjust pH to 6.0. Autoclaving Solutions A and B helps to extend shelf life.

5. Primary antibodies:

    a. p53 (CM5), rabbit polyclonal antibody, 1:1,000, antigen-retrieval technique applied (Vector Laboratories).

    b. Phospho-p53 (serine 15), rabbit polyclonal antibody, 1:250 (adult brain), 1:150 (embryo), antigen-retrieval technique applied (Cell Signaling Technology).

    c. Active caspase 3, rabbit monoclonal antibody, 1:500, antigen-retrieval technique applied (BD PharMingen).

    d. ssDNA rabbit polyclonal antibody, 1:300 (IBL Co).

    e. γH2AX (serine 139), rabbit polyclonal antibody, 1:500, antigen-retrieval technique applied (Abcam).

    f. 53BP1, rabbit polyclonal antibody, 1:500, antigen-retrieval technique applied (Bethyl).

6. Secondary antibodies:

    a. Biotinylated Goat anti-Rabbit IgG (H + L), 1:500 (Jackson ImmunoResearch).

    b. Cy3 (cyanine) conjugated Goat (or Donkey) anti-Rabbit IgG (H + L), 1:400 (Jackson ImmunoResearch), alternatively

     Alexa Fluor 555 Goat (or Donkey) anti-Rabbit IgG (H + L), 1:1,000 (Invitrogen).

    c. FITC (fluorescein) conjugated Goat (or Donkey) anti-Rabbit IgG (H + L), 1:200 (Jackson ImmunoResearch), alternatively Alexa Fluor 488 Goat (or Donkey) anti-Rabbit IgG (H + L), 1:1,000 (Invitrogen).

7. Blocking solution: 5% normal serum (Goat serum, Vector Laboratories or Donkey serum, Jackson ImmunoResearch)/1% bovine serum albumin (crystallized, Sigma) in PBST (0.4% Triton X-100, Sigma in PBS), and add thimerosal (Sigma) to make 0.01% solution for longer storage.

8. Biotin–Avidin System reagent: Vectastain Elite ABC kit (Vector Laboratories).

9. Substrate kits for colorimetric detection: Vector VIP (for violet color) or DAB (brown or dark blue color) substrate kit (Vector Laboratories). DAB substrate solution can be made; 4 mg of diaminobenzidine (DAB, Sigma) in 10 mL of 0.05 M Tris-HCl buffer, pH 7.2 and add 16.7 μL of 3% H_2O_2 before use for brown color development. Alternatively, 2 mg of DAB/250 mg Nickel sulfate (Sigma) in 10 mL of 0.175 M sodium acetate, pH 6.5 (adjusted with glacial acetic acid) and add 8.3 μL of 3% H_2O_2 before use for dark blue color development.

10. Mounting medium for fluorescence with DAPI (blue nucleus staining) or Propidium iodide (generates a red nuclear staining) (Vector Laboratories).

11. Terminal uridine deoxynucleotidyl transferase dUTP nick end labeling (TUNEL) kit: ApopTag in situ Apoptosis Detection Kit (Millipore).

3. Methods

The methods described below correspond to the apoptotic analyses shown in **Fig. 1**. As an example of the general approaches using these methods, DNA damage-induced apoptosis in the postnatal immature cerebellum is shown. However, the methods below are applicable to apoptotic analysis in all regions of the brain. It should be noted that the effects of DNA damage from ionizing radiation are widespread and can simultaneously affect large areas of immature neural cells. Depending on the type of insult or lesion being studied, the relative abundance of apoptotic cells will vary. Moreover, in some cases apoptosis may be sporadic, and so multiple approaches using different assays will ensure greater confidence in the evaluation of cell death. Because of the cellular heterogeneity and three-dimensional structure of the brain, it is also very important to compare carefully matched and defined neuroanatomical areas between control and experimental groups when performing apoptotic analyses via immunohistochemistry.

3.1. Neutral Red Staining (or Methyl Green) Staining to Detect Apoptotic Cells

Apoptotic cells will appear as fragmented and condensed individual cells or in groups alongside unaffected cells with normal morphology upon histological evaluation (e.g. **Fig. 1B, e, j**).

1. Wash slides containing tissue sections with PBS for 10 min (2 times).

2. Before staining, briefly rinse slides with tap water.

3. Stain slides with neutral red solution for 1–5 min (Neutral red) (*see* **Note 1**) or 5–10 min (Methyl green) (*see* **Note 2**).

4. Wash off extra staining solution in running tap water.

5. Dehydration step:

50% alcohol	~1 min
70% alcohol	~1 min (*see* **Note 3**)
95% alcohol	~1 min
100% alcohol	~1 min, 2 times
Xylene	~5 min, 3 times

6. Mount slides with nonaqueous mounting medium such as DPX for histology (Fluka).

3.2. Antigen Retrieval Treatment (*see Note 4*)

Antigen retrieval is a method that is often used to reveal the epitopes of an antigen that facilitates antibody binding and therefore a clear immunohistochemical signal *(14)*.

1. Incubate slides with Antigen-Retrieval working solution in Coplin jar made of polypropylene with screw cap for 10 min.

2. Tightly close screw cap and place 4 Coplin jars symmetrically on the microwave tray, and microwave for 10 minutes (total 420 kJ, 70% power of 1,000 watt microwave) (*see* **Note 5**). Check the level of working solution to cover all the slides and add more solution as necessary. Handle with caution since the solution will be very hot.

3. After microwaving, remove screw tops, and cool slides down in Antigen Retrieval working solution slowly to room temperature. It usually takes up to 1 h or longer.

3.3. Immunohistochemistry: Colorimetric Detection

1. Brain tissue fixation is done by transcardial perfusion with 4% buffered paraformaldehyde followed by routine cryopreparation and cryosectioning procedures.

2. Perform Antigen Retrieval Treatment (*see* **Subheading 3.2**), if it is necessary (*see* **Note 6**).

3. Wash slides with PBS for 10 min (2 times).

4. Incubate slides with H_2O_2 solution (2 mL of 30% H_2O_2 in 100 mL of methanol) for 30 min to quench endogenous peroxidase.

5. Rinse slides in PBS for 5 min (2 times).

6. Place slides in humid containers.

7. Gently spread blocking solution (*see* **Note 7**) to cover entire sections and incubate for 1–2 h on an orbital shaker at room temperature.

8. Decant blocking solution on absorbents.

9. Apply primary antibody solution diluted in blocking solution to slides in humid containers and incubate overnight on an orbital shaker at room temperature.

10. Rinse slides in PBS for 10 min (2 times).

11. Incubate slides with biotinylated IgG diluted in blocking solution for 1–2 h on an orbital shaker at room temperature.

12. Rinse slides in PBS for 10 min (2 times).

13. Add the ABC complex solution diluted in PBS (*see* **Note 8**) on the slides and incubate for 1–2 h on an orbital shaker at room temperature.

14. Rinse in PBS for 10 min (2 times).

15. Visualize positive signals with substrate kits; monitor the development of color under the microscope (*see* **Note 9**).

16. Stop enzyme reaction by rinsing slides with running tap water.

17. Counter staining with Methyl green (*see* **Subheading 3.1**).

18. Dehydration and mounting (*see* **Subheading 3.1**).

3.4. Immunohisto-chemistry: Fluorescence Detection

1. Perform Antigen Retrieval Treatment (*see* **Subheading 3.2**) if it is necessary.

2. Wash slides with PBS for 10 min (1 time).

3. Rinse slides in PBST for 15 min (2 times).

4. Place slides in humid containers.

5. Gently spread blocking solution (*see* **Note 7**) to cover entire sections and incubate for 1–2 h on an orbital shaker at room temperature.

6. Decant blocking solution on absorbents.

7. Apply primary antibody solution diluted in blocking solution to slides in humid containers and incubate overnight on an orbital shaker at room temperature.

8. Rinse slides in PBST for 15 min (2 times).

9. Add fluorescent secondary antibody diluted in blocking solution on the slides and incubate for 1–2 h in the dark.

10. Rinse slides in PBST for 15 min (2 times) in the dark.

11. Wash slides with PBS briefly.

12. Mount slides with mounting medium for fluorescence (*see* **Note 10**).

3.5. TUNEL or Double Staining with TUNEL

1. Perform Antigen Retrieval Treatment (*see* **Subheading 3.2**) for double staining. This is not required for TUNEL method itself (*see* **Note 11**).

2. Wash slides with PBS for 10 min (2 times).

3. Postfix sections in precooled fixative (2 ethanol/1 acetic acid) for 5 min at –20°C.

4. Rinse slides in PBS for 5 min (2 times).

5. Apply equilibration buffer to slides provided in the kit in a humid chamber for at least 10 s at room temperature.

6. Decant equilibration buffer.

7. Apply working strength terminal deoxynucleotidyl transferase (TdT) on the slides and incubate in a humid chamber for 1 h at 37°C (*see* **Note 12**).

8. Stop enzyme reaction in working strength stop/wash buffer provided in the kit for 10 min.

9. Rinse slides in PBS 10 min (2 times).

10. For double staining with TUNEL, perform **steps 3–8** of **Subheading 3.4**.

11. Add working strength anti-digoxigenin conjugate (fluorophore tagged) diluted in the blocking solution provided in the kit, for double staining add fluorescent secondary antibody in the same blocking solution together (*see* **Note 13**).

12. Incubate for 1 h at room temperature in the dark (*see* **Note 12**).

13. Wash slides in PBS 15 min (2 times) in the dark.

14. Mount slides with mounting medium for fluorescence (*see* **Note 10**).

---

4. Notes

1. One of the hallmarks of apoptosis is condensation of chromatin, called pyknosis, which can be visualized by Neutral red staining (*see* **Fig. 1B, e and j**). Round dense staining of nucleus or fragmented nuclear staining is the sign of cell death. However,

some brain areas contain granule cells that have small round nucleus, so careful reading of Neutral red staining requires not to mislabel this type of cells as pyknotic cells.

2. Proliferating cells and differentiating neurons tend to stain much stronger than fully mature neurons do in Methyl green nuclear staining.

3. 50~70% alcohol partially removes Neutral red or Methyl green staining.

4. Antigen Retrieval was originally developed for formalin-fixed, paraffin-embedded tissues *(14, 15)*; this method can also be applied to paraformaldehyde-fixed frozen tissues for better immunohistochemistry signals.

5. Using a pressure cooker or a steamer is an alternative method for heating.

6. Even though all exemplified primary antibodies in this protocol except ssDNA antibody require Antigen Retrieval treatment for better/specific signals, this treatment is not always applicable to all primary antibodies. It is necessary to test the applicability of Antigen Retrieval treatment to new antibodies.

7. The choice of species for normal serum is dependent on the species in which secondary antibody is raised. For example, use goat normal serum for the blocking solution, when biotinylated or fluorophore conjugated secondary antibody is raised in goat.

8. The ABC complex in PBS should be made at least 30 min prior to use.

9. It is important to cover the entire sections with substrate solution quickly for even development of positive signals. Adding Triton X-100 (0.04%) to substrate solution helps to break the surface tension of the solution.

10. Fluorophore Cy3 or/and FITC are compatible with DAPI counterstaining. The combination of FITC and PI is good, yet Cy3 signal is not distinguishable from PI counterstaining (*see* **Fig. 1B, C**).

11. This TUNEL protocol is based on ApopTag fluorescence in Situ Apoptosis Detection Kit, such as Fluorescein Detection Kit (Millipore) (*see* **Fig. 1C**).

12. Keep the timing of incubation. Longer incubation gives high nonspecific background staining.

13. The TUNEL kit provides fluorophore tagged Anti-Digoxigenin that is affinity purified sheep polyclonal antibody, so double staining is possible with antibodies raised in other species.

Acknowledgment

The authors thank McKinnon lab members for their input and advice. Support of this work was from the NIH and the ALSAC charities of SJCRH.

References

1. Dehay, C., and Kennedy, H. (2007). Cell-cycle control and cortical development, *Nat Rev* **8**, 438–450.

2. Huang, Z. J., Di Cristo, G., and Ango, F. (2007). Development of GABA innervation in the cerebral and cerebellar cortices, *Nat Rev* **8**, 673–686.

3. Jacobsen, M. (1991). *Dev Neurobiol*, Third Edition ed., Plenum Press, New York.

4. Wang, V. Y., and Zoghbi, H. Y. (2001). Genetic regulation of cerebellar development, *Nat Rev* **2**, 484–491.

5. Chong, M. J., Murray, M. R., Gosink, E. C., Russell, H. R., Srinivasan, A., Kapsetaki, M., Korsmeyer, S. J., and McKinnon, P. J. (2000). Atm and Bax cooperate in ionizing radiation-induced apoptosis in the central nervous system, *Proc Natl Acad Sci USA* **97**, 889–894.

6. Lee, Y., Chong, M. J., and McKinnon, P. J. (2001). Ataxia telangiectasia mutated-dependent apoptosis after genotoxic stress in the developing nervous system is determined by cellular differentiation status, *J Neurosci* **21**, 6687–6693.

7. Orii, K. E., Lee, Y., Kondo, N., and McKinnon, P. J. (2006). Selective utilization of non-homologous end-joining and homologous recombination DNA repair pathways during nervous system development, *Proc Natl Acad Sci USA* **103**, 10017–10022.

8. Lee, Y., and McKinnon, P. J. (2007). Responding to DNA double strand breaks in the nervous system, *Neuroscience* **145**, 1365–1374.

9. Norbury, C. J., and Zhivotovsky, B. (2004). DNA damage-induced apoptosis, *Oncogene* **23**, 2797–2808.

10. Roos, W. P., and Kaina, B. (2006). DNA damage-induced cell death by apoptosis, *Trends Mol Med* **12**, 440–450.

11. Jeffers, J. R., Parganas, E., Lee, Y., Yang, C., Wang, J., Brennan, J., MacLean, K. H., Han, J., Chittenden, T., Ihle, J. N., McKinnon, P. J., Cleveland, J. L., and Zambetti, G. P. (2003). Puma is an essential mediator of p53-dependent and -independent apoptotic pathways, *Cancer Cell* **4**, 321–328.

12. Rogakou, E. P., Pilch, D. R., Orr, A. H., Ivanova, V. S., and Bonner, W. M. (1998). DNA double-stranded breaks induce histone H2AX phosphorylation on serine 139, *J Biol Chem* **273**, 5858–5868.

13. Ward, I., and Chen, J. (2004). Early events in the DNA damage response, *Curr Top Dev Biol* **63**, 1–35.

14. Shi, S. R., Cote, R. J., and Taylor, C. R. (2001). Antigen retrieval techniques: current perspectives, *J Histochem Cytochem* **49**, 931–937.

15. Shi, S. R., Key, M. E., and Kalra, K. L. (1991). Antigen retrieval in formalin-fixed, paraffin-embedded tissues: an enhancement method for immunohistochemical staining based on microwave oven heating of tissue sections, *J Histochem Cytochem* **39**, 741–748.

Chapter 20

Genetic Mapping of Anti-Apoptosis Pathways in Myeloid Progenitor Cells

Dan Liu and Zhou Songyang

Summary

In mammalian cells, apoptotic and anti-apoptotic pathways may be investigated using a variety of biochemical, molecular, and genetic approaches. Retrovirus mediated genetic screens have proven a powerful tool in mapping out the network of players in a number of signaling pathways. We have developed the ERM (for enhanced retroviral mutagen) mutagenesis approach to identify novel players in the growth factor dependent survival pathways. ERM has been shown to be efficient and amenable to genome wide genetic screens in mammalian cells without the need of cDNA library construction. The advantages of the ERM method include regulatable expression, flexible design, and efficiency.

Key words: Enhanced retroviral mutagen, Genetic screen, Signaling pathway, Apoptosis

1. Introduction

To identify new players in the anti-apoptotic pathways, we developed the ERM (for enhanced retroviral mutagen) approach for genome-wide gain-of-function genetic screening in mammalian cells *(1–3)*. The ERM strategy takes advantage of the random nature of retroviral integration and improves upon traditional wild-type retrovirus mediated genetic screens. Features such as a regulatable promoter and epitope tags make the ERM strategy highly adaptable to high-throughput screens and a variety of signaling pathways. In particular, we were interested in illustrating the IL-3 (interleukin-3) signaling and anti-apoptosis pathway and the genes that mediate IL-3-independent survival *(1, 2)*. To accomplish this,

Peter Erhardt and Ambrus Toth (eds.), *Apoptosis*, Methods in Molecular Biology, vol. 559
DOI 10.1007/978-1-60327-017-5_20
© Humana Press, a part of Springer Science + Business Media, LLC 2004, 2009

we constructed ERM vectors and utilized the IL-3 dependent cell line 32D cells as a model system. The ERM viruses were introduced into 32D cells, and the cells were cultured in the absence of IL-3. The survival clones are then analyzed to identify the genes targeted by ERM.

2. Materials

2.1. Vectors and Cell Lines

1. A set of 3 pBabe-based ERM screening vectors (in three reading frames).
2. The retroviral packaging cell line BOSC23 *(4)*, cultured in DMEM supplemented with 10% fetal calf serum (*see* **Note 1**).
3. The IL-3-dependent mouse myeloid progenitor cell line 32D cells, cultured in RPMI supplemented with 10% heat-inactivated fetal calf serum and 10–20% of conditioned media from WEHI-3B cells (or commercially available IL-3 supplement) (*see* **Note 2**).

2.2. Reagents

1. Make polybrene (Sigma) stock solution in 1× PBS at 10 mg/mL, filter through a 0.2 μm filter, aliquot and freeze at –20°C.
2. Make puromycin stock solution in 1× PBS at 2 mg/mL, filter through a 0.2 μm filter, aliquot and freeze at –20°C.
3. Make tetracycline or doxycycline stock solution in 1× PBS at 5 mg/mL, filter through a 0.2 μm filter, aliquot and freeze at –20°C.

2.3. Oligonucleotide Primers

1. RT-1: 5′-*GCAAATACGACTCACTATAGGG*ATCCNNNN(GC)ACG-3′, N = AGCT, T7 primer sequences are in *italic*.
2. Myr1: 5′-ACCATGGGGAGCAGCAAGAGCAAACCAAAA-GACCCCAGCCAACGC-3′.
3. T7: 5′-GCAAATACGACTCACTATAGGG-3′

3. Methods

The genetic screen procedure can be roughly divided into three stages. Stage 1 involves preparing the cell lines and generating the ERM vectors needed for mutagenesis screens. Stage 2 involves the actual screening process where mutant clones are generated and expanded. Stage 3 involves analysis of the clones to identify the targeted loci and secondary analysis and confirmation of the function of the genes identified (*see* **Fig. 1**).

3.1. Generation of ERM Vectors and 32D-tTA Cells

1. Clone the ERM cassette into the *Nhe*I site of the U3 region within the 3′ LTR of the retroviral pBabe-puro vector (**Fig. 2**) *(5)* (*see* **Note 3**). Several desired features are cloned into the pBabe backbone to obtain the ERM vector (*see* **Note 4**). Three sets of vectors corresponding to the three reading frames of

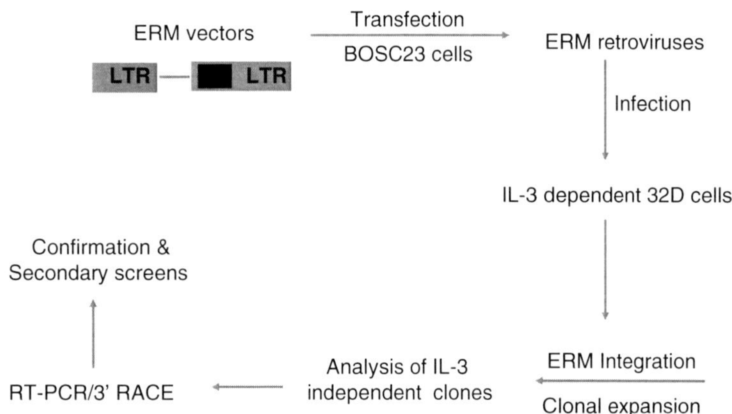

Fig. 1. ERM-mediated genetic screen in mammalian cells.

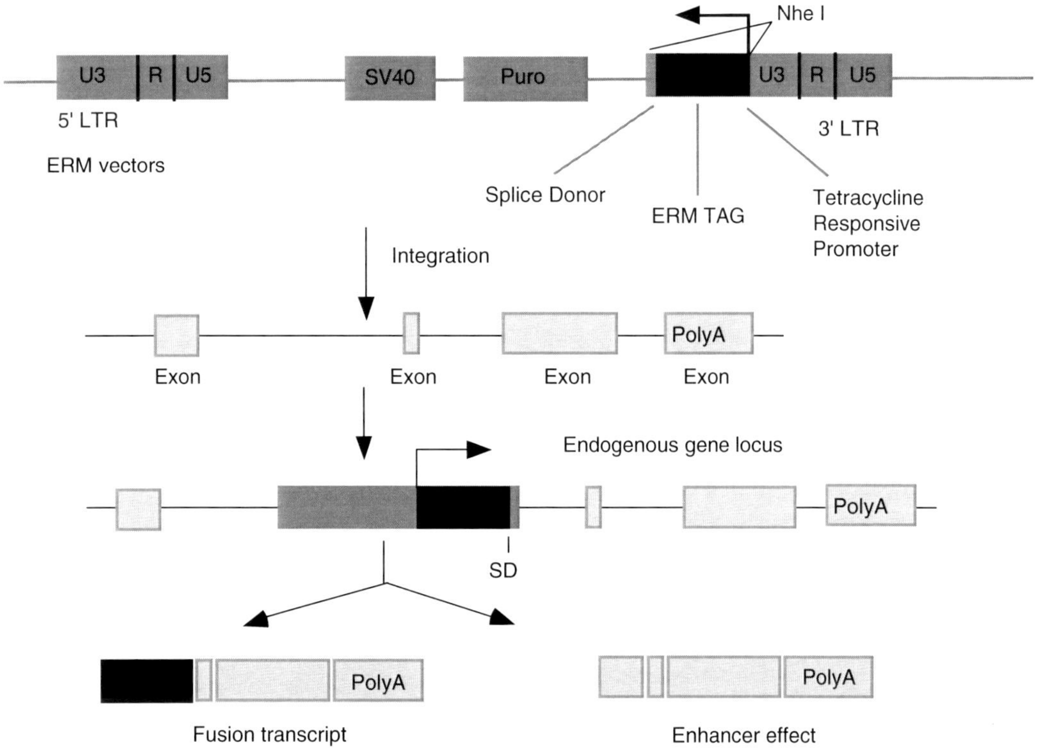

Fig. 2. Insertional tagging and activation of endogenous genes by ERM retroviruses. SD, splice donor.

the ERM Tag are needed to target endogenous genes in all possible reading frames (RF1, RF2, and RF3).

2. Generate the assay cell line by establishing 32D cells stably expressing tTA (for *tet-off* promoter) *(3)*. This was accomplished by introducing retroviruses (using the MSCV-neo vector) encoding tTA into 32D and then selecting the cells in neomycin for subsequent screens (*see* **Note 5**).

3.2. Carrying out ERM-Mediated Genetic Screen

1. Day 0: Maxiprep the DNA to be used for transfection and obtain the O.D. to ensure quality (*see* **Note 6**). Make sure the 32D-tTA cells are healthy and growing exponentially. Seed exponentially growing BOSC23 cells at $2.5 \times 10^6/60\,mm$ plate for transfection (*see* **Note 7**).

2. Day 1: Carry out the transfection. For 60 mm plates, ~10–20 µg of ERM vectors may be used, depending on the transfection reagents (e.g., Lipofectamine^{2000} from Invitrogen).

3. Day 2: Change media (4–5 mL) for BOSC23 cells. Feed 32D-tTA cells to ensure exponential growth and that enough cells will be available for screening (*see* **Note 8**).

4. Day 3: Harvest the viruses by collecting the supernatant from BOSC23 cells, and filtering it through a 0.45 µm syringe filter (*see* **Note 9**). The viruses are then used to spin infect 32D-tTA cells in the presence of polybrene (4 µg/mL) with a MOI (multiplicity of infection) of ≤1, in 24 or 6-well plates. Spin infection can be carried out at 2,500 rpm for 1–2 h at room temperature using a desktop centrifuge with microtiter plate holders (*see* **Note 10**).

5. Day 4: Change media and feed the infected cells.

6. Day 5–6: Replate the ERM-infected cells into 96-well (or 384-well) plates for genetic screens (*see* **Note 11**). The cells are first maintained in the presence of IL-3 to allow recovery and expansion.

7. Day 7–21: The cells are then cultured in the absence of IL-3 to screen for IL-3 independent survival genes. Once the clones have established, puromycin may be added to allow selection to occur (*see* **Note 12**). This process takes an average of 2 weeks. Further expansion may be carried out in puromycin-containing media.

3.3. Identification of ERM-Targeted Genes by RT-PCR and 3' RACE and Confirmation of Gene Function

1. Carry out tetracycline responsiveness tests to authenticate that the clones isolated are a direct result of ERM integration. Individual clones are cultured without IL-3 and in the presence or absence of tetracycline or doxycycline (1–5 µg/mL). If addition of the drug results in increased apoptosis (*tet-off*), the clone may be further analyzed (*see* **Note 13**).

2. Expand the tet-responsive clones in culture and collect enough cells ($\geq 10^5$–10^6) for RNA extraction, using a commercially available kit (e.g., with the RNeasy Mini Kit from QIAGEN). The amount of RNA isolated can be estimated based on the number of cells used.

3. Carry out reverse transcription using the isolated RNA, the random primer RT-1, and reverse transcriptase (e.g., Super-script[III] from Invitrogen) (*see* **Note 14**).

4. Perform PCR reactions using the cDNA, and the Myr1 and T7 primer set. The PCR products are then fractionated, gel purified, and directly sequenced with T7 primer (*see* **Note 15** and **16**).

5. Confirm the survival function of the genes identified. Once the identities of the gene loci targeted by ERM are known, further analyses of these genes become possible.

    a) If sequence analysis of the integration site reveals that a fusion transcript is generated, expression vectors encoding the gene fused to the ERM Tag should be generated (*see* **Note 17**). Ectopic expression of the fusion product in 32D cells should result in IL-3-independent cell survival. The cells are plated in the absence of IL-3 containing conditioned media or IL-3 supplement, and the survival rate is determined at different time points after IL-3 withdrawal compared with control cells. Apoptotic cells may be scored by Trypan Blue exclusion. Alternatively, the cells may be stained with PI and analyzed on a flow cytometer to determine the percentage of cells in various cell cycle stages.

    b) If the ERM integration occurs out of frame in a particular clone, it would indicate that the phenotypes observed may be a result of ERM enhancer effect (*see* **Note 16**). The integrated locus may either encode growth factors/growth factor receptors, or is in proximity to such genes. It is possible that gene expression of growth factors/receptors may be upregulated, resulting in an autocrine loop and preventing IL-3 withdrawal induced apoptosis. To test this, the supernatant from these ERM clones may be collected, filtered through a 0.2 μm filter, and used to culture the parental 32D cells in the absence of IL-3 containing conditioned media or IL-3 supplement. Growth and expansion of the cells would suggest an autocrine mechanism for the isolated clones.

6. Carry out secondary screens. For example, endogenous gene function may be inhibited by RNAi in 32D cells (either through stable expression of a shRNA vector or transient expression of siRNA oligos). These cells are then cultured in the absence of IL-3, and scored for survival at different time points. Apoptosis may be assayed by dye-exclusion or PI/Annexin V staining.

Inhibition of survival genes is expected to result in accelerated apoptosis compared with control cells.

4. Conclusions

The ERM technology is not limited only to screens that search for genes important for particular signaling pathways. Incorporation of different ERM tag sequences can enable screening a variety of phenotypes. For example, using fluorescence tags can allow the cells to be analyzed using high-throughput fluorescence microscopy to study the changes in protein trafficking and translocation. ERM may, therefore, help to fill the gap in this area that has long been difficult to study genome wide due to technical limitations.

5. Notes

1. Retroviral packaging cell lines need to be maintained and regularly passaged to avoid overcrowding. It is highly recommended that cells of low passage numbers be used, because extensively cultured cells tend to lose the ability to generate high-titer viruses. BOSC23 cells are used for generating ecotropic viruses and 293-ampho cells (Clontech) may be used for generating amphotropic viruses.

2. Conditioned WEHI-3B media is obtained by culturing WEHI-3B in RPMI supplemented with 10% heat-inactivated fetal calf serum to at least 1–2×10^6/mL. The supernatant is then collected (once there is close to 70–80% cell death), filtered through a 0.2 μm filter, aliquoted, and frozen at –80°C. Depending on the signaling pathways to be analyzed, many different cell lines can be used. The protocols are essentially the same for human vs. murine cells (making sure the correct packaging cell lines are used), or suspension vs. adherent cells.

3. The ERM vectors were designed to increase efficiency and ease of identification. Unlike conventional vectors, ERM does not rely on the retroviral LTR promoters that are relatively weak and sometimes suppressed.

4. The ERM cassette encompasses sequences for the ERM Tag, and a consensus splice donor sequence (AAGGTAAGT) under the control of a tetracycline regulatable promoter (*tet-off* in this case) *(6, 7)*. This design circumvents potential problems

arising from the cryptic splice acceptor sequence in the 3′ end of the 5′ LTR. In the ERM Tag, we included an epitope tag sequence (AU1) as well as the membrane targeting myristoylation signal sequences (**Fig. 2**). A fusion transcript may be generated as a result of ERM integration and subsequent splicing events (**Fig. 2**). The fusion transcript (5′ ERM Tag fused to the endogenous exons) may be ascertained by RT-PCR and 3′ RACE. The tetracycline-regulated promoter of the ERM vectors would also enable the separation of authentic integration events from spontaneous mutants.

5. To fully utilize the tetracycline responsive promoter, the tet operon binding protein (tTA) needs to be expressed in the cells for the *tet-off* system. For *tet-on*, rtTA will need to be used *(1, 2)*.

6. The quality of DNA is important for generating high-titer viruses. While miniprep DNA may be used, we generally find better results with maxiprep DNA.

7. The state of confluence of the packaging cells depends on the type of transfection reagents to be used and the culturing conditions of the cells. The optimal conditions need to be empirically determined and consistent with the manufacturer's suggestions.

8. In theory, 1×10^6 cells ERM-infected should be enough to cover the entire genome, because human or mouse genome contains about 30,000 genes *(8)*. Adherent cells may be plated at 30–40% confluence (in 60 mm plate) 12–16 h before infection. Suspension cells such as 32D cells may be plated in 6 or 24-well plates (0.5–1×10^6 cells/well) a few hours before infection.

9. It is not advisable to use 0.2 μm filters because it drastically decreases the viral titer. This clarification step is primarily to get rid of cells and debris. With one 60 mm plate of 4–5 mL media, the titer obtained ranges between 0.5 and 2×10^6 CPU/mL. The collected virus may be frozen and reused. Generally, we observe a two to fivefold drop in viral titer after freeze and thaw. Repeated freeze and thaw is not advisable. If necessary, freeze the collected viruses in aliquots.

10. Virus-containing media may be changed immediately after spin infection, or left overnight if compatible with the cell growing media (especially for adherent cells). Infection of suspension cells is best carried out in smaller multi-well plates, and the cells will need to be expanded soon after infection to minimize crowding.

11. Adding puromycin (1–2 μg/mL) to select infected cells is optional at this stage. In many cases, the selection should be held off until later because of potential stress on the cells.

12. The rate of expansion and recovery of the clones will vary, especially if the genes affected by ERM play a role in cell growth or proliferation. At this point, the analysis of the clones may be staggered so that fast growing clones can be analyzed first.

13. The presence of the *tet-off* promoter in the ERM vectors allows for this authentication step. The tet-responsive promoter may also become useful during the screening process, especially in cases where ERM-mediated gene activation needs to be regulated temporally.

14. For the RT-1 primer, the ACG triplet at the 3′ end helps to reduce the number of PCR products generated. Furthermore, the myristylation signal sequence does not contain the CGT sequence, primer binding therefore should occur outside of the ERM Tag region. In all the analysis following the screen, control cells in which no ERM has been introduced should be included to help rule out PCR artifact.

15. PCR conditions need to be optimized empirically. Usually, 1/3 of the RT products can be used for 30–35 cycles of PCR. It may be necessary to carry out two rounds of nested PCR (e.g., 5–10 cycles of first round PCR followed by another 30 cycles).

16. ERM retrovirus integration may result in the generation of transcripts that fuse the ERM Tag sequences with endogenous gene sequences (**Fig. 2**). These transcripts may be uncovered by the procedures delineated above. Alternatively, the presence of exogenous ERM enhancer elements may result in upregulated transcription at gene loci in proximity to ERM integration sites, through endogenous promoters. Authentication of such integration events may be possible by isolating genomic DNA for inverted PCR *(9)*.

17. Fusion transcripts generated from ERM integration most likely involve the fusion of ERM Tag to the full-length or truncated gene product. As a result, this version of the gene needs to be ectopically expressed (together with the ERM Tag) to determine whether IL-3-independent survival can be recapitulated.

Acknowledgments

This work is supported by the Welch Foundation (to Z.S.) and NIH grants CA84208 and GM69572. D.L. is supported in part by the American Heart Association. Z.S. is a Leukemia and Lymphoma Society Scholar.

References

1. Liu, D., Yang, X., and Songyang, Z. (2000). Identification of CISK, a new member of the SGK kinase family that promotes IL-3-dependent survival. *Curr Biol* **10**, 1233–6.

2. Ding, Z., Liang, J., Lu, Y., Yu, Q., Songyang, Z., Lin, S. Y., and Mills, G. B. (2006). A retrovirus-based protein complementation assay screen reveals functional AKT1-binding partners. *Proc Natl Acad Sci USA* **103**, 15014–9.

3. Liu, D., Yang, X., Yang, D., and Songyang, Z. (2000). Genetic screens in mammalian cells by enhanced retroviral mutagens. *Oncogene* **19**, 5964–72.

4. Pear, W. S., Nolan, G. P., Scott, M. L., and Baltimore, D. (1993). Production of high-titer helper-free retroviruses by transient transfection. *Proc Natl Acad Sci USA* **90**, 8392–96.

5. Morgenstern, J. P., and Land, H. (1990). Advanced mammalian gene transfer: high titre retroviral vectors with multiple drug selection markers and a complementary helper-free packaging cell line. *Nucleic Acids Res* **18**, 3587–96.

6. Gossen, M., Bonin, A. L., and Bujard, H. (1993). Control of gene activity in higher eukaryotic cells by prokaryotic regulatory elements. *Trends Biochem Sci* **18**, 471–5.

7. Paulus, W., Baur, I., Boyce, F. M., Breakefield, X. O., and Reeves, S. A. (1996). Self-contained, tetracycline-regulated retroviral vector system for gene delivery to mammalian cells. *J Virol* **70**, 62–7.

8. Brent, M. R. (2005). Genome annotation past, present, and future: how to define an ORF at each locus. *Genome Res* **15**, 1777–86.

9. Suzuki, T., Shen, H., Akagi, K., Morse, H. C., Malley, J. D., Naiman, D. Q., Jenkins, N. A., and Copeland, N. G. (2002). New genes involved in cancer identified by retroviral tagging. *Nat Genet* **32**, 166–74.

Chapter 21

Analysis of Apoptosis in Isolated Primary Cardiac Myocytes

Adel Mandl, Ambrus Toth, and Peter Erhardt

Summary

Acute myocardial infarction represents the leading cause of morbidity and mortality in the western societies. Importantly, both apoptosis and necrosis of cardiomyocytes have been implicated in the pathomechanism of myocardial infarction. The simplest way to analyze apoptosis in cardiac cells is the application of isolated neonatal primary cardiac myocytes, in which ischemia/reperfusion can be mimicked *in vitro* by exposing them to hypoxia and serum starvation, followed by restored oxygen and serum conditions, referred to as hypoxia/reoxygenation. In this chapter, we describe protocols routinely applied in our lab for investigating cardiomyocyte apoptosis. In summary, a better understanding of the apoptotic pathways and their regulation in the heart will potentially yield novel therapeutic targets for cardiac infarction.

Key words: Apoptosis, Primary cardiomyocyte, Western blot, Fractionation, Immunocytochemistry

---

1. Introduction

Thrombotic occlusion of a coronary artery triggers acute myocardial infarction, which is the leading cause of morbidity and mortality in the western societies. In the region supplied by the occluded coronary artery, cardiomyocytes undergo massive cell death, both necrosis and apoptosis, contributing to functional decline of the myocardium *(1–5)*.

Many patients with moderate infarct size recover from acute cardiac infarction, whereas those with large infarcts rather progress to heart failure due to remodeling *(6–9)*. Heart failure is the ultimate outcome of cardiac infarction representing a condition

Peter Erhardt and Ambrus Toth (eds.), *Apoptosis*, Methods in Molecular Biology, vol. 559
DOI 10.1007/978-1-60327-017-5_21
© Humana Press, a part of Springer Science+Business Media, LLC 2004, 2009

when the heart is unable to provide sufficient blood flow to supply the peripheral tissues with nutrients and oxygen *(6–9)*. Heart failure is preceded by ventricular remodeling that among others includes apoptosis, fibrosis, and inflammation.

Most adult cardiomyocytes are terminally differentiated cells and withdraw from cell cycle during ontogeny, losing the ability to divide *(10)*. Therefore, in general, two different therapeutic approaches can be considered to cure cardiac infarction and prevent heart failure. These include prevention of cardiac cell death, ideally both apoptosis and necrosis (through gene or drug therapy) or replacement of the lost and dead myocytes restoring cardiac function (through cell therapy). Unlike necrosis, which is thought to be an essentially irreversible process, the highly regulated, step-by-step nature of apoptosis suggests that it may provide targets for therapeutic intervention *(1–5)*.

Ischemia/reperfusion, which represents occlusion of a coronary artery followed by restoration of the blood flow, is widely used as an animal model to investigate cardiac infarction. In these in vivo studies, it has been demonstrated that cardiac myocytes undergo apoptosis in response to both ischemia and ischemia followed by reperfusion *(1, 3, 9)*. The simplest way to analyze apoptosis in cardiac cells, however, is the application of isolated neonatal primary cardiac myocytes. Ischemia/reperfusion can be mimicked in vitro by exposing them to hypoxia and serum starvation, followed by restored oxygen and serum conditions, referred to as hypoxia/reoxygenation *(11–13)*.

In the current chapter, first we describe the isolation of rat and mouse neonatal cardiomyocytes. These cells, when cultured for 5 days or less, either undergo their last division or do not divide any more *(10)*. Although there are differences between adult and neonatal cells, they are very similar to each other in many aspects, and therefore neonatal myocytes provide the most widely used cardiac cell system to study fundamental processes of cell life such as apoptosis, hypertrophy, or contractility. To use high quality cardiomyocytes in in-vitro studies is particularly important because accumulation of dead cells increase the background level of apoptosis and may mask the otherwise significant induction. In our hand, inclusion of a Percoll gradient was critical for obtaining sufficient purity by avoiding fibroblast contamination.

In the second part of this chapter, we describe a variety of different measurements of apoptosis with emphasis on specific steps used in assays performed with cardiomyocytes. These focus on measurements reflecting caspase activity (TUNEL assay, caspase 3 cleavage) and mitochondrial involvement (cytochrome c release). Since ischemia leads to ATP depletion, under ischemic conditions necrosis is the predominant form of cell death, whereas during reperfusion apoptosis is more significant *(14)*. Therefore, we also included an assay of necrosis based on detection of membrane damage, which is a characteristic feature of early necrosis but not apoptosis. In addition, several other apoptosis assays described elsewhere in this book can easily be adapted to the cardiomyocyte system to further expand the battery of methods detecting apoptosis in cardiac cells, such as Annexin staining and induction of endoplasmic reticulum (ER) stress.

In summary, a better understanding of the apoptotic pathways and their regulation in the heart will potentially yield novel therapeutic targets for cardiac infarction and other cardiovascular diseases.

2. Materials

2.1. Isolation of Rat Neonatal Cardiomyocytes

1. 15 and 50 mL conical tubes, 10 cm Petri dish, 25 mL flasks.

2. 24-well plate and/or 35-mm dish and/or 60-mm dish and/or 100-mm dish.

3. 0.22 μm filter.

4. 70% ethanol.

5. Cotton balls.

6. Curved scissors, curved forceps, dissecting scissors.

7. Mini stir bar. Magnetic stirrer in a 37°C incubator.

8. Calf serum.

9. Fibronectin from bovine plasma (Sigma; 1 mg/mL solution): 1:100 dilution in 1× PBS.

10. Centrifuge for 15 and 50 mL tubes.

11. 10× isolation buffer: 68 g NaCl, 47.6 g HEPES, 1.2 g NaH$_2$PO$_4$, 10 g glucose, 4 g KCl, 1 g MgSO$_4$ up to 1,000 mL distilled water, pH to 7.35, filter through 0.22 μm, store at 4°C for up to 1 year.

12. 1× isolation buffer: 100 mL of 10× isolation buffer and 900 mL distilled water, filter through 0.22 μm, store at 4°C. Prepare freshly every 2 weeks.

13. 1× red isolation buffer: dissolve 10 mg phenol red (Sigma) in 500 mL of 1× isolation buffer, filter through 0.22 μm and keep at 4°C. Prepare freshly every 2 weeks.

14. Working enzyme solution: for 5–8 hearts prepare 50 mL buffer. Dissolve 25 mg collagenase (Worthington) and 40 mg pancreatin (Sigma) in 1× isolation buffer separately, make up to 50 mL, filter through 0.22 μm into one 50 mL tube. Leave at room temperature, always prepare freshly.

15. Percoll discontinuous gradient: prepare it during the last digestion cycle. The densities to be used were published earlier by Kenneth Chien's laboratory *(15)*. Use five 15-mL tubes, one for the Percoll stock solution, one for top and another one for bottom gradient solutions, and two for the samples (*see* **Note 1**).

a) Percoll stock (final density: 1.110 g/mL)

Percoll (1.130 g/mL) (Amersham Biosciences)	9.9 mL
10× Isolation buffer	1.1 mL

b) Top solution (density: 1.059 g/mL)

Percoll stock	4.5 mL
1× Isolation buffer	4.9 mL

c) Bottom solution (density: 1.082 g/mL)

Percoll stock	4.7 mL
1× Red isolation buffer	2.5 mL

16. Maintenance medium for rat cardiac myocytes: 331.6 mL DMEM (4 volumes), 82.9 mL Medium 199 (1 volume), 50 mL horse serum (10%), 25 mL fetal bovine serum (5%), 5 mL antibiotics – Pen/Strep (10,000 U/mL/10,000 µg/mL), 5 mL glutamine (stock 200 mM), 0.5 mL BrdU (stock 100 mM) (to prevent cardiac fibroblast proliferation).

17. Nutridoma medium *(11)*: 149.4 mL DMEM (4 volumes), 37.4 mL Medium 199 (1 volume), 10 mL horse serum (5%), 1 mL Nutridoma reagent (0.5%) (Roche), 2 mL glutamine (200 mM), 0.2 mL BrdU (stock 100 mM) (to prevent cardiac fibroblast proliferation).

18. Hemocytometer grid to count cells.

19. Normoxic incubator (37°C).

2.2. Isolation of Mouse Neonatal Cardiomyocytes (in Addition to Materials Listed for Rat Cardiomyocyte Isolation)

1. Eppendorf tube, 35 mm dish, 60 mm dish, 24-well plate.

2. Materials are the same as for rat cardiac myocyte isolation. If genotypes of the mice are not known, no Percoll reagent is used (*see* **Subheading 3.2**).

2.3. Apoptosis Induction by Hypoxia/ Reoxygenation

1. Serum and glucose free DMEM medium (Invitrogen) (for hypoxia).
2. Hypoxic incubator (37°C).
3. Maintenance medium (*see* **Subheading 2.1, item 16**) (for reoxygenation).
4. Normoxic incubator (37°C).

2.4. Western Blot Analysis of Cleaved Caspase 3

1. Cell scraper, Eppendorf tubes, ice bucket, ice cold 1× PBS, Eppendorf centrifuge at 4°C, vortex mixer.
2. Lysis buffer stock: 50 mM Tris-HCl, pH 7.4, 100 mM NaCl, 0.1 mM EGTA, 0.1 mM EDTA, 1% Triton X-100. Add freshly: 1 mM sodium orthovanadate, 20 mM sodium fluoride, 30 mM β-glycerophosphate, 1 mM phenylmethyl-sulfonyl fluoride, and protease inhibitor cocktail complete (Roche).
3. BCA-protein concentration reagent (Pierce), 96-well plate, microplate reader or other methods for determining protein concentration.
4. 6× sample buffer (*see* **Note 2**), 1× sample buffer (1:6 dilution of 6× sample buffer with lysis buffer (*see* **Subheading 2.4, item 2**)).
5. Bio-Rad precast gel, running buffer (TGS buffer, Bio-Rad), dual color standard.
6. Nitrocellulose membrane, blotting pads, transfer buffer (TG buffer, Bio-Rad), methanol, stir bar, magnetic stirrer in cold room.
7. Ponceau stain (Sigma).
8. Rocking platform.
9. Blocking buffer: 5% non-fat dry milk and 1 mM sodium ortho-vanadate in 1× TBS. Always prepare freshly.
10. TBS + 0.2% Tween-20.
11. Primary antibody (cleaved caspase 3, 1:500, (Cell Signaling Technology), anti-actin, 1:5,000 (Calbiochem) as loading control.
12. Alexa Fluor 680-labeled secondary antibodies, 1:10,000 dilution (Molecular Probes).
13. IRDye 800-labeled secondary antibodies, 1:10,000 dilution (Rockland Immunochemicals).
14. Odyssey Infrared Imaging System (Li-Cor).

2.5. Cell Fractionation (Cytochrome c Release)

1. Mitochondria isolation kit for cultured cells (Pierce): Mitochondria isolation reagent A, Mitochondria isolation reagent C.

2. Cell scraper, Dounce homogenizer, 2 mL Eppendorf tubes.

3. EDTA-free Protease inhibitor cocktail (Pierce).

4. High speed bench top microcentrifuge (4°C).

5. Vortex mixer.

6. 1× PBS, CHAPS (Pierce), TBS.

7. BCA protein assay.

8. Anti-LDH antibody, 1:500 (Santa Cruz) and anti-cytochrome oxidase antibody, 1:500 (Molecular Probes) as loading controls for mitochondrial leakage.

9. Anti-cytochrome-c, 1:500 (BD Biosciences), and anti-Smac, 1:500 (Santa Cruz) primary antibodies, etc.

10. Alexa Fluor 680-labeled secondary antibodies (Molecular Probes).

11. IRDye 800-labeled secondary antibodies (Rockland Immunochemicals).

2.6. Immunocytochemistry (Cytochrome c Release)

1. 24-well tissue culture dishes, 12 mm circle glass coverslips stored in 95% ethanol.

2. Flame, broad-tipped forceps.

3. Fibronectin from bovine plasma (Sigma 1 mg/mL solution): 1:100 dilution in 1× PBS.

4. Paraformaldehyde (Electron Microscopy Sciences).

5. PBS, pH 7.4.

6. 0.3% (v/v) Triton X-100 in PBS.

7. Goat serum (Invitrogen).

8. Anti-cytochrome-*c*, 1:500 (BD Biosciences), we use different cytochrome-*c* antibodies for Western and IC.

9. Fluorophore-conjugated secondary antibodies.

10. Microscope slides.

11. 20G needle (*see* **Note 3**).

12. 300 ng/mL DAPI (Invitrogen) or 2 µg/mL Hoechst 33258 (Invitrogen) nucleic acid stains.

13. Fluorsave reagent (Calbiochem).

2.7. TUNEL Staining

1. In Situ Cell Death Detection Kit (Roche).

2. Glass coverslips, broad-tipped forceps.

3. Fibronectin from Bovine Plasma (Sigma 1 mg/mL solution): 1:100 dilution in 1× PBS.

4. Paraformaldehyde (Electron Microscopy Sciences).

5. PBS, pH 7.4.

6. 0.3% (v/v) Triton X-100 in PBS.

7. Goat serum (Invitrogen).

8. Micrococcal nuclease or DNase I, grade I (Roche).

9. Microscope slides.

10. 300 ng/mL DAPI (Invitrogen) or 2 μg/mL Hoechst 33258 (Invitrogen) nucleic acid stains.

11. Fluorsave reagent (Calbiochem).

2.8. Vital Staining for Irreversible Membrane Damage

1. Ethidium homodimer-2 (EH-2) and SYTO-10 staining kit (LIVE/DEAD® Reduced Biohazard Viability/Cytotoxicity Kit, Invitrogen).

2. HEPES-buffered saline solution (HBSS): 135 mM NaCl, 5 mM KCl, 1 mM $MgSO_4$, 1.8 mM $CaCl_2$, 10 mM HEPES, pH 7.4.

3. 4% Glutaraldehyde in HBSS, freshly prepared from recently acquired, reagent-grade glutaraldehyde (50%).

3. Methods

3.1. Isolation of Rat Neonatal Cardiomyocytes

1. Sterilize the pup's (1–2 days old) chest with a cotton ball soaked in 70% ethanol, decapitate the pup with curved scissors, open the chest and remove the heart with curved forceps, put it into ice cold 1× isolation buffer. Repeat this step for each heart.

2. Wash the hearts with 30 mL cold 1× isolation buffer (in the tissue culture hood).

3. Transfer the hearts into a 100-mm tissue culture dish, keep them in cold 1× isolation buffer, trim atria and connective tissue with dissecting scissors on ice. Cut the hearts into 4–6 pieces and put into a 25-mL flask with a small magnetic stirrer rod. Remove all remaining isolation buffer from the hearts.

4. Mince 4–8 hearts (not more) into one flask (usually 2 flasks/ rat mother).

5. Add 6 mL working enzyme solution into each flask and incubate 20 min in a 37°C incubator on a magnetic stirrer (start stirring at low frequency, around rate 3, slightly increase at each cycle).

6. In the meantime, thaw calf serum (used for inactivating the enzymes). When thawed, aliquot 2 mL serum into each

15-mL tubes (depending on how many digestion cycles are needed).

7. After 20 min stirring, transfer the supernatants of the heart chunks into the tubes containing serum. Store these in 37°C incubator (5% CO_2) during subsequent digestion cycles and loosen the lid of the tubes.

8. Repeat steps 5–7 until all the heart chunks are digested (usually 6–8 cycles).

9. During the digestion cycles, coat culture dishes and/or cover slips with fibronectin solution (1:100 dilution in 1× PBS from a stock) and incubate for 2–3 h at 37°C or overnight at 4°C (*see* **Note 4**).

10. After the last digestion cycle, pool all samples into 50-mL tubes and centrifuge at $1,000 \times g$ for 5 min.

11. During the last digestion cycle prepare the Percoll gradient into 2 × 15-mL tubes. First put 4 mL top solution into both tubes, and then very slowly and carefully add 3 mL bottom solution UNDER the top solution. A fine interface should be seen between the two solutions (**Fig. 1**).

12. Resuspend your pelleted cells in 4 mL of 1× isolation buffer and pipette 2 mL cell suspension onto the top of each Percoll gradient very slowly and carefully! Now three zones should be clearly separated from each other: bottom solution, top solution, and cell suspension (**Fig. 1**).

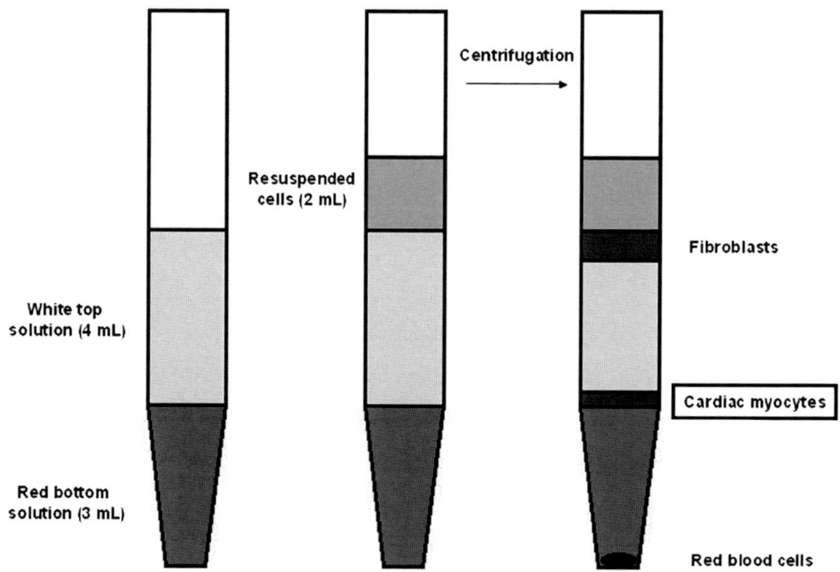

Fig. 1. Position of different cell types before and after the discontinuous Percoll gradient centrifugation.

13. Centrifuge the 2 tubes at $4,400 \times g$ for 32 min, adjusting the acceleration to 1 (slowest) and the brake to 0. Therefore, the acceleration and deceleration of the samples will be at the minimum, preventing the gradients from mixing randomly.

14. After centrifugation, two interfaces should be seen, both filled with cells. Cardiac fibroblasts are located in the upper interface. Carefully discard the fibroblasts by pipetting them off. Cells in the lower interface are cardiac myocytes. Collect cardiac myocytes from both tubes by a pipette into 30 mL of 1× isolation buffer (in a 50-mL tube). Red blood cells can be seen at the very bottom of the tube and can be discarded (*see* **Note 5**) (**Fig. 1**).

15. Centrifuge the isolation buffer-cardiac myocyte suspension at $330 \times g$ for 10 min.

16. Discard supernatant and resuspend cells in 30 mL of 1× isolation buffer to wash out the Percoll reagent.

17. Centrifuge at $330 \times g$ for 10 min again.

18. Discard supernatant and resuspend cells in 10 mL maintenance medium.

19. Count cells in the hemocytometer grid.

20. Plate cells into dishes or glass cover slips coated with fibronectin (*see* **Note 6**); use maintenance medium to dilute cell suspension:

    a. Cover slips in 24-well plate: $0.1–0.5 \times 10^5$ cells/well

    b. 35-mm dish: $0.5–2 \times 10^6$ cells

    c. 60-mm dish: $2–4 \times 10^6$ cells

    d. 100-mm dish $1–2 \times 10^7$ cells

21. Change medium the following day.

3.2. Isolation of Mouse Neonatal Cardiomyocytes

1. This protocol is based on the assumption that the genotypes of the mouse pups are not known at the time of cardiac myocyte isolation (*see* **Note 7**).

2. Use 1- to 2-day-old pups and sterilize their chest with a cotton ball soaked in 70% ethanol, decapitate the pup with curved scissors, open the chest, and remove the heart with curved forceps, put it into labeled 35-mm tissue culture dish with 5 mL ice cold 1× isolation buffer. Perform each step on ice.

3. Remove the tip of tail for genotyping and place it into labeled Eppendorf tube. Ensure that the tail corresponds to the heart for future genotyping and store at –20°C.

4. Repeat **steps 2** and **3** for each heart and tail.

5. Wash the hearts with 5 mL cold 1× isolation buffer (in the tissue culture hood) on ice.

6. Trim atria and connective tissue with dissecting scissors on ice and thoroughly mince them into small pieces using small dissecting scissors. Remove all remaining isolation buffer from the hearts.

7. Add 2 mL working enzyme solution to each heart.

8. Transfer each suspension into a 25-mL flask and add a small stirring rod.

9. Incubate for 10 min in a 37°C incubator on a magnetic stirrer (start stirring at low frequency).

10. In the meantime, thaw calf serum (used to inactivate the enzymes). When thawed, aliquot 2 mL serum into 15-mL tubes (depending on how many digestion cycles you need).

11. After 10 min, transfer the supernatants of the heart chunks into the tubes containing serum. Store these in 37°C incubator (5% CO_2) during subsequent digestion cycles and loosen the lid of the tubes.

12. Repeat digestion steps until all heart tissue is completely dissociated (~3–4 digestion cycles).

13. During the digestion cycles, coat 24-well plates with fibronectin solution (1:100 dilution in 1× PBS from a stock) and incubate for 2–3 h at 37°C or overnight at 4°C (*see* **Note 4**).

14. Pool samples for each individual heart and centrifuge at $330 \times g$ for 4 min.

15. Rinse once with 5 mL of 1× isolation buffer.

16. Centrifuge again and discard the supernatant.

17. Resuspend the cells in 2 mL maintenance medium.

18. Preplate cells into 60-mm dishes for 1.5 h in 37°C incubator (5% CO_2).

19. After 1.5 h, remove nonadherent cells (mostly cardiomyocytes) and plate into 2 wells of a 24-well plate (if required count cells, plate $150–300 \times 10^3$ cells per well). The adherent cells remaining in the preplating dish are mainly cardiac fibroblasts, because fibroblasts tend to attach to the surface faster than cardiac myocytes.

20. The following day, examine cells and change media as required (supplement new maintenance media with BrdU to inhibit fibroblast growth).

21. Process the tails as soon as possible to identify the genotypes of the mice.

3.3. Apoptosis Induction by Hypoxia/Reoxygenation

1. To induce hypoxia, change medium on cardiac myocytes to serum and glucose-free DMEM two days after isolation.

2. Incubate the cells in a humidified environment at 37°C in a three-gas incubator maintained at 5% CO_2 and 1% O_2

(oxygen expelled by nitrogen) for various times (hypoxic phase) (*see* **Note 8**). Ensure that your experiment is designed to include control (normoxic), hypoxic and hypoxic/reoxygenated samples as well. Normoxic samples are continued to be maintained in serum and glucose-containing maintenance medium in normoxic incubator.

3. Following various duration of hypoxia, collect hypoxic samples depending on the experimental goals (mRNA expression, immunoblot or immunocytochemistry analysis, etc.).

4. After hypoxia, generally half of the samples are reoxygenated for 16–24 h in serum and glucose-containing maintenance medium in normoxic incubator.

5. After reoxygenation, the samples are collected depending on the experimental goals (mRNA expression, immunoblot or immunocytochemistry analysis, etc.). Normoxic control samples are also collected at the same time as the reoxygenated samples.

3.4. Western Blot Analysis of Cleaved Caspase 3

1. Preparation of cell lysates (*see* **Note 9**).

    a) Collect cells by scraping into a 15-mL tube (leave the media on the cells to be able to collect detached apoptotic cells) (*see* **Note 10**).

    b) Transfer into an Eppendorf tube, spin at $2{,}000 \times g$ for 5 min.

    c) Discard the supernatant and wash the cells with 1× PBS.

    d) Spin at $2{,}000 \times g$ for 5 min, discard the supernatant.

    e) Lyse the pellet in 60 µL lysis buffer on ice for 20 min by vortexing it every 5 min.

    f) Spin at $14{,}000 \times g$ in an Eppendorf microcentrifuge for 10 min at 4°C.

    g) Transfer the supernatant into a new tube and discard the pellet.

2. Protein concentration measurement

    a) Determine the protein concentration with BCA protein assay using 5 µL sample each (always use duplicates).

    b) Add 10 µL of 6× sample buffer to all samples.

    c) Normalize the samples to each other using 1× sample buffer.

    d) Boil them for 5 min.

    e) Cool at room temperature for 5 min.

    f) Spin down condensation in each sample prior to loading the gel. Store remaining samples at –80°C.

3. SDS-PAGE

    a) Assemble 15% Tris-HCl precast gel in gel ring (Ready Gels from Bio-Rad) in running buffer.

    b) Load equal amount of samples (~10 µg) into the wells.

    c) Use markers.

    d) Run at 80–120 V (constant voltage).

4. Membrane transfer

    a) Cut a piece of nitrocellulose membrane and wet in transfer buffer.

    b) Prewet the sponges and filter papers (slightly bigger than the gel) in 1× transfer buffer.

    c) Assemble "sandwich": sponge – filter paper – gel – membrane – filter paper – sponge. Make sure that the gel is closer to the anode in the transfer apparatus. Place a stir bar to the bottom of the apparatus.

    d) Transfer for 1.5 h at 70 V at 4°C (e.g., in cold room) on a magnetic stirrer.

    e) When finished, optionally stain with Ponceau to check the quality of protein transfer onto the membrane.

    f) Remove Ponceau by washing the membrane with distilled water (removes excess staining).

    g) Block membrane in blocking buffer for 1 h at room temperature on a rocking platform.

5. Antibodies and detection

    a) Incubate with primary antibody diluted in TBST overnight at 4°C with constant agitation (on a rocking platform).

    b) Wash three times for 10 min with TBST.

    c) Incubate with fluorescently labeled secondary antibody diluted in TBST for 1 h at room temperature in a dark box on a rocking platform.

    d) Wash three times for 10 min with TBST.

6. Detect and quantify bands using Odyssey Infrared Imaging System (Li-Cor). Odyssey is equipped with two infrared channels for direct fluorescence detection on membranes. With two detection channels, two separate targets can be probed in the same experiment.

3.5. Cell Fractionation (for Cytochrome c Release Analysis)

1. Induce apoptosis in cardiac myocytes by hypoxia/reoxygenation or any other method. Concurrently, incubate a control, noninduced culture as well (*see* **Note 11**).

2. Collect cells by scraping (leave the media on the cells to be able to collect detached apoptotic cells) (*see* **Note 10**).

3. Transfer them into a 15 mL tube, spin at $850 \times g$ for 2 min.

4. Discard the supernatant and wash the cells with ice cold 1× PBS and, subsequently, transfer them into a 2 mL Eppendorf tube.

5. Spin at $850 \times g$ for 2 min, discard the supernatant.

6. From this point, all steps should be carried out at 4°C, including centrifugation. Process one sample at a time.

7. Pipette the required amount of Reagent A and C into two tubes and freshly add protease inhibitors to both. Some protease inhibitors lose their activity within 24 h of dilution. Therefore, it is recommended to add inhibitors to the reagents immediately prior to starting the experiment.

8. Add 800 μL of Mitochondria Isolation Reagent A. Vortex at medium speed for 5 s and incubate tube on ice for exactly 2 min.

9. Transfer cell suspension to a prechilled Dounce homogenizer.

10. Homogenize cells on ice. Apply enough strokes (about 20–30) to effectively lyse the cells (*see* **Note 12**).

11. Return lyzed cells to a prechilled 2-mL Eppendorf tube and add 800 μL of Mitochondria Isolation Reagent C.

12. Rinse Dounce homogenizer with 200 μL of Mitochondria Isolation Reagent A and add to the tube containing the sample.

13. Invert tube several times to mix (do not vortex).

14. Centrifuge tube at $700 \times g$ for 10 min at 4°C. The pellet contains cellular debris, nuclei and intact cells, while the supernatant contains cytosol and mitochondria.

15. Transfer the supernatant to a prechilled 2-mL Eppendorf tube and centrifuge at $3,000 \times g$ for 10 min at 4°C.

16. Transfer the supernatant to a new, prechilled 2-mL Eppendorf tube and centrifuge at $12,000 \times g$ for 15 min at 4°C.

17. The supernatant is the cytosolic fraction and the pellet contains the isolated mitochondria. Transfer the supernatant to a new, prechilled Eppendorf tube, and centrifuge again at $12,000 \times g$ to remove any residual mitochondria, transfer the supernatant into a new tube, then store at 4°C (this will be the cytosolic sample for the ensuing immunoblot analysis, *see* **Subheading 3.5, step 21.**).

18. Add 500 μL Mitochondria Isolation Reagent C to the pelleted mitochondria, and centrifuge at $12,000 \times g$ for 5 min. Discard the supernatant.

19. To lyse mitochondria and release mitochondrial content, add 2% CHAPS in TBS to the mitochondrial pellet and vortex for 1 min.

20. Centrifuge at $12,000 \times g$ for 5 min. The supernatant contains the soluble mitochondrial proteins (this will be the mitochondrial sample for the ensuing immunoblot analysis, *see* **Subheading 3.5, step 21**).

21. Measure protein concentration of both cytosolic and mitochondrial fraction, normalize and run samples on SDS-PAGE (Immunoblot analysis, *see* **Subheading 3.4, steps 2–6**).

22. Aliquot the remaining samples and freeze at –80°C; avoid freeze-thaw cycles.

23. For testing the success of fractionation, probe your blot with LDH and cytochrome oxidase primary antibodies. LDH reflects the cytosolic, while cytochrome oxidase the mitochondrial fraction (*see* **Note 13**).

24. Use cytochrome *c* (1:500) and Smac (1:500) primary antibodies to see the release of these proteins from the mitochondria.

3.6. Immunocyto-chemistry of Apoptotic Proteins

1. Prepare coverslips:
   a) Soak coverslips in 1 M NaOH for 2 h (use a large surface so that all coverslips are treated).
   b) Rinse coverslips well with sterile H_2O (three times for 5 min each until NaOH is washed out).
   c) Rinse with 70% ethanol.
   d) Store them in 95% ethanol.

2. Plate cells on coverslips:
   a) Using a forceps, remove excess ethanol by touching the coverslip on a tissue and briefly dry each using the flame.
   b) Place the sterile coverslip into the well.
   c) Coat coverslips with fibronectin solution (1:100 dilution in 1× PBS from a stock) and incubate for 2–3 h at 37°C or overnight at 4°C (*see* **Note 4**).
   d) Gently press the coverslip to the bottom of the well using sterile forceps or pipette.
   e) Plate the cell suspension onto the coverslip.
   f) Induce apoptosis in cardiac myocytes by hypoxia/reoxygenation or any other method. Concurrently, incubate a control, noninduced culture as well.

24-well plate:	$0.1–0.5 \times 10^5$ cells/well

3. Fixation:
   a) Remove the medium, avoid washing to save loosely attached apoptotic cells.

b) Fix the samples in 4% paraformaldehyde in PBS for 15 min at room temperature in the hood.

c) Rinse the samples three times with 1× PBS for 5 min each.

4. Permeabilization and blocking: incubate the samples for 1 h with 5% normal goat serum in PBS containing 0.3% Triton X-100.

5. Primary and secondary antibody incubation:

a) While blocking, dilute primary antibody in PBS-0.3% Triton X-100.

b) Remove blocking solution and apply diluted primary antibody (*see* **Note 14**).

c) Incubate cells overnight at 4°C or 1 h at room temperature.

d) Place the coverslips back into the 24-well plate and wash the cells three times in PBS, 5 min each wash.

e) Incubate cells with fluorochrome-conjugated secondary antibody at their appropriate dilutions in PBS-0.3% Triton X-100 for 1 h at room temperature in dark.

f) Decant the secondary antibody solution and wash three times with PBS for 5 min each in dark.

6. Counterstaining:

a) Incubate cells in Hoechst 33258 or DAPI (DNA stain) for 10 min.

b) Wash three times with PBS for 5 min.

7. Mounting:

a) Mount coverslip with a drop of Fluorsave Reagent (*see* **Note 15**).

b) Observe using fluorescent microscope.

c) Store samples in dark at 4°C.

3.7. TUNEL Staining

1. Always plate enough cells for negative and positive controls.

2. Follow **steps 1–4** described in **Subheading 3.6** (*see* **Note 16**).

3. TUNEL staining:

a) Incubate your positive control with *DNase*I or micrococcal nuclease for 10 min at room temperature to induce DNA strand brakes.

b) Dilute Enzyme Solution in Label Solution 1:5, mix them well, always prepare freshly. Use 20 μL solution for each coverslips.

c) Incubate your negative control with Label Solution only (no TdT enzyme).

d) Apply working strength terminal deoxynucleotidyl transferase (TdT) on the slides and incubate in a humid, dark chamber for 1 h at 37°C (*see* **Note 14** for technical details).

e) Rinse slides three times with 1× PBS.

4. Counterstaining:

a) Incubate cells in Hoechst 33258 or DAPI (DNA stain) for 10 min.

b) Wash three times with PBS for 5 min.

5. Mounting:

a) Mount coverslip with a drop of Fluorsave Reagent (*see* **Note 15**).

b) Observe using fluorescent microscope.

c) Store samples in dark at 4°C.

3.8. Vital Staining for Irreversible Membrane Damage

1. Grow adherent cells on sterile glass coverslips by following **steps 1–2** described in **Subheading 3.6**.

2. Remove the culture medium covering the cells and replace it with an equal volume of HBSS.

3. Prepare a diluted mixture of Component A (SYTO-10, green) and Component B (ethidium homodimer-2, red) dyes by pipetting 2 µL of each into a common 1 mL volume of HBSS (1:500 dilution of each). Mix thoroughly by vortexing or by pipetting up and down several times.

4. Remove the HBSS covering the cells and replace it with the diluted dye mixture; 200–500 µL should be sufficient. Incubate in dark for 15 min at room temperature.

5. Remove the dye solution, and wash the cells with fresh HBSS.

6. Add 4% glutaraldehyde in HBSS (freshly prepared), and incubate for at least 15 min before observation.

7. For improved and more persistent differential staining of the live and dead cell populations, leave the cells in 4% glutaraldehyde fixative for 1 h, then remove the fixative and cover the cells with HBSS. Observe the slides immediately after staining using fluorescent microscope (*see* **Note 17**). SYTO-10, a green fluorescent nucleic acid stain, is a highly membrane-permeant dye and labels all cells, including those with intact plasma membranes. Ethidium homodimer-2 is a cell-impermeant red fluorescent nucleic acid stain that labels only cells with compromised membranes (which is characteristic of necrotic and late apoptotic cells).

4. Notes

1. The exact density of the Percoll gradient is critical for satisfactory separation of red blood cells, cardiomyocytes, and fibroblasts from each other. If the separation is not sufficient, a new batch of Percoll usually solves the problem.

2. By using 6× sample buffer, higher protein concentration can be achieved.

3. Use 20G needle and bend it at the end forming a small hook, which will help handling the coverslips.

4. We have tried several different coating materials and fibronectin has proven to be the most effective. Coated dishes can be stored for a couple months at 4°C, if sealed by parafilm.

5. For proper removal of all fibroblasts, use a 2 mL sterile pipette, then discard it and use a new one and collect the cardiomyocytes very slowly. Collect as many cardiomyocytes as you can, but avoid contamination with red blood cells.

6. Variable proportion of the originally plated cardiac myocytes will only survive; therefore, the final density varies from isolation to isolation.

7. If mating of homozygote mice is feasible, the hearts of the pups can be combined and cardiac myocytes can be isolated based on the rat isolation protocol.

8. Under these conditions, a significant proportion of cardiac myocytes will die in 8 h. If the hypoxic incubator is not used constantly but only periodically, make sure to turn it on 16–24 h before use to reach the required temperature and reduced oxygen levels for the experiment.

9. To evaluate caspase 3 cleavage by immunoblot analysis, 35-mm dishes usually provide sufficient amount of protein.

10. Apoptotic cells tend to attach to the dish loosely; therefore, collect cells with a scraper without removing the media.

11. Always use freshly harvested cells (1–2×10^7 cells plated in a 100-mm dish seem to give the best result). Never freeze the samples before they are completely processed.

12. To check the efficiency of cell lysis, pipette 5 µL of cell lysate onto a glass slide, add coverslip, and view under a microscope. Compare with 5 µL of the nonlysed cells. A shiny ring around the cells indicates that the cells are still intact. If more than 20–30% of the cells have the shiny ring, continue homogenizing and check again.

13. If you experience mitochondrial protein leakage to the cytosol, the homogenization might have been too vigorous. Try to decrease the number of strokes used.

14. The amount of diluted primary antibody or TdT enzyme used can be decreased to 20 μL per coverslip. Put a small drop of primary antibody solution or TdT enzyme solution onto a piece of parafilm and then place the coverslip with cells facing down carefully onto the antibody solution or TdT enzyme solution with the help of a bent needle. Fill up an empty pipette box with water (to avoid the coverslip from drying up) and lay the parafilm with the coverslips onto it, and then close the box.

15. Put one drop of Fluorsave reagent onto the microscope slide. Using a bent needle and the forceps place the coverslip carefully to the edge of the drop, slowly let the coverslip lowered onto the slide completely. Press the coverslip to the slide with the other end of the forceps and remove the excess of the mounting medium using an aspirator.

16. If you wish to combine immunocytochemistry and TUNEL staining on the same samples, use primary and secondary antibodies first followed by TUNEL and finish with nuclear staining.

17. If multiple samples are processed on the same day, stagger the vital staining of coverslips to allow sufficient time for taking images by fluorescent microscope. Once exposed to fluorescent light, samples fade within minutes. Even if kept in dark, samples should be imaged within 30–60 min after staining.

References

1. Crow, M. T., Mani, K., Nam, Y. J., and Kitsis, R. N. (2004). The mitochondrial death pathway and cardiac myocyte apoptosis. *Circ Res* **95**, 957–70.

2. Gill, C., Mestril, R., and Samali, A. (2002). Losing heart: the role of apoptosis in heart disease – a novel therapeutic target? *Faseb J* **16**, 135–46.

3. Logue, S. E., Gustafsson, A. B., Samali, A., and Gottlieb, R. A. (2005). Ischemia/reperfusion injury at the intersection with cell death. *J Mol Cell Cardiol* **38**, 21–33.

4. Regula, K. M., and Kirshenbaum, L. A. (2005). Apoptosis of ventricular myocytes: a means to an end. *J Mol Cell Cardiol* **38**, 3–13.

5. Whelan, R. S., Mani, K., and Kitsis, R. N. (2007). Nipping at cardiac remodeling. *J Clin Invest* **117**, 2751–3.

6. Kitsis, R. N., and Narula, J. (2008). Introduction-cell death in heart failure. *Heart Fail Rev* **13**, 107–9.

7. Mani, K., and Kitsis, R. N. (2003). Myocyte apoptosis: programming ventricular remodeling. *J Am Coll Cardiol* **41**, 761–4.

8. Sutton, M. G., and Sharpe, N. (2000). Left ventricular remodeling after myocardial infarction: pathophysiology and therapy. *Circulation* **101**, 2981–8.

9. Yaoita, H., Ogawa, K., Maehara, K., and Maruyama, Y. (1998). Attenuation of ischemia/reperfusion injury in rats by a caspase inhibitor. *Circulation* **97**, 276–81.

10. Soonpaa, M. H., Kim, K. K., Pajak, L., Franklin, M., and Field, L. J. (1996). Cardiomyocyte DNA synthesis and binucleation

during murine development. *Am J Physiol* **271**, H2183–9.

11. Bueno, O. F., De Windt, L. J., Tymitz, K. M., Witt, S. A., Kimball, T. R., Klevitsky, R., Hewett, T. E., Jones, S. P., Lefer, D. J., Peng, C. F., Kitsis, R. N., and Molkentin, J. D. (2000). The MEK1-ERK1/2 signaling pathway promotes compensated cardiac hypertrophy in transgenic mice. *Embo J* **19**, 6341–50.

12. Shiraishi, I., Melendez, J., Ahn, Y., Skavdahl, M., Murphy, E., Welch, S., Schaefer, E., Walsh, K., Rosenzweig, A., Torella, D., Nurzynska, D., Kajstura, J., Leri, A., Anversa, P., and Sussman, M. A. (2004). Nuclear targeting of Akt enhances kinase activity and survival of cardiomyocytes. *Circ Res* **94**, 884–91.

13. Yue, T. L., Wang, C., Gu, J. L., Ma, X. L., Kumar, S., Lee, J. C., Feuerstein, G. Z., Thomas, H., Maleeff, B., and Ohlstein, E. H. (2000). Inhibition of extracellular signal-regulated kinase enhances Ischemia/Reoxygenation-induced apoptosis in cultured cardiac myocytes and exaggerates reperfusion injury in isolated perfused heart. *Circ Res* **86**, 692–9.

14. Gottlieb, R. A., Burleson, K. O., Kloner, R. A., Babior, B. M., and Engler, R. L. (1994). Reperfusion injury induces apoptosis in rabbit cardiomyocytes. *J Clin Invest* **94**, 1621–8.

15. Sheng, Z., Pennica, D., Wood, W. I., and Chien, K. R. (1996). Cardiotrophin-1 displays early expression in the murine heart tube and promotes cardiac myocyte survival. *Development* **122**, 419–28.

Chapter 22

Cell Death in Myoblasts and Muscles

Lawrence M. Schwartz, Zhengliang Gao, Christine Brown, Sangram S. Parelkar, and Honor Glenn

Summary

One of the hallmarks of development is that many more cells are produced than are ultimately needed for organogenesis. In the case of striated skeletal muscle, large numbers of myoblasts are generated in the somites and then migrate to take up residence in the limbs and the trunk. A subset of these cells fuses to form multinucleated skeletal muscle fibers, while a second group, known as satellite cells, exits the cell cycle and persists as a pool of lineage-restricted stem cells that can repair damaged muscle. The remaining cells initiate apoptosis and are rapidly lost. Primary myoblasts and established satellite cell lines are powerful tools for dissecting the regulatory events that mediate differentiative decisions and have proven to be important models. As well, muscle diseases represent debilitating and often fatal disorders. This chapter provides a general background for muscle development and then details a variety of assays for monitoring the differentiation and the death of muscle. While some of these methods are specialized to address the phenotypic properties of skeletal muscle, others can be employed with a wide variety of cell types.

Key words: Apoptosis, Autophagy, Myotube, Caspase, MTT, FACScan, Cell culture, C_2C_{12} cell, Differentiation, Necrosis

1. Introduction

Skeletal muscle accounts for about 40% of the body's total weight and approximately 75% of its cellular mass. It represents the body's primary amino acid reservoir and is subjected to protracted periods of nonpathological atrophy and hypertrophy in response to starvation, activity, and development. Muscle is also affected by a variety of pathological conditions that can result in motor defects, reduced quality of life, and death.

Peter Erhardt and Ambrus Toth (eds.), *Apoptosis*, Methods in Molecular Biology, vol. 559
DOI 10.1007/978-1-60327-017-5_22
© Humana Press, a part of Springer Science+Business Media, LLC 2004, 2009

Muscle precursor cells, myoblasts, typically arise in the somites during embryogenesis in response to a variety of inductive and repressive signals emanating from the neural tube and the floor plate/notochord, as well as from the lateral ectoderm and mesoderm *(1, 2)*. Precursor cells from the epaxial region of the somites migrate to form the back musculature, while the cells from the hypaxial region migrate to the newly formed limb buds.

This developmental scheme for the generation of muscle presents a few fundamental problems for the embryo. Importantly, progenitor tissues such as the dermomyotome cannot "know" how many myoblasts will actually be needed for organogenesis. Consequently, some mechanism must exist to insure that sufficient numbers of myoblasts are generated and migrate to the correct location, while surplus cells are rapidly removed. This latter issue becomes important when one considers that myoblasts are mitotically competent migratory cells that pose a risk of neoplasm or organ dysgenesis should their growth become deregulated. In fact, myoblast-derived rhabdomyosarcomas are the most common form of soft tissue tumors in children *(3)*.

Evolution has favored a developmental scheme that provides the plasticity required to maintain valuable cells while simultaneously eliminating either surplus or potentially deleterious ones. This scheme, which was popularized in a review by Martin Raff in 1992 *(4)*, suggested that all developing cells are on the verge of committing suicide. Those cells that receive appropriate trophic signals from interacting neighbors activate survival programs, while cells failing to receive these signals initiate programmed cell death, typically by apoptosis.

Much of our insight into the regulation of muscle survival and differentiation comes from in vitro studies of either primary myoblasts or established lines like C_2C_{12} cells *(5, 6)* (*see* **Fig. 1**). In fact, C_2C_{12} cells can produce functional muscle fibers in vivo following transplantation, although these fibers tend to be smaller than those produced with primary cells *(7)*. C_2C_{12} cells are a stable, nontransformed muscle stem cell that can be maintained indefinitely in vitro. Expression of the two basic helix–loop–helix (bHLH) transcription factors, MyoD and Myf5, cycle out of phase with one another and may mark these cells as predisposed to form muscle fibers or satellite cells, respectively *(8, 9)*.

Following transfer to a low serum differentiation medium (DM), a subset of cells expresses high levels of MyoD and induces a variety of survival proteins including p21 and AKT (or PI 3-kinase/AKT), which confer an apoptosis-resistant phenotype *(10, 11)*. These cells align with one another and fuse to form multinucleated striated skeletal myotubes, the in vitro counterpart of muscle fibers. Myotubes, like muscle fibers, are highly resistant to physiological insults and rarely undergo apoptosis. Instead, they preferentially initiate autophagy, a program that

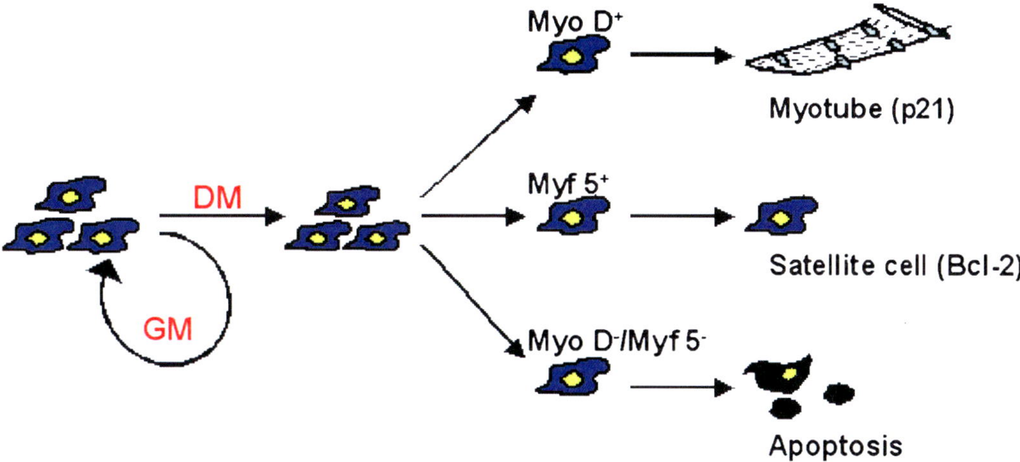

Fig. 1. Myoblast differentiative decisions. In growth medium (GM), myoblasts continue to cycle. When transferred to low serum differentiation medium (DM), cells make one of three choices. Some express MyoD and p21 as differentiation and survival factors, respectively, and fuse to form myotubes. Others activate Myf5 and Bcl-2 and arrest as reserve/satellite cells. When re-exposed to GM, these cells cycle. Some cells fail to activate either differentiation or survival proteins and undergo apoptosis.

allows the cell to consume organelles and bulk cytoplasm *(12)*. Autophagic cells are highly apoptosis resistant and can survive for long periods with suboptimal nutrition *(13)*.

A second population of myoblasts does not express detectable levels of MyoD and instead upregulates the bHLH myogenic protein Myf5 and the antiapoptotic protein Bcl-2 *(14)*. These cells arrest as a pool of lineage-restricted stem cells that remain quiescent until they are re-exposed to growth factors. These "reserve cells" are the in vitro counterparts of muscle "satellite cells," and many researchers use the terms interchangeably. When appropriately stimulated, these cells re-enter the cell cycle and their progeny recapitulate all of the developmental decisions displayed by the parental cells, including myotube formation.

A final group of myoblasts fails to upregulate either bHLH myogenic factors or survival proteins and instead undergoes apoptosis. These cells presumably represent the pool of "surplus" cells that were not required for proper myogenesis. Following transfer to DM these cells begin to die by 24 h and then the cell death rate returns to baseline by about 36 h *(15, 16)*. At present the regulatory factors that control this default to apoptosis are poorly understood, although members of the tumor necrosis factor family have been implicated *(17)*.

One of the virtues of using primary myoblasts for in vitro studies is that they can be derived from mutant animals and allow for the study of both cellular responses and developmental decisions. This is particularly important for animals carrying germline deletions that result in embryonic lethality.

In this chapter, we present protocols for the isolation and analysis of myoblasts, including differentiation and cell death. Details of the differentiation assays are included because the diversity of cells that can be produced in vitro following the loss of growth factors can confound some death assays. For example, myotubes are metabolically more active than myoblasts, so assays that measure mitochondrial function as a surrogate for cell number, like MTT and XTT (see below), may lead to misinterpretations.

2. Materials

2.1. Primary Mouse Myoblast Isolation

1. 70% ethanol in a squirt bottle.
2. Phosphate-buffered saline (PBS), pH 7.2 (10×, Invitrogen) diluted to 1× with water.
3. Standard tissue culture dishes (100, 60, and 35 mm, Falcon).
4. Low power dissecting microscope.
5. Collagenase/dispase/CaCl$_2$ solution: 1.5 U/mL collagenase D (Roche), 2.4 U/mL dispase II (Roche) in 2.5 mM CaCl$_2$.
6. Medium surgical scissors and microscissors.
7. Two pairs of fine forceps.
8. Sterile razor blade.
9. 100-µm and 40-µm cell strainers (Falcon).
10. Ten-micrometer filter units: made from a small piece of 10-µm nitex mesh (Sefar America) taped to the small end of a plastic funnel, wrapped in foil, and autoclaved.
11. Collagen solution: Add 1 mL of concentrated acetic acid to 179-mL water. Sterile filter through 0.2-µm filtration unit, then add 20 mL of 0.1% calf skin collagen in 0.1 N acetic acid (Sigma). Store at 4°C.
12. Collagen-coated tissue culture dishes: incubate dishes in the collagen solution overnight at room temperature. Remove and save the used collagen for future use. Rinse the plates with sterile water, allow them to dry, and store at room temperature for >6 months.
13. F-10-based primary myoblast growth medium: 20% fetal bovine serum (FBS) (Atlanta Biologicals) in Hams F-10 nutrient mixture (Invitrogen) plus 2.5 ng/mL basic fibroblast growth factor (human; Promega), 100 U/mL penicillin and 100 µg/mL streptomycin (Invitrogen).

14. F-10/DMEM-based primary myoblast growth medium: same as F-10 medium above, but substitute half the volume of Hams F-10 with DMEM (Invitrogen)

15. Differentiation medium: 2–5% horse serum (HyClone) in DMEM plus 100 U/mL penicillin and 100 µg/mL streptomycin.

16. Humidified 37°C, 5% CO_2 incubator.

17. Inverted microscope.

2.2. C_2C_{12} Mouse Myoblast Subculture And Differentiation

1. C_2C_{12} cells (ATCC, American Type Culture Collection)

2. Phosphate-buffered saline (PBS), pH 7.2 (10×), liquid (Invitrogen).

3. Trypsin, 0.25% (1×), liquid (Invitrogen).

4. Growth medium (GM): DMEM (high glucose, Invitrogen) plus 10% FBS (Atlanta Biologicals) 100 U/mL penicillin and 100 m g/mL streptomycin. (Invitrogen), pH 7.2 (*see* **Note 1**).

5. Differentiation medium (DM): DMEM plus 0.1% FBS, 5 µg/mL insulin (Sigma) and 5 µg/mL transferrin (Sigma), 100 U/mL penicillin and 100 m g/mL streptomycin, pH 7.2 (*see* **Note 2**).

6. Freezing medium: FBS with 7% dimethyl sulfoxide (Sigma).

7. 100-mm BD Falcon tissue culture dish (BD)

8. Costar/Corning 6-, 12-, 24-well plates (Corning).

2.3. Myosin Heavy Chain Staining

1. Cultures of C_2C_{12} cells or primary myoblasts cultured in DM for the duration desired.

2. Glass microscope coverslips.

3. Laminin-coated glass coverslips (optional).

4. Monoclonal antisera against myosin heavy chain (MHC) (1:100; MF20, Developmental Studies Hybridoma Bank).

5. FITC- or Cy3^{Tm}-conjugated rabbit anti-mouse antiserum (Jackson Laboratories).

6. DAPI (4′,6-diamidino-2-phenylindole, Sigma), or To-Pro-3 (Invitrogen).

7. 2% paraformaldehyde: 1 g electron microscopy grade paraformaldehyde per 50-mL PBS in small bottle with a stir bar. Add a few drops of NaOH and heat in hood at 60–70°C to dissolve. Cool to room temperature and adjust pH to 7.4.

8. PBST: PBS plus 0.1% Tween-20.

9. 0.5% Triton-X 100 detergent in PBS.

10. 1% bovine serum albumin (BSA) in PBST.

2.4. Vital Dye/ Membrane Exclusion Assays

2.4.1. Trypan blue

1. Hemacytometer with cover glass.
2. 15-mL centrifuge tubes.
3. Microfuge tubes.
4. Trypan blue stain 0.4% (Sigma).
5. Phosphate-Buffered Saline (PBS), pH 7.2 (10×), liquid (Invitrogen).
6. Trypsin, 0.25% (1×), liquid Invitrogen.

2.4.2. Live/Dead Assay

1. LIVE/DEAD® Viability/Cytotoxicity Assay Kit (Molecular Probes/Invitrogen).

2.5. Annexin V Staining

1. Annexin-binding buffer: 10 mM HEPES, 140 mM NaCl, 2.5 mM $CaCl_2$, pH 7.4.
2. Propidium iodide 100× stock: 0.1 mg/mL water.
3. Annexin V conjugate (e.g., Annexin V-Alexa Fluor 488, Molecular Probes).
4. Round-bottom 5-mL tubes.

2.6. Caspase 3/7 Activation Assay

1. Opaque-walled 96-well plates (Fisher Scientific).
2. Lysis solution (9% v/v Triton X-100).
3. CytoTox-ONE reagent (Promega).
4. Stop solution (Promega).
5. Caspase-Glo 3/7 reagent (Promega).
6. Multichannel pipettor.
7. Plate shaker.
8. Plate reader with luminescence detection and fluorescence detection with 560-nm excitation and 590-nm emission filters.

2.7. Mitochondrial Activity Assays

2.7.1. MTT

1. 96-well tissue culture treated microplate (BD Falcon).
2. Multichannel pipettor.
3. Hemacytometer with cover glass.
4. Plate reader with 570-nm wavelength absorbance detection.
5. Plate shaker.
6. MTT stock: 5 mg/mL stock of MTT (3-(4,5-dimethylthiazol-2-yl)2,5-diphenyltetrazolium bromide) (Invitrogen) in 1× PBS. This solution should be sterilized by passing through a 0.22-μm filter in a sterile tissue culture hood. This MTT stock solution may be kept at 4°C for up to a week.
7. MTT working solution: Add one-tenth of the volume of MTT stock solution to the required amount of DMEM growth medium to yield a final concentration of 500 μg/mL.
8. DMSO (dimethyl sulfoxide) (Sigma).

2.7.2. XTT

1. XTT stock: 1 mg/mL stock of XTT (sodium 3'-[1-(pheny laminocarbonyl)-3,4-tetrazolium]-bis(4-methoxy-6-nitro)benzene sulfonic acid hydrate) solution in phenol red free DM along with a suitable amount of electron coupling reagent (provided and recommended by the manufacturer). This XTT stock solution should be sterilized by passing through a 0.22-µm filter in a sterile tissue culture hood. The XTT stock solution may now be aliquoted and stored at −20°C.

2. XTT working solution: Add 20% of the volume of XTT stock solution to the required amount of DM to yield a final concentration of 200 µg/mL.

3. Methods

3.1. Primary Mouse Myoblast Isolation

1. Sacrifice mice by decapitation (neonates) or CO_2 (adults).

2. Rinse off limbs with 70% ethanol and remove from body. Using dissecting scope, dissect muscle away from skin and bone and place muscle in 35-mm dish containing 0.5-mL PBS on ice.

3. In a tissue culture hood, mince up tissue with microscissors and razor blade. Continue the rest of the procedure in the hood.

4. Add about 2 mL dispase/collagenase solution per gram of tissue and continue mincing until slurry can be taken up into a 10-mL pipette.

5. Transfer tissue to 15-mL tube and incubate at 37°C for about 20 min, with occasional triturating to break up tissue.

6. Place 100-µm cell strainer in a 50-mL centrifuge tube, then add the slurry to the strainer. Rinse with a few mL of F-10-based growth medium. Repeat with the 40-µm and 10-µm cell strainers.

7. Centrifuge cells for 5 min at $350 \times g$, aspirate supernatant, resuspend in 10-mL F-10 medium, and plate on uncoated 10-cm tissue culture dish.

8. Incubate at 37°C and 5% CO_2 for 2 h.

9. Remove medium and floating cells (fibroblasts will be left stuck to plate) and transfer to collagen-coated dish. The resulting population is about 80% myoblasts (round, compact, desmin-positive cells) and 20% fibroblasts (flat, spread, desmin-negative cells.)

10. After incubating for 2–3 days, aspirate medium and add a small volume of PBS to the dish. Tap the side of the dish to dislodge the weakly-adherent myoblasts, then transfer these cells to a fresh collagen-coated dish of F-10-based growth medium. This medium gives myoblasts a growth advantage over fibroblasts. This procedure can be repeated every few days until all fibroblasts are absent from the culture.

11. When cells reach about 60% confluency, trypsinize and split as described for passaging C 2 C 12 cells below, at 1:3, replating onto collagen-coated dishes.in DMEM/F-10 medium, to allow cells to grow more quickly

12. Myoblasts can be frozen for storage using standard cell culture protocols or differentiated as described for C_2C_{12} cells below (*see* **Fig. 2**).

3.2. Culturing C_2C_{12} Cells

1. Thaw frozen C_2C_{12} cells rapidly in a 37°C water bath and then dilute them into 10 mL of GM in a 100-mm tissue culture dish.

2. Incubate the cells in GM at 37°C in 10% CO_2.

3. Change the medium after all cells have adhered, around 4–10 h (*see* **Note 3**).

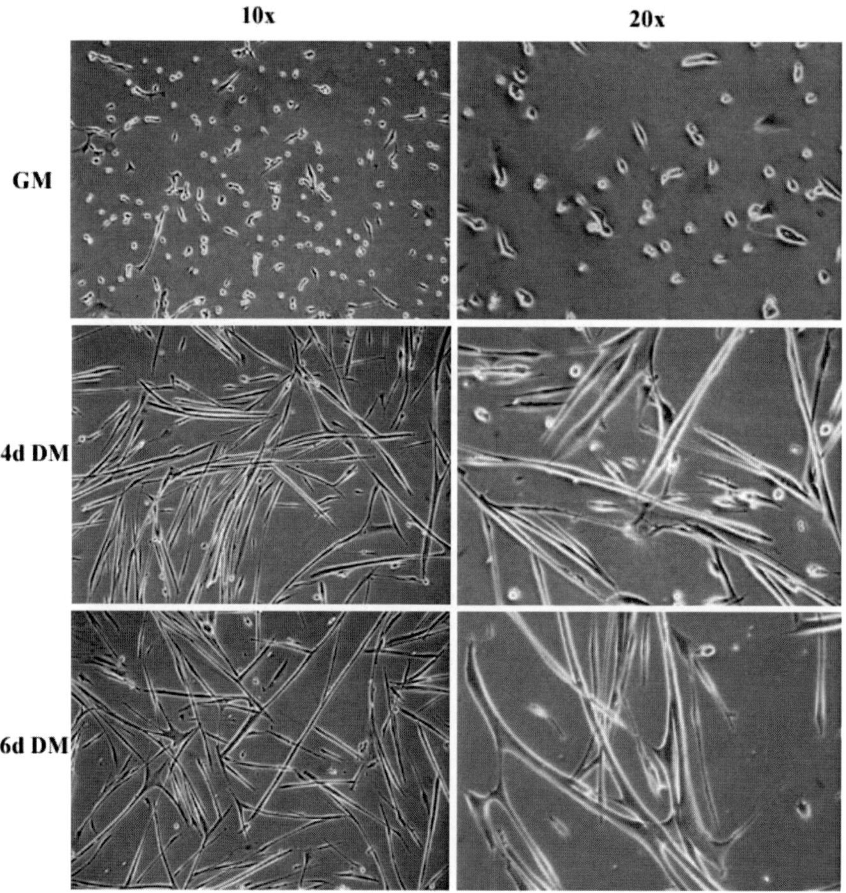

Fig. 2. Differentiation of primary myoblasts. Primary mouse myoblasts were grown on collagen-coated dishes and photographed in growth medium (GM) and after 4 and 6 days in differentiation medium (DM). Removal of growth factors results in most myoblasts fusing to form myotubes, which are often spontaneously contractile. A few mononucleated reserve cells remain.

3.2.1. Passaging
C$_2$C$_{12}$ cells

C$_2$C$_{12}$ cells double every 16–24 h and should also be passaged regularly, typically every 2–4 days, to insure that they do not reach more than 70–80% confluency. Failure to do this will result in the selection of differentiation-incompetent cells, which will dramatically alter the results.

1. Aspirate the medium and wash the plate once with 1× PBS.

2. Apply ~2 mL trypsin solution, briefly and gently swirl the plate in the shape of a cross and return the plates to incubator for 5–10 min or until the cells detach from the substrate (no more than 15–20 min).

3. Shake or bang the plates on a hard surface to facilitate detachment. The cells will appear round and phase-bright under phase microscopy.

4. Terminate the digestion by adding 5–10 mL of GM.

5. Use a sterile Pasteur pipette to "blow" the cells off the substrate.

6. Transfer the suspended cells into a 15-mL Falcon tube.

7. Pellet the cells at $200 \times g$ for 3–5 min.

8. Discard the supernatant and resuspend the cells in 1–2 mL GM. The cells can be split at 1:3–1:20 ratios as needed and replated in fresh dishes with GM (*see* **Note 4**).

3.2.2. Storing C$_2$C$_{12}$ cells

1. To make a liquid nitrogen cell stock, collect fast growing (70–80% confluent) healthy cells as is done with passage.

2. Immediately resuspend the pellets in proper amount of freezing medium (1–1.5 mL per plate) and then add the cell suspension to cryogenic vials with proper labels at 0.5 mL per vial.

3. Transfer the vials to a freezing container (such as a Cryo 1°C Freezing Container, "Mr. Frosty," Nalgene) and place in a –80° C freezer overnight. The cells can remain at –80°C for short-term storage, but cell survival will diminish over time.

4. Place the vials in a liquid nitrogen Dewar for long-term storage.

3.2.3. Differentiating
C$_2$C$_{12}$ cells

1. Allow cells to grow in GM to about 100% confluence and then switch to DM.

2. Change the medium every day or when significant amount of cell death/floating cell debris is present (*see* **Note 5**).

3. In about 1–2 days small myotubes will start to appear, and by 2–4 days after transfer to DM large multinucleated myotubes should be visible (*see* **Note 6**).

3.3. Characterization of Myoblast Cultures

While there are several methods for determining the percentage of myoblasts that have fused to form myotubes in vitro, the easiest is to simply examine the cultures under phase contrast optics. The satellite/reserve cell population adheres tightly to the plate and will appear as flat cells that are phase dark while the myotubes will be raised and phase bright. The clusters of nuclei should be readily apparent.

Immunocytochemistry or immunofluorescence with antisera directed against MHC can be used to generate publication grade images. These are particularly attractive and informative when coupled with nuclear dyes like DAPI (4'-6-diamidino-2-phenylindole) or To-Pro-3 and rhodamine-labeled phalloidin, a fungal toxin that specifically binds polymeric actin (*see* **Fig. 3**). The data from these images can be used to determine the fusion index: the percentage of nuclei within multinucleated myotubes divided by the total number nuclei in the field.

3.3.1. MHC Staining

1. Culture and differentiate C_2C_{12} cells or primary myoblasts on plates or glass microscope slide cover slips. At desired time points, carefully aspirate the medium and wash cells with 1× PBS up to three times. Take care not to dislodge the myotubes, which tend to be weakly adherent.

2. Fix the cells with 2% paraformadehyde in PBS for 10–30 min at room temperature, followed by three washes with PBS.

3. Permeabilize cells by exposure to 0.5% Triton X-100 in PBS for 10–20 min. Wash three times with PBST.

4. Block for 10–20 min with 1% BSA in PBST. Normal goat serum (5%) or a combination of both NGS and BSA can be used as well.

5. Incubate in anti-MHC antibody in 1% BSA in PBST overnight at 4°C or at room temperature for 1–2 h.

Actin   MHC   Nuclei

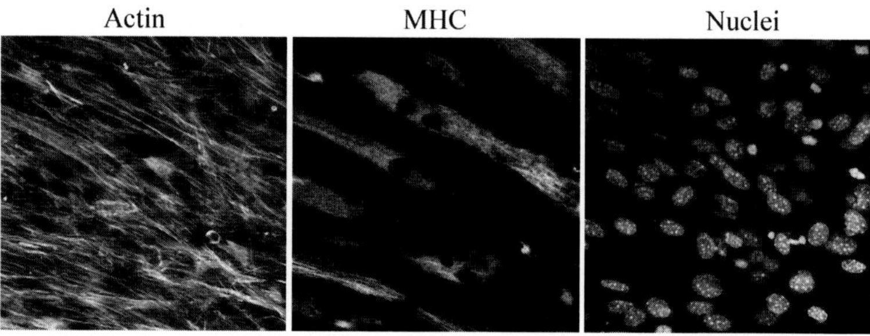

Fig. 3. Immunocytochemical analysis of myosin heavy chain expression in C_2C_{12} myoblasts. Cells were exposed to DM for 3 days then fixed and tripled labeled. Actin was labeled with rhodaminde-phalloidin. Myosin heavy chain was detected by primary antibody followed by FITC-conjugated secondary. Nuclei were counterstained with To-Pro 3.

6. Wash three times with PBST each for 3–5 min.

7. Incubate in a fluorescently labeled secondary antibody at room temperature for 20–60 min. Alternatively, an ABC kit can be used and traditional DAB (diamino benzidine tetrahydrochloride) staining performed for nonfluorescent imaging.

8. Wash three times with PBST each for 5–10 min.

9. Either wash with100 ng/mL DAPI in PBS or use mounting medium containing DAPI to label nuclei.

10. Examine at least three random fields and count the total number of nuclei and the number of nuclei within MHC-positive cells.

11. Calculate the differentiation index [(the number of MHC-positive cells/the total number nuclei in the field) × 100] and the fusion index [(the number of nuclei within multinucleated myotubes/the total number of nuclei in the field) × 100].

3.4. Vital Dye/ Membrane Exclusion Assays

Trypan blue is an old standard assay for cell death in vitro. It is a vital dye that is excluded by intact cells, but can readily enter those with disrupted membranes. Most cells that die in vitro do so via apoptosis, a process that maintains membrane integrity even as the cell dies (*18*). Apoptotic cells will remain trypan blue negative until they undergo a secondary necrosis, presumably because the small apoptotic bodies become ATP depleted and can no longer prevent the influx of water and ions that leads to necrosis. Consequently, while this is an easy in vitro assay for cell death, it measures a very late endpoint in the process and does not accurately measure the number of apoptotic cells in a culture.

A more "modern" approach for measuring cell viability in vitro is the LIVE/DEAD® assay, which is based on the use of two different fluorescent vital dyes that can selectively label cells with intact or disrupted membranes. The nonfluorescent calcein AM dye readily enters cells where it encounters nonspecific esterases that convert it to a nonpermeable carboxylic acid derivative that fluoresces green (excitation/emission ~495 nm/~515 nm) and marks "living" cells. The second dye, ethidium homodimer (EthD-1), is excluded by cell membranes and cannot stain living cells. However, it readily enters cells with disrupted membranes and intercalates into DNA, which enhances its fluorescence ~40-fold. Dead cells fluoresce red (excitation/emission ~495 nm/~635 nm).

Because the LIVE/DEAD® assay is based on fluorescence, it can be analyzed via microscopy, FACScan analysis, or on a fluorescent plate reader, although we have only had limited success with the latter approach. As with trypan blue, this assay is more a measure of secondary necrosis than apoptosis.

3.4.1. Trypan Blue

1. Grow C_2C_{12} cells in growth medium or differentiation medium in 12- and/or 24-well plates in triplicates for a given amount of time.

2. At desired time points, trypsinize the plates and collect the cells in 15-mL tubes. In most cases, it is desirable to include the culture medium and PBS buffer as well in order to collect any detached cells.

3. Pellet the cells and resuspend them in 50–200 μL of PBS (cell density can be adjusted as desired).

4. Transfer 20–50 μL of the cells into a microfuge tube and add an equal volume of 0.4% trypan blue solution (dilution 1:1). Mix gently but thoroughly.

5. Add 10 μL of the solution to a hemocytometer and count both the total number of cells and the subset that are trypan blue positive. An optimal cell density may produce 100–200 cells per count (*see* **Note 7**).

6. % Viability = (Total number of viable cells/Total number of viable and nonviable cells) × 100.

3.4.2. Live/Dead Assay

1. Culture C_2C_{12} or its derivatives on glass cover slips (for high-quality microscopic analyses) or in 12- or 24-well plates.

2. At desired time points, wash the cells gently with PBS one or two times (*see* **Note 8**).

3. Warm up the LIVE/DEAD reagent stock solutions to room temperature and mix the 2 mM EthD-1 solution (Component B) and 4 mM calcein AM stock solution (Component A) as provided by the manufacturer to create a final working solution of 2 μM calcein AM and 4 μM EthD-1. Vortex thoroughly and add directly to cells. Incubate for 30–45 min at room temperature (*see* **Note 9**).

4. Calcein AM and EthD-1 can be viewed simultaneously with conventional fluorescein long-pass filters or observed separately with calcein by a standard fluorescein band-pass filter and EthD-1 via filters for propidium iodide or Texas Red®. Fluorescent images can be quantified automatically with some software programs like Metavue (Universal) or Image J (http://rsb.info.nih.gov/ij/) (*see* **Note 10**).

3.5. Annexin V Staining

While trypan blue and LIVE/DEAD® staining represent convenient tools for measuring cell death in vitro, they are not apoptosis-specific. Instead, they are measuring either primary necrotic cell death or the secondary necrosis that occurs in vitro when apoptotic bodies are not phagocytosed and subsequently become ATP-depleted. One of the early events in apoptosis is the translocation of the membrane phospholipid phosphatidylserine from the inner leaflet of the plasma membrane to the outer layer by activated scramblase *(19)*. The protein Annexin V binds to phosphatidylserine with high affinity and can be used in a variety of assays to identify apoptotic cells *(20)*. When coupled with the DNA intercalating dye propidium iodide, labeled Annexin V can provide a sensitive and powerful tool for distinguishing apoptotic and necrotic cells, respectively, via fluorescence microscopy or FACScan analysis. It should be noted that myoblasts become transiently Annexin V positive when they undergo fusion early in myogenesis, but this happens at points of cell–cell contact that are unlikely to persist when cells are prepared for FACS can analysis *(21)*.

3.5.1. Annexin V staining

1. Grow cells in two 10-cm plates.

2. Add death-inducing agent to one plate (*see* **Table 1**) and let the other serve as a control (*see* **Table 2**)

3. Remove supernatant to 15-mL tube, then trypsinize adherent cells.

4. Add cells to supernatant.

5. Centrifuge cells and resuspend in 1-mL PBS.

Table 1
Toxic treatments for C_2C_{12}cells

Class of drug	Treatment	Optimal concentration
Growth factor deprivation	Incubate in serum-free medium	Serum free medium for 24–48 h
Adenosine receptor agonist	2-Chloroadeno-sine	100 μM
Kinase Inhibitor	Staurosporine	0.5–10 μM
Proteasome inhibitor	MG132 (Z-Leu-Leu-Leu-al)	0.25–5 μM
Oxidative stress	H_2O_2	12.5–100 μM in sodium pyruvate free medium
Mitochondrial toxin	N-Acetyl-D-sphingosine (C2 Ceramide)	50–125 μM
Mitochondrial toxin	Rotenone	1–40 μM
Protein misfolding	L-Azetidine-2-carboxylic acid	0.5–4 mM

Table 2
Experimental design for annexin V experiments

Tube	Cell treatment	Annexin	PI
Negative control	None	–	–
Experimental control	None	+	+
Annexin V-only control	Death-inducing	+	–
PI-only control	Death-inducing	–	+
Experimental sample	Death-inducing	+	+

6. Remove a 10 μL sample of cells, add 10 μL trypan blue, and count the number of both living and dead cells with a hemo-cytometer.

7. Centrifuge cells and resuspend in annexin binding buffer to approximately 1×10^6 cells/mL.

8. Transfer 100 μL of cell suspension to a 5-mL tube compatible with flow cytometer.

9. Add 5 μL annexin V and incubate at room temperature for 15 min.

10. Add 400 μL annexin binding buffer

11. Add 5 μL propidium iodide stock

12. Put tubes on ice and run on flow cytometer. Run the negative control tube first, followed by the Annexin V and PI control tubes (see **Table 2**) to set machine voltages and compensation.

13. Cells positive for PI are dead, cells negative for annexin V are living, and cells positive for annexin V are apoptotic.

3.6. Caspase 3/7 Activation Assay

Unlike the vital dye staining methods described above, caspase 3/7 activity is a relatively true measure of apoptosis rather than necrosis. As well, it also measures an early event in the process rather than the final end result.

This protocol makes use of a luciferase-based assay for caspase-3/7 activity. Briefly, DEVD, the peptide sequence recognized by activated caspases 3 and 7, is covalently linked to a proluminescent substrate. Cleavage of the target peptide releases aminoluciferin, which is a substrate for luciferase, resulting in a luminescent signal that is directly proportional to caspase activity.

In order to correct for any differences in total cell numbers between samples, we normalize the caspase signal to activity of the constitutively expressed lactate dehydrogenase (LDH). We use an LDH activity assay that relies on LDH-dependent reduction of resazurin to produce the fluorescent product resorufin.

1. Culture cells in opaque-walled 96-well plate (Plate #1). Each treatment or cell line should be plated in triplicate in 105 μL of medium. Also prepare blank wells with each medium used but without cells.

2. Prepare Caspase-Glo 3/7 reagent and CytoTox-One according to manufacturer's protocol (*see* **Note 11**). Equilibrate reagents to room temperature.

3. Remove culture plate from incubator and equilibrate to room temperature (approximately 20 min).

4. Add 2 μL of lysis solution to each well. Shake for 30 s.

5. Pipette wells up and down once to mix.

6. Transfer 50 μL of lysate from each well into a second opaque-walled 96-well plate (Plate #2).

7. Add 50 μL of Casp-Glo reagent to Plate #2 including blanks. Shake for 30 s.

8. Incubate for 60 min at room temperature.

9. Add 50 μL of CytoTox-ONE reagent to each well of Plate 1, including the blank wells. Shake for 30 s.

10. Incubate at room temperature for 10 min.

11. Add 25 μL of stop solution to Plate 1. Shake for 10 s.

12. Read fluorescence at 560/590.

13. Following the 60-min incubation **(step 8)**, read luminescence of Plate 2.

14. The values from the corresponding blank wells should be subtracted from both the luminescence (caspase activity) and the fluorescence (LDH activity) readings. The fluorescence values can then be used to normalize the caspase 3/7 data in order to correct for any variation in total cell numbers between samples (*see* **Note 12**).

3.7. Mitochondrial Activity

Both MTT (3-(4,5-dimethylthiazol-2-yl)2,5-diphenyltetrazolium bromide) and XTT (sodium 3′-[1-(phenylaminocarbonyl)-3,4-tetrazolium]-bis (4-methoxy-6-nitro) benzene sulfonic acid hydrate) are hydrolyzed by the mitochondrial succinatetetrazolium reductase system in metabolically active cells. MTT forms an insoluble precipitate within cells, while the XTT product is soluble. Each is used to quantify relative cell numbers in cultures and can therefore be used as either proliferation or death assays, depending on the treatment. Obviously, care needs to be taken when interpreting the data since treatments that enhance proliferation or reduce cell death may yield comparable numbers.

3.7.1. Mitochondrial Activity: MTT

1. Culture proliferating C_2C_{12} cells in a 10-cm plate until they reach about 60–70% confluency.

2. Trypsinize the cells as in section 3.2.1.

3. After the cells have lifted off, neutralize the trypsin by adding a few mL of GM.

4. Resuspend the cells using a Pasteur pipette to prevent cell aggregation.

5. Count a small aliquot of the cell suspension with a hemocytometer while the cells are being pelleted by centrifugation at about $150 \times g$. Calculate the number of cells/mL.

6. Resuspend the cells in fresh GM to a concentration of $5\text{-}10 \times 10^4$ cells/mL and then use a multichannel pipette to transfer about 100 µL to each well of a sterile tissue culture grade 96-well plate. The cell density to be seeded depends on the incubation time of the treatments. For a 24-h toxic insult treatment a 5×10^4 cells/mL seeding density is ideal (*see* **Note 13**).

7. Leave at least one column of wells on the 96-well plate empty to measure background absorbance (blank values) during the spectrophotometric analysis.

8. Incubate the plates for 24 h. The cell density should be at about 45% depending upon the treatment time to be used.

9. Incubate cells with death inducing agent (e.g., **Table 1**) and after the appropriate treatment time, typically 12 h, aspirate off the medium and add 100 µL/well of MTT working solution. This yields a final concentration of about 50 µg/well of MTT.

10. Incubate the plates at 37°C for at least 1½– 4 h until the purple water-insoluble formazan crystals develop within surviving metabolically active cells.

11. Remove the MTT containing medium and wash all of the wells once with PBS.

12. Remove the PBS and dissolve the intracellular crystals using 100 µL/well DMSO. Keep the plate on a shaker for at least 10 min at room temperature.

13. Measure the absorbance using a suitable plate reader at 570-nm wavelength.

14. Deduct the values for the blank from the remaining wells and calculate the cell viabilities.

3.7.2. Mitochondrial Activity: XTT

The same general approach used with MTT is also used with XTT although there are some modest differences as outlined below:

1. Following treatment with the apoptosis inducer replace the medium with 100 µL/well of XTT working solution. This yields a final XTT concentration of 20 µg/well.

2. Incubate the plates at 37°C for at least 2–4 h until distinct soluble, orange colored formazan crystals develop.

3. Gently mix on a shaker for at least 10 min to disperse the crystals.

4. Measure the absorbance using a suitable plate reader at 450-nm wavelength. In addition, it is recommended to measure the absorbance at a reference wavelength of 690 nm and subtract these values from those obtained at 450 nm.

5. Deduct the values for the blank from the remaining wells and calculate the cell viabilities.

4. Notes

1. Some laboratories use growth medium with 20% FBS, while others use 15% calf serum plus 5% FBS due to the high cost of FBS. Inclusion of high levels of FBS or calf serum in the growth medium will result in faster and more extensive differentiation when the cells are switched to differentiation medium. For most studies, a GM that includes 10% FBS is adequate.

2. Some laboratories use a differentiation medium composed of DMEM with 2% horse serum. Media of either formulation produce comparable results.

3. Alternatively, dilute the thawed cells into 10–15 mL of GM in a Falcon tube and pellet the cells at $200 \times g$ for 3–5 min with a bench top tissue culture centrifuge. Resuspend the pellet in 1 mL of GM and transfer the cells to a 100-mm dish with 10 mL of GM. After an overnight incubation, change the medium to remove floating cells and debris. Change the medium every 1–2 days.

4. It is best to split the cells well before they become 100% confluent. Pasteur pipettes can be used to facilitate passage with 1 mL of cell suspension roughly equal to 20 drops; plate the cells evenly and do not allow cells to reach confluence (even locally), otherwise they will begin to differentiate in growth media. Cells are usually grown to about "70–80%" confluence. Be careful about the percent confluency as it is very subjective; it is recommended to develop a consistent sense of confluence of one's own from experience In addition, 0.05% trypsin/0.02% EDTA can be used in place of the 0.25% trypsin.

5. About 5–10% of the cells die when cultured in GM. This percentage transiently increases to 15–20% when the cells

are switched to DM. This wave of apoptotic death peaks at about 24 h and then falls to baseline by 36 h. When using DM containing horse serum cells should be switched to DM when they reach ~80% confluency rather than the 100% that is used for the FBS/insulin/transferrin DM or the cells will differentiate poorly. Differentiation is about a day slower in the horse serum-containing DM but in general results in reduced cell death and myotubes that are larger and more firmly attached to the substrate (less phase bright). C_2C_{12} cells will also adhere better if the plastic plates are coated with laminin, the preferred extracellular matrix protein for myotubes (*22*). This is not an issue with short-term cultures but can be a factor for studies where myotubes will be maintained for a long experiment.

6. It should be noted that C_2C_{12} cells derived from different sources may have different growth and differentiation properties, making comparisons between independent studies more complicated. When we work with new primary myoblasts or different C_2C_{12} derivatives, we routinely test three different confluencies (70–80%, 90–100%, and 100%) for their ability to differentiate in DM, since the lines can display different predispositions to differentiate.

7. After prolonged incubation, viable cells will begin to take up the dye as well, so it is important to work quickly, preferably 15–30 min after the initial staining of the cells. Alternatively, samples can be counted in batches to minimize the variation resulting from prolonged incubation. Proper focus of the microscope is important to insure that stained cells are adequately distinguished from nonstained ones. A consistent standard should be maintained for scoring cells so that cellular debris is not inappropriately scored as individual cells.

8. Washing with PBS can remove or dilute serum esterase activity, which is usually present in serum-supplemented growth media and thus reduce extracellular fluorescence background. It can also wash away cell debris, which may be an issue depending on the experimental design.

9. For 12- or 24-well plates, a volume of 150–300 µL EthD-1 and 100–200 µL calcein AM is usually needed per well. The aqueous solution of calcein AM is labile to hydrolysis and should be made fresh and used within a day.

10. C_2C_{12} cells are prone to die when deprived of culture media and maintained in PBS containing staining solution for more than 30–60 min. When using software to quantify cell death, it should be noted that a single dead cell might produce multiple apoptotic bodies, while a myotube of multiple nuclei may constitute a single counting event. This can be somewhat compensated by making appropriate changes

to the software settings/parameters. Alternatively, a microplate reader can be used for quantification. Calcein AM can be excited by standard fluorescein filters of 485 ± 10 nm while EthD-1 is compatible with typical rhodamine filters of 530 ± 12.5 nm. The fluorescence emissions should then be acquired separately with calcein AM at 530 ± 12.5 nm, and EthD-1 at 645 ± 20 nm.

11. The CytoTox-ONE reagent should be protected from light and once reconstituted is stable for 1 week at 4°C. The working reagent may be frozen and thawed without loss of signal. The Caspase-Glo reagent is stable for 3 days after reconstitution at 4°C. Freezing and thawing the reconstituted reagent will cause significant loss of signal.

12. The LDH assay **(Section 3.6, steps 9–12)** can be performed during the incubation period for the caspase assay **(Section 3.6, step 8)**.

13. The cells can also be incubated in phenol red free differentiation medium for up to 4 days to generate myotubes if they are the focus of the study.

Acknowledgments

We thank Jacques Tremblay for sharing his protocols for primary myoblast isolation. We are also very grateful to Helen Blau (Stanford University) for all the resources and protocols she makes available on her laboratory Web site (http://www.stanford.edu/group/blau/reagents.htmL). This work was supported by grants from the NIH, the Collaborative Biomedical Research Program, and the Center of Excellence in Apoptosis Research (CEAR) to LMS.

References

1. Ehrhardt, J., Brimah, K., Adkin, C., Partridge, T., and Morgan, J. (2007). Human muscle precursor cells give rise to functional satellite cells in vivo. *Neuromuscul. Disord.* **17**, 631–638.

2. Peault, B., Rudnicki, M., Torrente, Y., Cossu, G., Tremblay, J. P., Partridge, T., Gussoni, E., Kunkel, L. M., and Huard, J. (2007). Stem and progenitor cells in skeletal muscle development, maintenance, and therapy. *Mol. Ther.* **15**, 867–877.

3. McDowell, H. P. (2003). Update on childhood rhabdomyosarcoma. *Arch. Dis. Child.* **88**, 354–357.

4. Raff, M. C. (1992) Social controls on cell survival and cell death. *Nature* **356**, 397–400.

5. Yaffe, D., and Saxel, O. (1977). A myogenic cell line with altered serum requirements for differentiation. *Differentiation* **7**, 159–166.

6. McArdle, A., Maglara, A., Appleton, P., Watson, A. J., Grierson, I., and Jackson, M. J. (1999). Apoptosis in multinucleated skeletal muscle myotubes. *Lab. Invest.* **79**, 1069–1076.

7. Wernig, A., Irintchev, A., Hartling, A. et al. (1991). Formation of new muscle fibres and tumours after injection of cultured myogenic cells. *J. Neurocytol.* **20**, 982–997.

8. Kitzmann, M., Carnac, G., Vandromme, M., Primig, M., Lamb, N. J., and Fernandez, A. (1998). The muscle regulatory factors MyoD and myf-5 undergo distinct cell cycle-specific expression in muscle cells. *J. Cell Biol.* **142**, 1447–1459.

9. Lindon, C., Montarras, D., and Pinset, C. (1998). Cell cycle-regulated expression of the muscle determination factor Myf-5 in proliferating myoblasts. *J. Cell Biol.* **140**, 111–118.

10 Shen, X., Collier, J. M., Hlaing, M., Zhang, L., Delshad, E. H., Bristow, J., and Bernstein, H. S. (2003). Genome-wide examination of myoblast cell cycle withdrawal during differentiation. *Dev. Dyn.* **226**, 128–138.

11. Jiang, B. H., Aoki, M., Zheng, J. Z., Li, J., and Vogt, P. K. (2003). Myogenic signaling of phosphatidylinositol 3kinase requires the serinethreonine kinase Akt/protein kinase B. *Proc. Natl Acad. Sci. USA* **96**, 2077–2081.

12 Wang, X., Blagden, C., Fan, J., Nowak, S. J., Taniuchi, I., Littman, D. R., and Burden, S. J. (2005). Runx1 prevents wasting, myofibrillar disorganization, and autophagy of skeletal muscle. *Genes Dev.* **19**, 1715–1722.

13. Mizushima, N. (2007). Autophagy: process and function. *Genes Dev.* **21**, 2861–2873.

14. Dominov, J. A., Dunn, J. J., and Miller, J. B. (1998). Bcl-2 expression identifies an early stage of myogenesis and promotes clonal expansion of muscle cells. *J. Cell Biol.* **142**, 537–544.

15. Hu, Y., Cascone, P. J., Cheng, L., Sun, D., Nambu, J. R., and Schwartz, L. M. (1999). Lepidopteran DALP, and its mammalian ortholog HIC-5, function as negative regulators of muscle differentiation. *Proc. Natl Acad. Sci. USA* **96**, 10218–10223.

16. Gao, Z. L., Deblis, R., Glenn, H., and Schwartz, L. M. (2007). Differential roles of HIC-5 isoforms in the regulation of cell death and myotube formation during myogenesis. *Exp. Cell Res.* **313**, 4000–4014.

17. O'Flaherty, J., Mei, Y., Freer, M., and Weyman, C. M. (2006). Signaling through the TRAIL receptor DR5/FADD pathway plays a role in the apoptosis associated with skeletal myoblast differentiation. *Apoptosis* **11**, 2103–2113.

18. Arends, M. J., and Wyllie, A. H. (1991). Apoptosis: mechanisms and roles in pathology *Int. Rev. Exp. Pathol.* **32**, 223–254.

19. Sahu, S. K., Gummadi, S. N., Manoj, N., Aradhyam, G. K. (2007). Phospholipid scramblases: an overview. *Arch. Biochem. Biophys.* **462**, 103–114.

20. Fadok, V. A., Bratton, D. L., Rose, D. M., Pearson, A., Ezekewitz, R. A. B., and Henson, P. M. (2000) A receptor for phosphatidylserine-specific clearance of apoptotic cells. *Nature* **405**, 85–90.

21. van den Eijnde, S. M., van den Hoff, M. J., Reutelingsperger, C. P., van Heerde, W. L., Henfling, M. E., Vermeij-Keers, C., Schutte, B., Borgers, M., Ramaekers, F. C. (2001). Transient expression of phosphatidylserine at cell–cell contact areas is required for myotube formation. *J. Cell Sci.* **114**, 3631–3642.

22. Gullberg, D., Tiger, C. F., and Velling, T. (1999). Laminins during muscle development and in muscular dystrophies. *Cell. Mol. Life Sci.* **56**, 442–460.

Part VI

Analysis of Apoptosis in Model Organisms

Chapter 23

Reliable Method for Detection of Programmed Cell Death in Yeast

Xinchen Teng and J. Marie Hardwick

Summary

Accumulating evidence suggests that yeasts are capable of undergoing programmed cell death (PCD) to benefit long-term survival of the species, and that yeast and mammals may share at least partially conserved PCD pathways. In our experience, mammalian apoptosis assays have not been readily applicable to yeast. Therefore, to take advantage of yeast as a genetic tool to study PCD, we developed a yeast cell death assay that can reliably reveal viability differences between wild-type strains and strains lacking the mitochondrial fission genes *DNM1*/Drp1 and *FIS1*, orthologs of mammalian cell death regulators. Cell viability following treatment with acetic acid is quantified by colony formation and vital dye (FUN1) staining to reproducibly detect dose-dependent, genetically programmed yeast cell death.

Key words: Yeast, Programmed cell death, Apoptosis, Fis1, Dnm1, Acetic acid, Colony forming assay, FUN1, Mitochondria, Fission

1. Introduction

The debate over the existence of programmed cell death (PCD) machinery in yeast stems in part from uncertainty about the benefit of a cell suicide program in unicellular species and the mechanisms for selecting such a program during evolution. However, purposeful death of a subset of yeast cells in the population has been suggested to benefit the population as a whole during nutrient limitation *(1)*, failed mating attempts *(2)*, and exposure to killer viruses *(3)*. Furthermore, deliberate cell death was suggested to be a critical consequence of cell-to-cell communication by colonial organisms, resulting in the elimination of older members to benefit younger members *(4)*. The controversy over

Peter Erhardt and Ambrus Toth (eds.), *Apoptosis*, Methods in Molecular Biology, vol. 559
DOI 10.1007/978-1-60327-017-5_23
© Humana Press, a part of Springer Science+Business Media, LLC 2004, 2009

yeast PCD also stems from conflicting terminologies. The yeast community often uses the term "apoptosis" to mean programmed death, while mammalian apoptosis is usually, but not always, reserved for caspase-dependent cell death to distinguish this pathway from other emerging pathways. Perhaps a more legitimate application of the term apoptosis for yeast is based on the evidence that dying yeast exhibit apoptosis-like features, fitting the original morphological definition of the term *(5)*.

Although yeast appear to lack obvious orthologs of Bcl-2 family proteins that regulate apoptosis in metazoans, mammalian Bcl-2 proteins can retain their anti- and pro-death functions when exogenously expressed in yeast *(6–10)*. While it is not known if the biochemical mechanisms of Bcl-2 family proteins in yeast and mammalian cells are the same or distinct, these observations by many laboratories are consistent with the existence of conserved PCD pathways between yeast and mammals. In contrast to Bcl-2 proteins, yeasts encode homologs of many other mammalian cell death regulators, and these have been suggested to fulfill a similar role in yeast, such as Mca1/Yca1, which is designated a metacaspase based on sequence similarity to mammalian caspases *(11)*, the dynamin-like mitochondrial fission protein Dnm1 (human Drp1) *(9)*, Bir1, which contains two N-terminal BIR repeats found in baculovirus and cellular IAPs (inhibitor of apoptosis proteins) *(12)*, the serine protease Nma111, which is homologous to mammalian HtrA2/Omi *(13)*, and an ortholog of human AIF *(14)*. Yeast knockouts and protein overexpression models have demonstrated that these factors influence cell death in yeast, but evidence that these factors arose during evolution in part to serve as cell suicide factors in yeast is experimentally challenging to demonstrate. Nevertheless, we anticipate that further analysis of their cell death functions in yeast will be revealing about mammalian processes. However, we have found it difficult to adapt published protocols to quantify cell death phenotypes in yeast with satisfying reproducibility.

Mammalian cell death assays that have been applied to yeast include the exposure of phosphatidylserine on the outer plasma membrane leaflet (Annexin V staining), nuclear DNA fragmentation (TUNEL labeling), activation of caspase-like activity (FITC-VAD-FMK cleavage), and release of cytochrome *c* from mitochondria *(15)*. Questions have been raised about whether or not these markers in yeast truly reflect a purposeful death pathway analogous to caspase-dependent apoptosis or any other death pathway in mammalian systems *(15, 16)*. For example, hundreds of caspase substrates have been identified in mammals, and the roles for a subset of these in facilitating apoptosis have been partially delineated, but similar evidence in yeast is largely lacking. While acetic acid *(17)*, pheromones *(2)*, and exogenous Bax *(7)* induce cytochrome *c* release from yeast mitochondria, the physiological role of cytochrome *c* release in yeast, or in any nonmammalian species, remains in question.

Given the difficulties in defining PCD in yeast, we have turned to a genetic definition, and then applied this definition to the development of a reliable assay to detect yeast cell death/survival following treatment with acetic acid. Our definition of a condition that induces yeast PCD is one that can detect the increased survival of yeast lacking the pro-death factor Dnm1/Drp1, a dynamin-like GTPase with pro-death function that is conserved in mammals, flies, and worms *(18–21)*. Conversely, the increased sensitivity to cell death by yeast *FIS1* knockout strains from the YKO collection defines the opposite phenotype *(21)*. Therefore, experimental conditions that distinguish the viabilities of these two knockout strains from the wild-type control strain are deemed to be reflective of PCD in yeast *(9)*. We have now further developed our earlier acetic acid assay so that other investigators can readily apply this strategy with high reproducibility.

Several cell death stimuli have been suggested to induce yeast PCD, including hydrogen peroxide *(22)*, acetic acid *(23, 24)*, high salt *(25)*, and others, but regardless of the specific death stimulus applied, the dose is critical. Any death stimulus can induce non-PCD in yeast or mammalian cells when administered in high doses. Analogous to a clonogenic assay in mammalian models, the yeast community generally defines cell survival as the ability to form a colony on solid medium. This is a stringent test of cell viability in mammalian models, as cultured mammalian cells often exhibit only delayed cell death, such as with Bcl-2 overexpression following a variety of death stimuli. In cases such as this, vital dyes or related strategies are required to reliably quantify cell death *(26, 27)*. A comparable assay for yeast is FUN1 staining to identify metabolic viability, a convenient alternative assay to quantify yeast viability.

2. Materials

2.1. Yeast Cultures and Acetic Acid Treatment

1. Yeast strains:

   Wild-type BY4741 (MATa *his3Δ1 leu2Δ0 met15Δ0 ura3Δ0*)

   Δ*fis1* (MATa *fis1*::kanMX4 *Δhis3Δ1 leu2Δ0 met15Δ0 ura3Δ0*)

   Δ*dnm1* (MATa *dnm1*::kanMX4 *his3Δ1 leu2Δ0 met15Δ0 ura3Δ0*).

2. YPD liquid medium: 2% peptone, 1% yeast extract, 2% glucose.

3. Roller drum (TC-7, New Brunswick Scientific).

4. Glass yeast culture tubes (18 × 150 mm) with loose fitting caps.

5. Acetic acid stock solution (17.3 M) (J. T. Baker).

2.2. Colony Forming Assay

1. 96-well microplates (200 μL).
2. Omni trays.
3. YPD agar plates (YPD + 2% agar).
4. 12-place multichannel pipette.

2.3. FUN1 Staining Assay

1. Live/Dead Yeast Viability Kit (Molecular Probes) containing FUN1 cell stain [10 mM solution in anhydrous dimethylsulfoxide (DMSO)] and Calcofluor White M2R (5 mM solution in water). Stored at –20°C.

2. FUN1 staining solution: 2% glucose/dextrose and 10 mM Na-HEPES (pH 7.2), sterilized with 0.2-μm filter and stored at room temperature.

3. Fluorescence microscope equipped with multipass filter sets appropriate for viewing DAPI, fluorescein, and rhodamine.

4. Microslides (25 × 75 × 1 mm) and micro cover glass (18 × 18 mm) (VWR scientific).

3. Methods

3.1. Culture Growth and Acetic Acid Treatment

1. WT and mutant yeast strains are streaked out from frozen –80°C glycerol stock (without thawing) onto YPD agar plates and incubated at 30°C for 2 days.

2. Single colonies are picked from each plate, inoculated into 2-mL YPD in yeast culture tubes, and incubated at 30°C on a roller drum overnight (~35 rotations/min).

3. Overnight cultures are diluted to OD_{600} = 0.2 in 5-mL YPD (or at least 1 mL more than the amount required) and further incubated at 30°C until mid-log phase, OD_{600} = ≤0.5. Before treatment, all cultures are adjusted to the same OD_{600} (for example the OD of the least dense culture). For best results, adjust the OD_{600} 30 min before reaching 0.5 by diluting the slightly overgrown samples in YPD, then continue incubation until OD_{600} = ~0.5. Further adjustments may be necessary to make all samples equal. If OD_{600} exceeds 0.6, results may be less reliable (*see* **Note 1**).

4. Dispense 2 mL of each mid-log phase culture into each of the two new tubes for treatment with two doses of acetic acid and the remainder is held as the untreated control. Addition of

23-µL and 28-µL acetic acid stock (17.3 M) to 2-mL cultures yields a final concentration of 199 mM and 242 mM, respectively. Vortex immediately after addition and incubate at 30°C for 4 h on rotator (*see* **Note 2**).

3.2. Viability Measured as Colony Forming Units (cfu)

1. For the untreated control, transfer 100 µL of culture to a 96-well plate and perform five 1:5 serial dilutions (20-µL culture is added to 80-µL ddH$_2$O and mixed well by pipetting 5–6 times). Then with clean tips plate 5 µL from each dilution (highest to lowest) onto YPD agar plates.

2. For samples treated for 4 h with acetic acid, perform serial dilutions and plating the same as in **step 1** (*see* **Note 3**).

3. Incubate agar plates at 30°C for 2 days to read the results (**Fig. 1**). Note, the difference between WT and Δ*fis1* is greater at 199 mM, while the difference between WT and Δ*dnm1* is greater at 242 mM acetic acid (*see* **Note 4**).

3.3. Viability Measures by FUN1 Staining Assay

1. After 4 h of acetic acid treatment, 400 µL of 199 mM acetic acid-treated cultures are mixed with 1-mL of sterile FUN1 staining solution in a 1.5-mL microcentrifuge tube and cells are pelleted at 6000 × g for 5 min. Cultures treated with 242 mM acetic acid are not usually suitable for FUN1 staining as too many cells are killed to yield reliable viable cell counts.

2. Discard the supernatant and resuspend the cell pellet in 1-mL sterile FUN1 staining solution. Add 0.5 µL FUN1 cell stain and 5 µL Calcofluor White M2R to the yeast suspension with final concentrations of 5 µM and 25 µM, respectively. Mix thoroughly.

3. Incubate at 30°C in the dark for 30 min.

4. Spin down the yeast pellet and remove all of the supernatant except for ~50 µL. Resuspend the yeast pellet and trap 1.5-µL stained yeast suspension between a slide and cover glass and observe on a fluorescence microscope.

5. Dead cells stain diffusely green, while healthy cells stain blue and contain a red bar inside (**9**). Count at least 200 cells from representative fields.

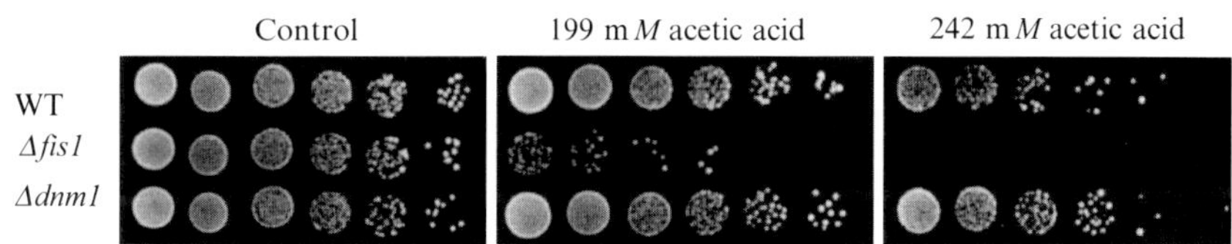

Fig. 1. Colony formation assay of WT, Δ*fis1*, and Δ*dnm1* strains treated with the indicated doses of acetic acid and plated in 5-fold serial dilutions (left to right).

4. Notes

1. Not only are equal cell numbers between samples important for calculating cell survival/death, but the metabolic state of the yeast culture to be treated is also critical. We have found that yeast cells in postdiauxic phase are more resistant to death stimuli than log phase cells (not shown). Therefore, all the samples must be grown under the same conditions, diluted from overnight cultures and recovered in fresh media for the same period. When performing a series of experiments, it is recommended to use the same batch of medium, as different lots can produce different growth characteristics. Because individual strains can contain genetic variants, it is also advisable to test several different colonies.

2. Although this protocol was developed for BY4741 strains and for YPD medium, other strains and media can be used successfully with some adjustments to the protocol. For synthetic dropout medium (0.67% yeast nitrogen base, 30 mg/L amino acids mixture, 2% glucose/dextrose), the yeast cells will grow somewhat slower and will require a longer time to reach mid-log phase following dilution from overnight cultures. The same strains grown in synthetic medium are somewhat more sensitive to the death stimulus, and usually require a lower dose of acetic acid (e.g., 130–180 mM).

3. Serial dilutions used here are a convenient method to reveal large viability differences, but may fail to detect real differences that are less than fivefold (for example, our standard for mammalian cells is only a 20% difference in survival for at least one time-point posttreatment with appropriate reproducibility), though smaller differences are frequently reported. For yeast strains with smaller differences in viability, colony numbers can be counted by plating larger volumes on 60-mm Petri dishes, by plating 1:2 dilutions in the critical range, or by adjusting the dose of acetic acid (**Fig. 1**).

4. Other death stimuli can also be used successfully to distinguish these three strains of yeast. Assays conditions that fail to distinguish the survival of *FIS1* and *DNM1* knockouts from their wild-type control may not yield reliable results. Other factors can affect cell death susceptibility, such as age, ploidy, and background strain. For example, older haploid strains are more sensitive than newly sporulated haploid strains.

Acknowledgments

The authors would like to thank Drs. Wen-Chih Cheng and Yihru Fannjiang who first developed this acetic acid-induced cell death assay for yeast. This work was supported by NIH grant RO1 GM077875.

References

1. Fabrizio, P., Battistella, L., Vardavas, R., Gattazzo, C., Liou, L. L., Diaspro, A., Dossen, J. W., Gralla, E. B., and Longo, V. D. (2004). Superoxide is a mediator of an altruistic aging program in *Saccharomyces cerevisiae*. *The Journal of Cell Biology* **166**, 1055–1067.

2. Severin, F. F., and Hyman, A. A. (2002). Pheromone induces programmed cell death in S. cerevisiae. *Current Biology* **12**, R233–R235.

3. Ivanovska, I., and Hardwick, J. M. (2005) Viruses activate a genetically conserved cell death pathway in a unicellular organism. *The Journal of Cell Biology* **170**, 391–399.

4. Vachova, L., and Palkova, Z. (2005). Physiological regulation of yeast cell death in multicellular colonies is triggered by ammonia. *The Journal of Cell Biology* **169**, 711–717.

5. Kerr, J. F., Wyllie, A. H., and Currie, A. R. (1972). Apoptosis: A basic biological phenomenon with wide-ranging implications in tissue kinetics. *British Journal of Cancer* **26**, 239–257.

6. Manon, S., Chaudhuri, B., and Guerin, M. (1997). Release of cytochrome c and decrease of cytochrome c oxidase in bax-expressing yeast cells, and prevention of these effects by coexpression of bcl-xL. *FEBS Letters* **415**, 29–32.

7. Priault, M., Camougrand, N., Kinnally, K. W., Vallette, F. M., and Manon, S. (2003). Yeast as a tool to study bax/mitochondrial interactions in cell death. *FEMS Yeast Research* **4**, 15–27.

8. Kane, D. J., Sarafian, T. A., Anton, R., Hahn, H., Gralla, E. B., Valentine, J. S., Ord, T., and Bredesen, D. E. (1993). Bcl-2 inhibition of neural death: Decreased generation of reactive oxygen species. *Science* **262**, 1274–1277.

9. Fannjiang, Y., Cheng, W. C., Lee, S. J., Qi, B., Pevsner, J., McCaffery, J. M., Hill, R. B., Basanez, G., and Hardwick, J. M. (2004). Mitochondrial fission proteins regulate programmed cell death in yeast. *Genes and Development* **18**, 2785–2797.

10. Vander Heiden, M. G., Choy, J. S., VanderWeele, D. J., Brace, J. L., Harris, M. H., Bauer, D. E., Prange, B., Kron, S. J., Thompson, C. B., and Rudin, C. M. (2002). Bcl-x(L) complements *Saccharomyces cerevisiae* genes that facilitate the switch from glycolytic to oxidative metabolism. *The Journal of Biological Chemistry* **277**, 44870–44876.

11. Madeo, F., Herker, E., Maldener, C., Wissing, S., Lachelt, S., Herlan, M., Fehr, M., Lauber, K., Sigrist, S. J., Wesselborg, S., and Frohlich, K.U. (2002). A caspase-related protease regulates apoptosis in yeast. *Molecular Cell* **9**, 911–917.

12. Yoon, H. J., and Carbon, J. (1999). Participation of bir1p, a member of the inhibitor of apoptosis family, in yeast chromosome segregation events. *Proceedings of the National Academy of Science of the United States of America* **96**, 13208–13213.

13. Fahrenkrog, B., Sauder, U., and Aebi, U. (2004) The *S. cerevisiae* htra-like protein nma111p is a nuclear serine protease that mediates yeast apoptosis. *The Journal of Cell Science* **117**, 115–126.

14. Wissing, S., Ludovico, P., Herker, E., Buttner, S., Engelhardt, S. M., Decker, T., Link, A., Proksch, A., Rodrigues, F., Corte-Real, M., Frohlich, K. U., Manns, J., Cande, C., Sigrist, S. J., Kroemer, G., and Madeo, F. (2004). An aif orthologue regulates apoptosis in yeast. *The Journal of Cell Biology* **166**, 969–974.

15. Hardwick, J. M., and Cheng, W. C. (2004). Mitochondrial programmed cell death pathways in yeast. *Developmental Cell* **7**, 630–632.

16. Cheng, W. C., Leach, K. M., and Hardwick, J. M. (2008). Mitochondrial death pathways in yeast and mammalian cells. *Biochim Biophys Acta* **1783**, 1272–1279.

17. Ludovico, P., Rodrigues, F., Almeida, A., Silva, M. T., Barrientos, A., and Corte-Real, M. (2002). Cytochrome *c* release and mitochondria involvement in programmed cell death induced by acetic acid in *Saccharomyces cerevisiae*.

Molecular Biology of the Cell **13**, 2598–2606.

18. Frank, S., Gaume, B., Bergmann-Leitner, E. S., Leitner, W. W., Robert, E. G., Catez, F., Smith, C. L., and Youle, R. J. (2001). The role of dynamin-related protein 1, a mediator of mitochondrial fission, in apoptosis. *Developmental Cell* **1**, 515–525.

19. Goyal, G., Fell, B., Sarin, A., Youle, R.J. and Sriram, V. (2007). Role of mitochondrial remodeling in programmed cell death in *Drosophila melanogaster*. *Developental Cell* **12**, 807–816.

20. Jagasia, R., Grote, P., Westermann, B., and Conradt, B. (2005). Drp-1-mediated mitochondrial fragmentation during egl-1-induced cell death in *C. elegans*. *Nature* **433**, 754–760.

21. Cheng, W. C., Teng, X., Park, H. K., Tucker, C. M., Dunham, M. J., and Hardwick, J. M. (2008) Deletion of mitochondrial fission gene Fis1 **deficiency** selects for a compensatory mutations responsible for cell death and growth control defects *Cell Death and Differentiation* **15**, 1838–1846.

22. Madeo, F., Frohlich, E., Ligr, M., Grey, M., Sigrist, S. J., Wolf, D. H., and Frohlich, K. U. (1999). Oxygen stress: A regulator of apoptosis in yeast. *The Journal of Cell Biology* **145**, 757–767.

23. Ludovico, P., Sansonetty, F., and Corte-Real, M. (2001). Assessment of mitochondrial membrane potential in yeast cell populations by flow cytometry. *Microbiology (Reading England)* **147**, 3335–3343.

24. Ludovico, P., Sousa, M. J., Silva, M. T., Leao, C., and Corte-Real, M. (2001). *Saccharomyces cerevisiae* commits to a programmed cell death process in response to acetic acid. *Microbiology (Reading England)* **147**, 2409–2415.

25. Huh, G. H., Damsz, B., Matsumoto, T. K., Reddy, M. P., Rus, A. M., Ibeas, J. I., Narasimhan, M. L., Bressan, R. A., and Hasegawa, P. M. (2002). Salt causes ion disequilibrium-induced programmed cell death in yeast and plants. *Plant J* **29**, 649–659.

26. Cheng, E. H., Levine, B., Boise, L. H., Thompson, C. B., and Hardwick, J. M. (1996). Bax-independent inhibition of apoptosis by bcl-xl. *Nature* **379**, 554–556.

27. Cheng, E. H., Kirsch, D. G., Clem, R. J., Ravi, R., Kastan, M. B., Bedi, A., Ueno, K., and Hardwick, J. M. (1997). Conversion of bcl-2 to a bax-like death effector by caspases. *Science* **278**, 1966–1968.

Chapter 24

Detection of Cell Death in *Drosophila*

Kimberly McCall, Jeanne S. Peterson, and Tracy L. Pritchett

Summary

Drosophila is a powerful model system for the identification of cell death genes and understanding the role of cell death in development. In this chapter, we describe three methods typically used for the detection of cell death in *Drosophila*. The TUNEL and acridine orange methods are used to detect dead or dying cells in a variety of tissues. We focus on methods for the embryo and the ovary, but these techniques can be used on other tissues as well. The third method is the detection of genetic interactions by expressing cell death genes in the *Drosophila* eye.

Key words: Drosophila, Apoptosis, Cell death, Embryo, Ovary

1. Introduction

Drosophila has unique genetic and cell biological advantages as a model system for the study of programmed cell death (reviewed in **ref.** *(1)*. Cell death occurs in diverse developmental processes in *Drosophila*, including the formation of the embryonic nervous system, the destruction of larval tissues during metamorphosis, the morphogenesis of the eye, and the generation of eggs in the ovary *(2–5)*. Additionally, cells die ectopically in response to external stimuli, such as X-rays *(3)*.

Genetic studies in *Drosophila* have uncovered three cell death activators, *reaper*, *hid*, and *grim* (collectively called RHG genes) *(6–8)*. Flies homozygous for the *H99* deletion, which removes

Peter Erhardt and Ambrus Toth (eds.), *Apoptosis*, Methods in Molecular Biology, vol. 559
DOI 10.1007/978-1-60327-017-5_24
© Humana Press, a part of Springer Science + Business Media, LLC 2004, 2009

these three genes, die during embryogenesis with a complete block in apoptosis *(6)*. A fourth related gene, *sickle*, lies outside the *H99* deficiency *(9–11)*. Although the RHG genes are essential for embryonic apoptosis, they are not required for the cell death of ovarian nurse cells, suggesting that there exists at least one other cell death pathway in flies *(12, 13)*. Overexpression of *reaper*, *hid*, or *grim* leads to wide-spread apoptosis which can be blocked by coexpression of the caspase inhibitor p35 or overexpression of inhibitor of apoptosis proteins (IAPs) *(7, 8, 14, 15)*. Reaper, Hid, and Grim bind to IAPs, displacing bound caspases (reviewed in **refs.16, 17)**. Caspase inhibition by IAPs is critical in *Drosophila* as homozygous loss-of-function *thread* (*diap1*) mutants die early in embryogenesis with extensive apoptosis *(18–20)*. Mammalian proteins such as Smac/Diablo and Htr/Omi share similar IAP binding motifs and function analogously to the RHG proteins (reviewed in **ref. 16)**.

The genome sequence of *Drosophila* has revealed fly homologs for most mammalian cell death genes. These cell death genes include seven caspases, one Apaf-1-related gene, two Bcl-2 family genes, four genes encoding BIR domain proteins, a TNF-related gene, a TNFR related gene, a FADD-like gene, a CAD-like gene, and a p53 ortholog (reviewed in **ref. 21)**. Loss-of-function phenotypes have been generated for several of these genes by classical genetics, transposon insertion, RNAi, or expression of dominant negative alleles (reviewed in **refs. 16, 17, 22)**. The Bcl-2 family members and TNF homolog appear to play relatively minor roles in *Drosophila*, and the major form of regulation occurs through the RHG proteins and Diap1. Dronc (caspase-9 ortholog) and Drice (caspase-3 ortholog) are the principal caspases required for *Drosophila* cell death.

One of the strengths of the *Drosophila* system is the ability to use cell biological techniques on a variety of developmental stages. The condensed nuclear morphology associated with apoptosis can be easily detected with DNA dyes such as DAPI and propidium iodide. Mountants with these dyes are available from Vector labs. Expression of cell death genes can be visualized by whole mount RNA in situ hybridization or immunocytochemistry. Commercial antibodies are rarely available for *Drosophila* proteins; however, antibodies useful for immunocytochemistry have been described for a small number of cell death proteins, such as Hid, Thread, and Cytochrome *c* *(20, 23–26)*. Fixation protocols for antibody staining of *Drosophila* tissues can be found in the references cited here or in *Drosophila* protocol books (such as **ref. 27)**. Antibodies against active human caspase-3 (Cell Signaling Technologies) cross-react with *Drosophila* caspases and/or their targets and provide a convenient readout for caspase activity in flies *(28–31)*. Additionally, transgenic lines have been described that express caspase reporters *(32, 33)*. The detection of macrophages is correlated with the amount of cell death and macrophages can be visualized by antibodies such as those against Peroxidasin and Croquemort *(3, 34–36)*. In addition to studying cell death in the intact fly, several *Drosophila* cell lines exist which can be subjected to the same analysis as mammalian cell lines *(37)*. RNAi works efficiently on these cell lines and is less complicated than in mammals because flies lack an interferon response to long dsRNAs. Genome-wide RNAi screens have been performed and used as a first step toward identifying mammalian genes *(38, 39)*.

Other forms of cell death exist in addition to apoptosis, particularly autophagic cell death and necrosis. Transmission electron microscopy can be used to define the type of cell death, and markers have been developed to detect autophagy. LC3-GFP (or Atg8-GFP) and Atg5-GFP become punctate in cells undergoing autophagy, and Lysotracker (Invitrogen) can be used to visualize an increase in lysosome size or number *(40, 41)*. Mitochondrial fission is associated with apoptosis in diverse organisms and mitochondrial-GFP markers or Mitotracker (Invitrogen) have been used to detect mitochondrial morphology changes in *Drosophila* *(25, 42, 43)*.

In this chapter, we describe three methods typically used for the detection of apoptosis in *Drosophila*. The chapter is updated from the previous chapters in the Methods of Molecular Biology series *(44, 45)*. The TUNEL and acridine orange methods are used to detect dead or dying cells in a variety of tissues. We focus on methods for the embryo and the ovary, but these techniques can be used on other tissues as well. The third method is the detection of genetic interactions by expressing cell death genes in the *Drosophila* eye.

2. Materials

The methods described below review general fly handling as well as specific apoptosis techniques. However, workers new to *Drosophila* should consult with a *Drosophila* laboratory before beginning to work with flies.

2.1. Fly Handling

1. Instant fly food (Fisher Scientific or Carolina Biological). Fly food should be poured into vials or bottles (available from Fisher Scientific or VWR). Other sources for *Drosophila* reagents and supplies can be found at the Bloomington *Drosophila* stock center Web site http://flystocks.bio.indiana.edu/.

2. Dissecting microscope with a 6.5×–50× magnification range.

3. Diffuser pad for CO_2 anesthesia or shaved ice and dry vials for cold anesthesia.

4. Baskets with 80μm nitex mesh (Sefar America) for dechorionating embryos. Baskets can be made by attaching a small piece of mesh to one end of a 1-in. long tube.

5. Fine forceps, tungsten needles, and glass plates or depression slides for dissection.

6. Apple juice agar plates: Mix 90 g of Difco agar with 3 L of water, autoclave for 50 min and cool in a 60°C water bath. Mix 1 L of apple juice with 100 g of sugar and heat to 60°C to dissolve. Combine agar/water and juice/sugar mixtures, stir, add 60 mL of a 10% solution of Tegosept (https://www.geneseesci.com) in ethanol and pour the plates. The tops of 35×10-mm plates (Falcon) will fit fly food bottles from Fisher. Alternatively, instant egg laying media may be purchased from https://www.geneseesci.com.

7. Egg laying chambers. These may be made by cutting a hole in the side of a dry fly food bottle and stuffing it with a cotton ball. The apple juice agar plate will fit on the mouth of the bottle.

8. Yeast paste: Mix granular yeast (Sci-Mart) and water in a 1:1 ratio into a smooth paste.

9. *Drosophila* Ringers (DR): 130 mM NaCl, 4.7 mM KCl, 1.9 mM $CaCl_2$, 10 mM Hepes, pH 6.9. Make 10× and store frozen.

2.2. TUNEL Materials

1. Phosphate buffered saline (PBS): 130 mM NaCl, 7 mM $Na_2HPO_4 \cdot 2H_2O$, 3 mM $NaH_2PO_4 \cdot 2H_2O$.

2. PBS with 0.1% Tween-20 (PBT).

3. Heptane.

4. TritonX-100.

5. Proteinase K (Fisher) stock solution: 20 mg/mL in dH_2O, stored frozen in 10 μL aliquots.

6. Bovine serum albumin (BSA, Fisher).

7. Fixative: 4% paraformaldehyde (Sigma) in PBS. Heat to dissolve.

8. 70% glycerol in PBS.

9. Methanol.

10. Household mercury-free bleach (Fisher Scientific; for dechorionating embryos).

For a colorimetric reaction you will need reagents in **items 11–14**. For fluorescent TUNEL, use reagents in **items 15** and **16**.

11. Normal goat serum (GIBCO, Invitrogen).

12. pH 9 buffer: 0.1 M Tris base, pH 9.5, 0.1 M NaCl, 50 mM $MgCl_2$, 0.1% Tween-20.

13. ApopTag reagents (Millipore): Equilibration buffer (EB), Reaction buffer containing nucleotides labeled with Digoxigenin

(RXB), Terminal deoxyribonucleotidyl transferase (TdT), and Stop-wash buffer (SWB).

14. Roche Applied Science reagents: Anti-Digoxigenin antibody complexed to Alkaline Phosphatase (Anti-Dig-AP), Nitroblue Tetrazolium(NBT), and 5-Bromo-4-Chloro-3-Indolyl Phosphate Toluidinium (X-Phos) stock solution.

15. ApopTag Fluorescein Direct In Situ Apoptosis Detection Kit (Millipore, Billerica, MA).

16. Vectashield mountant with DAPI (Vector Labs).

2.3. Acridine Orange Materials

1. Stock solution acridine orange (AO, Sigma) dissolved in dH_2O at 1 mg/mL, stored in dark at 4°C.

2. 0.1 M phosphate buffer, pH 7.0.

3. Heptane.

4. Halocarbon oil (series 700, Halocarbon Products Corporation).

5. Fluorescence microscope equipped with fluorescein, rhodamine and UV filters, and a camera.

2.4. Genetic Interactions Materials

1. pGMR vector *(14)* or other eye-specific vector. The UASt vector available from https://dgrc.cgb.indiana.edu/may also be used.

2. pπ25.7wcΔ2-3 plasmid *(27)*, available from many *Drosophila* laboratories.

3. Injection buffer: 5 mM KCl, 0.1 mM NaH_2PO_4 pH 6.8.

4. Fly microinjection facility.

5. Fly strains – *GMRrpr, GMRhid, GMRp35, GMRdiap1, GMRdiap2* are all available from the *Drosophila* stock center at Bloomington, IN, USA (*see* http://fly.bio.indiana.edu/).

6. A camera attachment for the dissecting microscope is necessary for making a photographic record of eye phenotypes.

3. Methods

3.1. Sample Preparation

For well-developed ovaries and/or good embryo production, start with equal numbers of 3–7-day-old male and female flies kept together in uncrowded conditions, transferring them to new food vials supplemented with wet yeast paste once or twice daily for 4 days or more before collecting samples.

3.1.1. Ovary Dissection

1. Anesthetize flies under CO_2 or on ice.

2. Dissect females in depression plates or slides in a drop of DR. Grasp fly between thorax and abdomen with forceps, and pull

at the terminal part of the abdomen with another pair of forceps to release the ovaries and other organs from the cavity.

3. Tease ovaries away from debris and separate ovarioles from each other with tungsten needles or forceps.

4. Transfer tissue in DR to microcentrifuge tubes (*see* **Notes 1 and 2**). Hold tissues on ice until all samples have been collected. Proceed to **Subheading 3.2.1** for TUNEL or in **Subheading 3.3** for AO staining.

3.1.2. Embryo Collection

1. Apply a dab of fresh yeast paste to an apple juice agar plate. Have the apple juice agar plate ready to fix against the mouth of the egg laying chamber. Transfer flies to the chamber, attach the plate with tape, invert the bottle, and allow flies to lay eggs for the desired time.

2. Use water and a fine brush to dislodge the embryos from the surface of the plate, and collect them with a Pasteur pipette or a large (1,000 µL) pipette tip (*see* **Note 2**).

3. Transfer the embryos to baskets and remove the water. Dechorionate embryos in baskets using 50% freshly diluted bleach for 2–5 min and wash several times with water. Proceed to **Subheading 3.2.2** for TUNEL or **Subheading 3.3** for AO staining.

3.2. TUNEL Staining

For TUNEL staining with a colorimetric reaction, ovaries or embryos are treated as for in situ hybridization and antibody staining. The protocol below was derived from the description of ovarian tissue preparation by Verheyen and Cooley *(46)*, from descriptions of embryo staining in protocols 54 and 95 in Ashburner *(27)* and from the description of TUNEL staining by White et al. *(6, 47)*. The fluorescent TUNEL staining protocol (*see* **Subheading 3.2.4**) was derived from protocols described in the product guide and in *(48)*.

3.2.1. Ovary Fixation

1. Remove DR from ovaries (*see* **Note 3**), add heptane/fix 5:1 and rotate for 30 min at room temperature (RT, except for fluorescent TUNEL protocol, *see* **Note 4**).

2. Remove heptane/fix and wash twice with excess PBT, taking care to remove all heptane droplets. Proceed to **Subheading 3.2.3**.

3.2.2. Embryo Fixation

1. Mix heptane and fix 1:1

2. Transfer embryos to the fixing solution (*see* **Notes 4** and **5**) and shake for 20 min at RT.

3. Remove fix (bottom layer) first and then remove heptane. Add fresh heptane and shake.

4. Add a double quantity of methanol and shake hard (vortex) for 2 min to remove the vitelline membrane.

5. Discard embryos at interface, remove heptane and then methanol, and wash twice with methanol.

6. Rehydrate through a series of 75%, 50%, and 25% methanol in PBT. Wash twice with PBT. Proceed to **Subheading 3.2.3** for colorimetric TUNEL or **Subheading 3.2.4** for fluorescent TUNEL.

3.2.3. Colorimetric TUNEL Staining Protocol

1. Treat fixed tissue with Proteinase K, 10 μg/mL in PBT (5 min for ovaries, 3 min for embryos) and wash twice with PBT (*see* **Note 6**).

2. Post-fix for 20 min in a solution of 4% paraformaldehyde in PBS, then wash five times, 5 min each, in PBT.

3. Equilibrate for 1 h at RT in EB.

4. Incubate overnight at 37°C in a reaction mix consisting of RXB and TdT in a 2:1 ratio, with 0.3% Triton-X 100.

5. Preabsorb anti-Dig-AP, diluted 1:2,000 in PBT, with fixed tissue for 2 h at RT or at 4°C overnight.

6. Remove RXB and TdT from tissue and incubate in SWB diluted to 1:34 in water at 37°C for 3–4 h. Remove SWB and wash three times, 5 min each, in PBT.

7. Block in a solution of 2 mg/mL BSA, 5% normal goat serum in PBT for 1 h at RT.

8. Incubate tissue in preabsorbed antibody for 2 h at RT or overnight at 4°C.

9. Wash four times for 20 min each in PBT and wash twice, 20 min each, in pH 9 buffer.

10. Add 18 μL of NBT/X-phos solution to 1 mL of pH 9 buffer, and incubate tissues, watching carefully for the color reaction.

11. Stop the reaction with 2–3 PBT washes and mount in 70% glycerol.

3.2.4. Fluorescent TUNEL Protocol

1. Treat fixed tissue with 20 μg/mL Proteinase K in PBS for 20 min (*see* **Note 6**).

2. Wash two times for 5 min each in PBT.

3. Post-fix for 5 min in fix.

4. Wash three times, 5 min each, in PBT.

5. Wash two times, 5 min each, in equilibration buffer (from kit).

6. Mix enzyme to working strength solution 30% TdT: 70% Reaction Buffer (from kit) and add to sample.

7. Incubate in the dark for 3 h in a 37°C waterbath, flicking tube every 30 min to mix contents.

8. Remove enzyme solution and wash quickly with working strength stop wash buffer in the dark (one 1-min wash and one 5-min wash).

9. Wash three times in PBT over 30 min in dark.

10. Mount in Vectashield with DAPI and view under epifluorescence.

3.3. Acridine Orange Staining

Acridine orange is a vital dye that differentially stains living and dying cells. The use of other vital dyes is described elsewhere *(3)*. The advantage of acridine orange is that it is performed quickly on live tissue. However, this also means that the tissue must be examined and photographed immediately after staining. These protocols are derived from protocols described in *(3, 12)*.

3.3.1. Embryo Protocol

1. Dilute AO stock solution to 5 µg/mL in 0.1 M phosphate buffer.

2. Collect embryos in mesh baskets as described in **Subheading 3.1.2** and wash only in water (*see* **Note 7**). Using a fine tipped paintbrush, transfer embryos from the mesh to tubes containing an equal volume of heptane and the 5 µg/mL AO solution. Microcentrifuge tubes or glass tubes with tight-fitting caps may be used.

3. Shake tubes vigorously by hand for 3–5 min. Shaking by hand improves the permeability of the embryos (*see* **Note 8**).

4. Pipette off the liquid and replace with heptane (*see* **Note 3**).

5. Pipette the embryos in heptane onto glass slides. Try to keep the embryos separated and soak up the heptane using a kimwipe twisted into a point (*see* **Note 9**). Quickly cover the embryos with halocarbon oil and a coverslip.

6. View the slide immediately under epifluorescence. AO staining is visible under both rhodamine and fluorescein filters. The rhodamine filter often looks better, as the fluorescein filter shows more background and smearing from residual heptane.

3.3.2. Ovary Protocol

1. Dilute AO stock solution to 10 µg/mL in 0.1 M phosphate buffer.

2. Transfer dissected ovaries to an eppendorf tube containing 15 µL heptane and 15 µL of 10 µg/mL AO solution.

3. Flick tube gently to mix and rotate for 5 min.

4. Transfer ovaries to slides and spread out the ovary tissue into individual egg chambers if possible. Pipette off the AO/heptane mixture or use a Kimwipe twisted into a point (*see* **Note 9**). Cover with halocarbon oil and a coverslip.

5. View the slide immediately under epifluorescence, using the fluorescein, rhodamine, or UV filter. Under UV, the apoptotic nuclei stain yellow or red *(12)*.

3.4. Detection of Genetic Interactions

The *Drosophila* eye is a commonly studied tissue for many cell biological processes. It has several advantages: it has a repeating structure made up of only a few cell types, the development of this tissue is highly organized and the eye is a nonessential tissue. Cell death normally occurs in the developing eye disc, and can easily be induced. Several eye-specific expression vectors exist and we will focus on one of these, the pGMR vector *(14)*, which is expressed in all cells in the differentiated portion of the eye disc. Many fly strains have been generated that ectopically express cell death genes in the eye **(Table 1)**. Several of these strains have been used to identify interacting genes through genetic modifier screens *(1)*.

Table 1
Drosophila transgenic strains overexpressing cell death genes in the eye

Gene class	Transgenic strain	References
Cell death activator	*GMRreaper*	*(15, 47)*
	GMRhid	*(7)*
	GMRgrim	*(8)*
	UASsickle[a]	*(9, 10)*
	GMRdebcl, UASdebcl[a]	*(49–51)*
	gl-p53	*(52, 53)*
	GMR-eiger	*(54)*
Cell death protector	*GMRp35*	*(14)*
	GMRdiap1	*(15)*
	GMRdiap2	*(15)*
	UASbuffy[a]	*(55)*
Caspase	*GMRdcp-1*	*(56)*
	GMRdrICE	*(56)*
	UASdronc[a]*, GMRdronc*	*(57, 58)*
	GMRstrica	*(59)*
	GMRdamm	*(60)*
Human genes	*GMRhuntingtin*	*(61)*
	UASbax[a]	*(62)*
	UASbcl-2[a]	*(62)*
	UASIce[a]	*(63)*
C. elegans genes	*UASced-3*[a]	*(63)*
	GMRced4	*(64)*

[a]UAS lines may be crossed to a variety of GAL4 drivers, such as GMRGAL4, to induce eye expression

3.4.1. Generation of Transgenic Flies

1. Subclone a gene of interest into the pGMR vector using standard molecular biology techniques.

2. Purify the final construct using CsCl banding or the Endofree Plasmid Kit (Qiagen). At this point, the construct may be sent to one of several companies that generates transgenic flies for a fee (see transgenic services at http://flybase.org/static_pages/allied-data/external_resources5.html). Larvae will be shipped and you may proceed to **step 6**. Alternatively, the transgenic flies may be made in the lab (*see* **steps 3–5**).

3. Combine 10 μg of the final pGMR construct with 4 μg of the pπ25.7wc 2-3 plasmid (purified with CsCl or Endofree kit), and ethanol precipitate. Resuspend in 20 μL of injection buffer.

4. Inject into the posterior of preblastoderm embryos that are *white⁻* (this requires a *Drosophila* microinjection facility). Further details on *Drosophila* transformation can be found in *(27)*.

5. Transfer larvae that survive the injection procedure to a vial of fly food. If standard cornmeal-molasses fly food is used, survival can be increased by supplementing with a small amount of soft instant flyfood.

6. Collect the adult flies as they emerge and mate them individually to *white⁻* flies. The pGMR vector carries the *white⁺* gene that will mark the transgenic flies. However, the injected flies will only carry the transgene in their germline, and expression in the eye will not be visible until the next generation of flies.

7. Examine the progeny for *white⁺* eyes (ranging from dark orange to red with the pGMR vector, or yellow to red with other vectors). Cross to *white⁻* flies to maintain the new transgenic strain and to generate homozygotes.

8. Examine the eyes of the heterozygous and homozygous flies. A smaller eye, diffuse pigmentation or irregular (rough) appearance may indicate ectopic cell death. An irregular appearance could also indicate a block in cell death, however (*see* **Note 10**).

3.4.2. Test for Genetic Interactions

1. Perform fly crosses with the new transgenic line and some of the strains are listed in **Table 1** (*see* **Note 11**). Examine progeny carrying both transgenes.

2. Inhibition of cell death can be accomplished by crossing to the *GMRp35* or *GMRdiap1* strain. The caspase inhibitor p35 is a potent inhibitor of cell death in many systems. However, not all caspases are inhibited by p35 so *GMRdiap1* should also be tested.

3. The transgenic line may be crossed to loss-of-function mutations as well. Such mutations are available for the *H99* genes, *diap1, dark, eiger*, and most of the caspases (reviewed in *16, 17, 22*). Additionally, RNAi lines have been described for

many of these genes *(65)*. Genetic interactions can often be visualized in heterozygous combinations (reviewed in **ref.*1***).

4. Photograph eyes under the dissecting microscope, or use scanning electron microscopy for more impressive images.

4. Notes

1. Use low retention microcentrifuge tubes to minimize tissue adhering to the side of the tube.

2. To move embryos or dissected tissue from plates to staining tubes, and from staining tubes to slides, use plastic pipette tips from which 3/16 in. of the end has been removed with a razor blade. Rinse the tip in PBT just before using, hold the pipette vertically at all times and pipette very slowly. If using Pasteur or transfer pipettes, keep the tissue within the narrow bottom portion of the pipette.

3. To change solutions in which embryos or tissues are incubating, use glass pipettes drawn out to a fine tip. Remove virtually all liquid from samples by touching the tip to the meniscus and moving it slowly toward the sample.

4. For the fluorescent TUNEL protocol, extensive fixation times can reduce the signal. We have had success with ovaries by fixing for 10 min in 1 part Buffer B (100 mM KH_2PO_4/ K_2HPO_4 (pH 6.8), 450 mM KCl, 150 mM NaCl, 20 mM $MgCl_2 \cdot 6H_2O$), 1 part 36% formaldehyde, 4 parts dH_2O, and an equal volume heptane *(46)* .

5. Embryos may be transferred directly from the mesh to the fix with a fine paintbrush. Alternatively, embryos may be washed in the baskets with 0.1% Triton-X 100 in water. The embryos are then pipetted to an empty tube and the embryos are allowed to settle at the bottom of the tube. Remove the 0.1% Triton-X 100 and replace with heptane/fix (*see* **Note 3**).

6. The incubation step with Proteinase K can be variable. When using a new solution of Proteinase K it is best to try several concentrations and incubation times to find the optimal balance between tissue degradation and signal intensity.

7. For acridine orange staining, it is critical that there is not a trace of detergent when embryos are washed (such as Triton-X 100). Detergent will completely abolish AO staining.

8. It is essential that the tubes containing embryos be shaken very hard by hand. Standard rotation of the tubes is not sufficient for the heptane/AO to permeabilize the vitelline membrane.

9. Do not allow embryos or ovary tissues to dry out when the heptane is removed, as they will shrivel up quickly. Heptane evaporates rapidly; blowing gently on the slide will speed up its evaporation.

10. A cell death phenotype can be confirmed by sectioning the eye, and examining with conventional light microscopy *(66)*. Imaginal eye discs may also be stained with acridine orange or TUNEL.

11. To identify genetic interactions, it is necessary that the flies are grown at the same temperature (usually 25°C) and the appropriate control strains are grown at the same time. The level of expression from the pGMR vector is highly dependent on the temperature.

Acknowledgments

We thank Daniela Drummond-Barbosa and current and previous lab members for help in devising these protocols. KM is supported by the NIH R01 GM60574.

References

1. Hay, B. A., Huh, J. R., and Guo, M. (2004). The genetics of cell death: approaches, insights and opportunities in *Drosophila*. *Nat. Rev. Genet.* **5**, 911–922.

2. Wolff, T. and Ready, D. F. (1991). Cell death in normal and rough eye mutants of *Drosophila*. *Development* **113**, 825–839.

3. Abrams, J. M., White, K., Fessler, L. I., and Steller, H. (1993). Programmed cell death during *Drosophila* embryogenesis. *Development* **117**, 29–43.

4. Jiang, C., Lamblin, A. F., Steller, H., and Thummel, C. S. (2000). A steroid-triggered transcriptional hierarchy controls salivary gland cell death during *Drosophila* metamorphosis. *Mol. Cell* **5**, 445–455.

5. McCall, K. (2004). Eggs over easy: cell death in the *Drosophila* ovary. *Dev. Biol.* **274**, 3–14.

6. White, K., Grether, M. E., Abrams, J. M., Young, L., Farrell, K., and Steller., H. (1994). Genetic control of programmed cell death in *Drosophila*. *Science* **264**, 677–683.

7. Grether, M. E., Abrams, J. M., Agapite, J., White, K., and Steller, H. (1995). The *head involution defective* gene of *Drosophila melanogaster* functions in programmed cell death. *Genes Dev.* **9**, 1694–1708.

8. Chen, P., Nordstrom, W., Gish, B., and Abrams, J. M. (1996). *Grim*, a novel cell death gene in *Drosophila*. *Genes Dev.* **10**, 1773–1782.

9. Srinivasula, S. M., Datta, P., Kobayashi, M., et al. (2002). *Sickle*, a novel *Drosophila* death gene in the *reaper/hid/grim* region, encodes an IAP-inhibitory protein. *Curr. Biol.* **12**, 125–130.

10. Wing, J. P., Karres, J. S., Ogdahl, J. L., Zhou, L., Schwartz, L. M., and Nambu, J. R. (2002). *Drosophila sickle* is a novel *grim-reaper* cell death activator. *Curr. Biol.* **12**, 131–135.

11. Christich, A., Kauppila, S., Chen, P., Sogame, N., Ho, S. I., and Abrams, J. M. (2002). The damage-responsive *Drosophila* gene *sickle* encodes a novel IAP binding protein similar to but distinct from *reaper, grim,* and *hid*. *Curr. Biol.* **12**, 137–140.

12. Foley, K. and Cooley, L. (1998). Apoptosis in late stage *Drosophila* nurse cells does not require genes within the *H99* deficiency. *Development* **125**, 1075–1082.

13. Peterson, J. S., Bass, B. P., Jue, D., Rodriguez, A., Abrams, J. M., and McCall, K. (2007). Non-

canonical cell death pathways act during *Drosophila* oogenesis. *Genesis* **45**, 396–404.

14. Hay, B. A., Wolff, T., and Rubin, G. M. (1994). Expression of baculovirus *p35* prevents cell death in *Drosophila*. *Development* **120**, 2121–2129.

15. Hay, B. A., Wassarman, D. A., and Rubin, G. M. (1995). *Drosophila* homologs of baculovirus inhibitor of apoptosis proteins function to block cell death. *Cell* **83**, 1253–1262.

16. Kornbluth, S. and White, K. (2005). Apoptosis in *Drosophila*: neither fish nor fowl (nor man, nor worm). *J. Cell Sci.* **118**, 1779–1787.

17. Hay, B. A. and Guo, M. (2006). Caspase-dependent cell death in *Drosophila*. *Annu. Rev. Cell Dev. Biol.* **22**, 623–650.

18. Wang, S. L., Hawkins, C. J., Yoo, S. J., Muller, H. A., and Hay, B. A. (1999). The *Drosophila* caspase inhibitor DIAP1 is essential for cell survival and is negatively regulated by HID. *Cell* **98**, 453–463.

19. Goyal, L., McCall, K., Agapite, J., Hartwieg, E., and Steller, H. (2000). Induction of apoptosis by *Drosophila reaper, hid* and *grim* through inhibition of IAP function. *EMBO J.* **19**, 589–597.

20. Lisi, S., Mazzon, I., and White, K. (2000). Diverse domains of THREAD/DIAP1 are required to inhibit apoptosis induced by REAPER and HID in *Drosophila*. *Genetics* **154**, 669–678.

21. Vernooy, S. Y., Copeland, J., Ghaboosi, N., Griffin, E. E., Yoo, S. J., and Hay, B. A. (2000). Cell death regulation in *Drosophila*: conservation of mechanism and unique insights. *J. Cell Biol.* **150**, F69–F75.

22. Kanda, H. and Miura, M. (2004). Regulatory roles of JNK in programmed cell death. *J. Biochem.* **136**, 1–6.

23. Haining, W. N., Carboy-Newcomb, C., Wei, C. L., and Steller, H. (1999). The proapoptotic function of *Drosophila* Hid is conserved in mammalian cells. *Proc. Natl Acad. Sci. USA* **96**, 4936–4941.

24. Varkey, J., Chen, P., Jemmerson, R., and Abrams, J. M. (1999). Altered cytochrome c display precedes apoptotic cell death in *Drosophila*. *J. Cell Biol.* **144**, 701–710.

25. Abdelwahid, E., Yokokura, T., Krieser, R. J., Balasundaram, S., Fowle, W. H., and White, K. (2007). Mitochondrial disruption in *Drosophila* apoptosis. *Dev. Cell* **12**, 793–806.

26. Yoo, S. J., Huh, J. R., Muro, I., et al. (2002). Hid, Rpr and Grim negatively regulate DIAP1 levels through distinct mechanisms. *Nat. Cell Biol.* **4**, 416–424.

27. Ashburner, M. (1989). *Drosophila, A Laboratory Handbook*, Cold Spring Harbor Laboratory, Cold Spring Harbor, NY.

28. Yu, S. Y., Yoo, S. J., Yang, L., et al. (2002). A pathway of signals regulating effector and initiator caspases in the developing *Drosophila* eye. *Development* **129**, 3269–3278.

29. Martin, D. N. and Baehrecke, E. H. (2004). Caspases function in autophagic programmed cell death in *Drosophila*. *Development* **131**, 275–284.

30. Baum, J. S., St George, J. P., and McCall, K. (2005). Programmed cell death in the germline. *Semin. Cell Dev. Biol.* **16**, 245–259.

31. Xu, D., Wang, Y., Willecke, R., Chen, Z., Ding, T., and Bergmann, A. (2006). The effector caspases drICE and dcp-1 have partially overlapping functions in the apoptotic pathway in *Drosophila*. *Cell Death Differ.* **13**, 1697–1706.

32. Takemoto, K., Nagai, T., Miyawaki, A., and Miura, M. (2003). Spatio-temporal activation of caspase revealed by indicator that is insensitive to environmental effects. *J. Cell Biol.* **160**, 235–243.

33. Mazzalupo, S. and Cooley, L. (2006). Illuminating the role of caspases during *Drosophila* oogenesis. *Cell Death Differ.* **13**, 1950–1959.

34. Nelson, R. E., Fessler, L. I., Takagi, Y., et al. (1994). Peroxidasin: a novel enzyme-matrix protein of *Drosophila* development. *EMBO J.* **13**, 3438–3447.

35. Franc, N. C., Dimarcq, J. L., Lagueux, M., Hoffmann, J., and Ezekowitz, R. A. (1996). Croquemort, a novel *Drosophila* hemocyte/macrophage receptor that recognizes apoptotic cells. *Immunity* **4**, 431–443.

36. Franc, N. C., Heitzler, P., Ezekowitz, R. A., and White, K. (1999). Requirement for *croquemort* in phagocytosis of apoptotic cells in *Drosophila*. *Science* **284**, 1991–1994.

37. Cherbas, L. and Cherbas, P. (2000). *Drosophila* cell culture and transformation, in *Drosophila Protocols* (Sullivan, W., Ashburner, M. and Hawley, R. S., Eds.), Cold Spring Harbor Laboratory, Cold Spring Harbor, NY, pp. 373–387.

38. Perrimon, N. and Mathey-Prevot, B. (2007). Applications of high-throughput RNA interference screens to problems in cell and developmental biology. *Genetics* **175**, 7–16.

39. Yi, C. H., Sogah, D. K., Boyce, M., Degterev, A., Christofferson, D. E., and Yuan, J. (2007). A genome-wide RNAi screen reveals multiple regulators of caspase activation. *J. Cell Biol.* **179**, 619–626.

40. Rusten, T. E., Lindmo, K., Juhasz, G., et al. (2004). Programmed autophagy in the *Drosophila* fat body is induced by ecdysone through regulation of the PI3K pathway. *Dev. Cell* **7**, 179–192.

41. Scott, R. C., Schuldiner, O., and Neufeld, T. P. (2004). Role and regulation of starvation-induced autophagy in the *Drosophila* fat body. *Dev. Cell* **7**, 167–178.

42. Cox, R. T. and Spradling, A. C. (2003). A Balbiani body and the fusome mediate mitochondrial inheritance during *Drosophila* oogenesis. *Development* **130**, 1579–1590.

43. Goyal, G., Fell, B., Sarin, A., Youle, R. J., and Sriram, V. (2007). Role of mitochondrial remodeling in programmed cell death in *Drosophila melanogaster*. *Dev. Cell* **12**, 807–816.

44. McCall, K., Baum, J. S., Cullen, K., and Peterson, J. S. (2004). Visualizing Apoptosis, in *Drosophila Cytogenetics Protocols* (Henderson, D., Ed.), Humana Press, Totowa, NJ, pp. 431–442.

45. McCall, K. and Peterson, J. (2004). Detection of apoptosis in *Drosophila*, in *Apoptosis: Methods and Protocols* (Brady, H., Ed.), Humana Press, Totowa, NJ, pp. 191–206.

46. Verheyen, E. and Cooley, L. (1994). Looking at oogenesis, in *Methods in Cell Biology* (Goldstein, L. S. B. and Fyrberg, E. A., Eds.), Academic, New York, NY, pp. 545–561.

47. White, K., Tsahaoglu, E., and Steller., H. (1996). Cell killing by the *Drosophila* gene *reaper*. *Science* **271**, 805–807.

48. Drummond-Barbosa D. and Spradling, A. C. (2001). Stem cells and their progeny respond to nutritional changes during *Drosophila* oogenesis. *Dev. Biol.* **231**, 265–278.

49. Igaki, T., Kanuka, H., Inohara, N., et al. (2000). Drob-1, a *Drosophila* member of the Bcl-2/CED-9 family that promotes cell death. *Proc. Natl Acad. Sci. USA* **97**, 662–667.

50. Brachmann, C. B., Jassim, O. W., Wachsmuth, B. D., and Cagan, R. L. (2000). The *Drosophila* Bcl-2 family member dBorg-1 functions in the apoptotic response to UV-irradiation. *Curr. Biol.* **10**, 547–550.

51. Colussi, P. A., Quinn, L. M., Huang, D. C., et al. (2000). Debcl, a proapoptotic Bcl-2 homologue, is a component of the *Drosophila melanogaster* cell death machinery. *J. Cell Biol.* **148**, 703–714.

52. Jin, S., Martinek, S., Joo, W. S., et al. (2000). Identification and characterization of a p53 homologue in *Drosophila melanogaster*. *Proc. Natl Acad. Sci. USA* **97**, 7301–7306.

53. Ollmann, M., Young, L. M., Di Como, C. J., et al. (2000). *Drosophila* p53 is a structural and functional homolog of the tumor suppressor p53. *Cell* **101**, 91–101.

54. Igaki, T., Kanda, H., Yamamoto-Goto, Y., et al. (2002). Eiger, a TNF superfamily ligand that triggers the *Drosophila* JNK pathway. *EMBO J.* **21**, 3009–3018.

55. Quinn, L., Coombe, M., Mills, K., et al. (2003). Buffy, a *Drosophila* Bcl-2 protein, has anti-apoptotic and cell cycle inhibitory functions. *EMBO J.* **22**, 3568–3579.

56. Song, Z., Guan, B., Bergman, A., et al. (2000). Biochemical and genetic interactions between *Drosophila* caspases and the proapoptotic genes *rpr*, *hid*, and *grim*. *Mol. Cell. Biol.* **20**, 2907–2914.

57. Meier, P., Silke, J., Leevers, S. J., and Evan, G. I. (2000). The *Drosophila* caspase DRONC is regulated by DIAP1. *EMBO J.* **19**, 598–611.

58. Quinn, L. M., Dorstyn, L., Mills, K., et al. (2000). An essential role for the caspase Dronc in developmentally programmed cell death in *Drosophila*. *J. Biol. Chem.* **275**, 40416–40424.

59. Xu, P., Vernooy, S. Y., Guo, M., and Hay, B. A. (2003). The *Drosophila* microRNA Mir-14 suppresses cell death and is required for normal fat metabolism. *Curr. Biol.* **13**, 790–795.

60. Harvey, N. L., Daish, T., Mills, K., et al. (2001). Characterization of the Drosophila caspase, DAMM. *J. Biol. Chem.* **276**, 25342–25350.

61. Jackson, G. R., Salecker, I., Dong, X., et al. (1998). Polyglutamine-expanded human huntingtin transgenes induce degeneration of *Drosophila* photoreceptor neurons. *Neuron* **21**, 633–642.

62. Gaumer, S., Guenal, I., Brun, S., Theodore, L., and Mignotte, B. (2000). Bcl-2 and Bax mammalian regulators of apoptosis are functional in *Drosophila*. *Cell Death Differ.* **7**, 804–814.

63. Shigenaga, A., Funahashi, Y., Kimura, K., et al. (1997). Targeted expression of *ced-3* and *Ice* induces programmed cell death in *Drosophila*. *Cell Death Differ.* **4**, 371–377.

64. Kanuka, H., Hisahara, S., Sawamoto, K., Shoji, S., Okano, H., and Miura, M. (1999). Proapoptotic activity of *Caenorhabditis elegans* CED-4 protein in *Drosophila*: implicated mechanisms for caspase activation. *Proc. Natl Acad. Sci. USA* **96**, 145–150.

65. Leulier, F., Ribeiro, P. S., Palmer, E., et al. (2006). Systematic in vivo RNAi analysis of putative components of the *Drosophila* cell death machinery. *Cell Death Differ.* **13**, 1663–1674.

66. Bonini, N. M. (2000). Methods to detect patterns of cell death in *Drosophila*. *Methods Mol. Biol.* **136**, 115–121.

Chapter 25

Detecting Apoptotic Cells and Monitoring Their Clearance in the Nematode *Caenorhabditis elegans*

Nan Lu, Xiaomeng Yu, Xiangwei He, and Zheng Zhou

Summary

Apoptosis is a genetically controlled process of cell suicide that plays an important role in animal development and in maintaining homeostasis. The nematode *Caenorhabditis elegans* has proven to be an excellent model organism for studying the mechanisms controlling apoptosis and the subsequent clearance of apoptotic cells, aided with cell-biological and genetic tools. In particular, the transparent nature of worm bodies and eggshells makes *C. elegans* particularly amiable for live cell microscopy. Here we describe a few methods for identifying apoptotic cells in living *C. elegans* embryos and adults and for monitoring their clearance during embryonic development. These methods are based on Differential Interference Contrast microscopy and on fluorescence microscopy using GFP-based reporters.

Key words: *C. elegans*, Apoptosis, Programmed cell death, Engulfment, Phagosome maturation, CED-1, PI(3)P, Time-lapse recording, GFP, mRFP, Differential interference contrast microscope

1. Introduction

During an animal's development and adult life, a large number of unwanted cells are eliminated by programmed cell death, or apoptosis. Apoptotic cells are rapidly engulfed (via phagocytosis) by phagocytes, or engulfing cells, from the body and are degraded inside a membrane-bound structure referred to as a "phagosome" (**Fig. 1**) *(1)*. Apoptosis plays important roles in sculpting structures; maintaining homeostasis, and eliminating abnormal, nonfunctional, or harmful cells *(2)*. Efficient removal of dying cells is the necessary last step of apoptosis; in addition, it actively prevents harmful inflammatory and autoimmune responses *(1)*.

Peter Erhardt and Ambrus Toth (eds.), *Apoptosis*, Methods in Molecular Biology, vol. 559
DOI 10.1007/978-1-60327-017-5_25
© Humana Press, a part of Springer Science + Business Media, LLC 2004, 2009

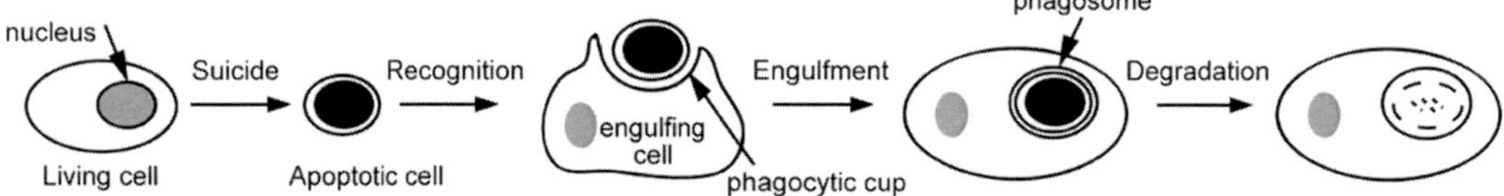

Fig. 1. A diagram describes the fate of an apoptotic cell in metazoans.

1.1. A Review of Published Methods for Detecting Distinct Features of Apoptotic Cells in C. elegans

The nematode *C. elegans*, a small free-living round worm, has been established as an excellent model organism for studying the mechanisms of apoptosis and the engulfment of apoptotic cells due to its simple anatomy, known cell lineage, well-established genetics, and easily distinguishable apoptotic cell morphology *(3, 4)*. During the development of the *C. elegans* hermaphrodite, 131 somatic cells and approximately 300–500 germ cells undergo apoptosis *(5–7)*. In the soma, due to the fixed cell lineage, both the identity of the cells that undergo apoptosis and the timing of death are invariable in *C. elegans (5, 6)*. Apoptotic cells are rapidly engulfed and degraded by neighboring cells *(5–7)*. Multiple types of cells can function as engulfing cells, including hypodermal cells, gonadal sheath cells, intestinal cells, and pharyngeal muscle cells *(6–8)*. One particularly useful feature of *C. elegans* is that animals at all developmental stages are transparent. Apoptotic cells are thus easily recognized within living animals under the Nomarski differential interference contrast (DIC) optics as highly reflective, button-like objects that are referred to as "cell corpses"(**Fig. 2a**) *(5–7)*. DIC microscopy is thus commonly used to detect cell corpses in *C. elegans (4)*. DIC microscopy, however, is unable to distinguish engulfed cell corpses from unengulfed ones because the plasma membrane of an engulfing cell is typically not visible under DIC microscope.

Besides DIC microscopy, a number of methods have been used to recognize apoptotic cells at all developmental stages in *C. elegans* based on their distinct cellular features. These include the transmission electron microscopy (TEM) for detecting cell corpses in larvae and adults, the TUNEL (*t*erminal transferase d*U*TP *n*ick *e*nd *l*abeling) assay that detects DNA ends generated during apoptosis in embryos, and the staining of larvae and adults with SYTO dyes. For an excellent review of these methods, please *see* **ref.9**. Recently, several methods have been developed to detect the exposure of phosphatidylserine (PS), a membrane phospholipids kept in the inner leaflet of the plasma membrane of living cells, on the outer surface of *C. elegans* apoptotic cells using PS-binding proteins, such as MFG-E8 and annexin V, as reporters *(10–12)*. In addition, apoptotic cells undergo chromatin condensation *(13)*. A chromatin-associated histone H3 reporter

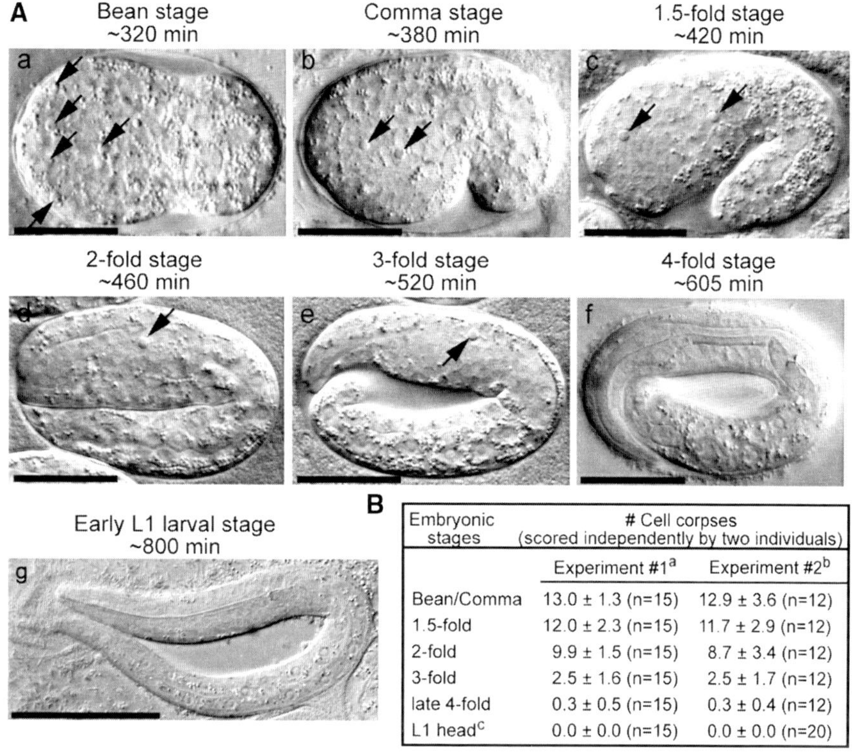

B

Embryonic stages	# Cell corpses (scored independently by two individuals)	
	Experiment #1[a]	Experiment #2[b]
Bean/Comma	13.0 ± 1.3 (n=15)	12.9 ± 3.6 (n=12)
1.5-fold	12.0 ± 2.3 (n=15)	11.7 ± 2.9 (n=12)
2-fold	9.9 ± 1.5 (n=15)	8.7 ± 3.4 (n=12)
3-fold	2.5 ± 1.6 (n=15)	2.5 ± 1.7 (n=12)
late 4-fold	0.3 ± 0.5 (n=15)	0.3 ± 0.4 (n=12)
L1 head[c]	0.0 ± 0.0 (n=15)	0.0 ± 0.0 (n=20)

Fig 2. Apoptotic cells display the same distinct morphology detectable by DIC microscopy in different embryonic development stages. (**A**) DIC images of a wild-type embryo at different developmental stages. Time (labeled in min) represents the time point the embryo enters the corresponding stage after the first cleavage. (**e**) and (**f**) were captured at mid-3-fold (~550 min) and late 4-fold (~770 min) stages, respectively. Dorsal is up and anterior is to the left. Scale bars: 20 μm in (**a–f**) and 50 μm in (**g**). *Arrows* indicate cell corpses. (**B**) The number of cell corpses in wild-type embryos and newly hatched L1 larvae scored under the DIC microscope by two different individuals in independent experiments. Data are presented as mean ± SD. *n*, number of animals scored. [a]Data published in *(15)*. [b]Data published in *(16)*. [c]Scored in the head of L1 larvae hatched within 1 h.

(HIS-72::GFP) *(14)*, which allows us to detect the distinct condensed chromatin morphology in apoptotic cells in *C. elegans* embryos, is another cell corpse-specific marker *(15)*.

Recently, we developed CED-1::GFP and 2xFYVE::mRFP, two fluorescent markers that label the surface of phagocytic cups and that of maturing phagosomes (**Fig. 1**). These two markers not only offer new methods for distinguishing cell corpses but also enable us to determine whether a cell corpse is engulfed *(8, 15–17)*. Using these markers, we further established time-lapse recording methods to monitor the processes of engulfment as well as degradation of individual apoptotic cells in developing embryos, a procedure that enables us to dissect the steps of apoptotic cell clearance on a subcellular level *(15–17)*.

1.2. The Basis for Using CED-1 and 2xFYVE to Identify Apoptotic Cells

CED-1 is a single-pass transmembrane protein expressed in engulfing cells and acts on cell surfaces as a phagocytic receptor for neighboring apoptotic cells *(8)*. CED-1 recognizes the cell-surface features of cell corpses, clusters on the phagocytic cup and then transiently on nascent phagosomes (**Figs. 3B** and **4B**) *(8, 17)*. This feature enables a CED-1::GFP reporter (**Fig. 3A**) to specifically label cell corpses that are in the process of being engulfed (**Fig. 4B**). In addition, CED-1::GFP is particularly useful for detecting unengulfed or partially engulfed cell corpses in engulfment-defective mutants (except the *ced-7* mutants), because the blockage or delay of pseudopod extension around

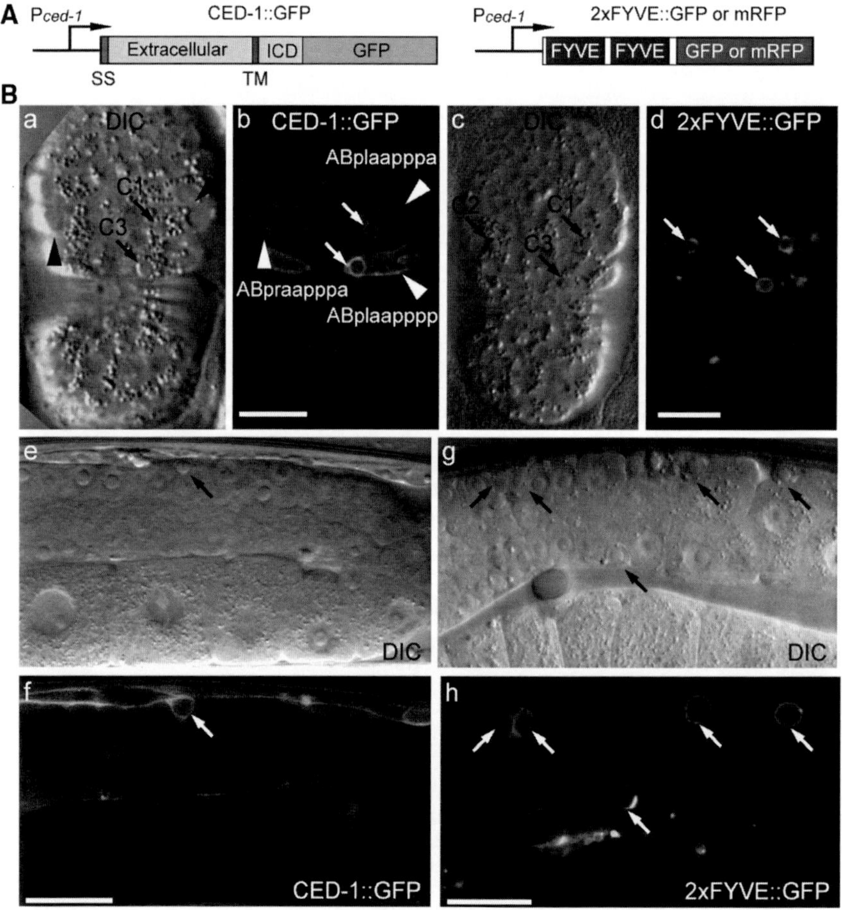

Fig. 3. CED-1 and 2xFYVE as markers for apoptotic cells. (**A**) Diagrams of the reporter constructs. $P_{ced-1}{}^{ced-1}$ promoter, *ICD* Intracellular domain, *SS* Signal sequence, *TM* transmembrane domain. (**B**) DIC and fluorescence images illustrating that CED-1:GFP and 2×FYVE:GFP are enriched on the engulfing cell membrane surrounding cell corpses. (**a–d**) ~330 min-stage wild-type embryos. Anterior is to the top. Ventral faces readers. Scale bars: 10 μm. *Arrows* indicate phagosomes containing cell corpses. *Arrowheads* label the three ventral hypodermal cells as engulfing cells for C1, C2, and C3. (**e–h**) Part of the gonad in wild-type adult hermaphrodites. Mid-body is to the left. Scale bars: 20 μm. *Arrows* indicate phagosomes containing germ cell corpses.

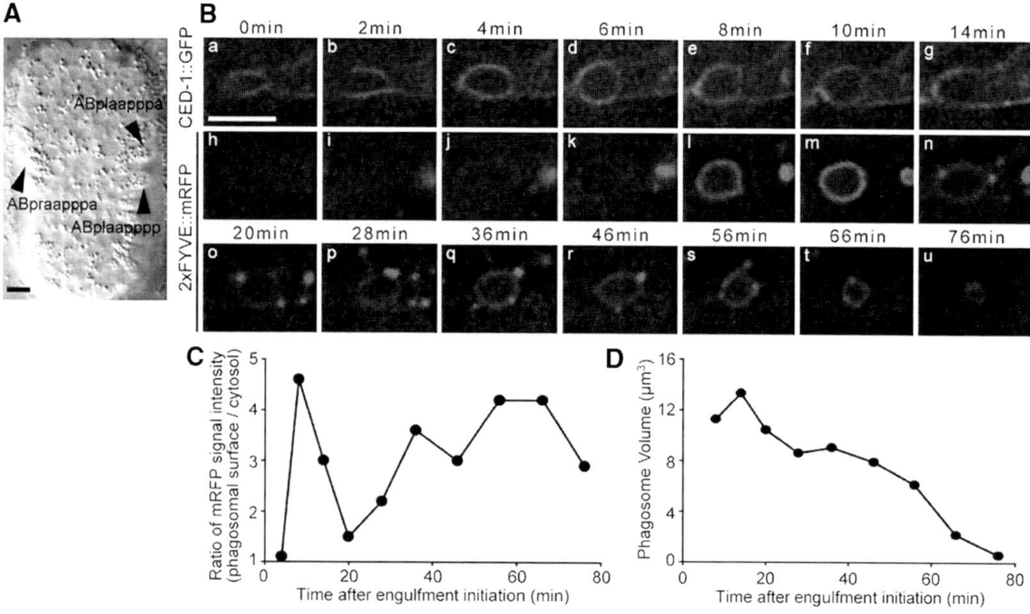

Fig. 4. Time-lapse recording of the engulfment and degradation of apoptotic cells in *C. elegans* embryos. (**A**) DIC image of a ~310 min-stage embryo at which stage the time-lapse recording should begin. Anterior is to the top. Ventral faces readers. *Arrowheads* indicate the three ventral hypodermal cells that will engulf C1, C2, and C3. C1, C2, and C3 have not displayed the distinct DIC morphology of cell corpses at this time point. (**B**) Time lapse images of the co-expressed CED-1:GFP (**a–g**) and 2xFYVE:mRFP (**h–u**) around cell corpse C3 in a wild-type embryo. 0 min: the time point when the engulfing cell extends pseudopods halfway around C3. The scale bar: 5 μm. (**C**) The relative PI3P signal intensity on the surface of the phagosome containing C3 measured from images in (B) (h–u) plotted over time. (**D**) The volume of the phagosome containing C3 measured from images in (**B**) (**h–u**) plotted over time.

cell corpses do not affect the ability of CED-1 to recognize cell corpses and cluster on phagocytic cups *(8, 10)*. As a consequence, in these mutants, CED-1::GFP is observed as bright, distinct partial circles around cell corpses, which represent not-enclosed phagocytic cups *(8, 10)*.

In wild-type *C. elegans* embryos, the clustering of CED-1::GFP around a cell corpse is detectable throughout the entire engulfment process (~5 min) and the first 9 min of phagosome maturation, which lasts 50–70 min in total *(17)*. As a result, at any given time point, only a small portion of cell corpses is labeled by CED-1::GFP in animals that display normal engulfment activity. Recently, we established 2xFYVE::GFP as a marker for cell corpses that remains on the phagosomal surface until the complete degradation of the cell corpse inside and therefore labels almost all cell corpses.

Phosphatidylinositol-3-phosphate (PI3P) is a phosphoinositide species that is specifically enriched on the surface of endosomes and phagosomes and that acts as a signaling molecule for vesicle trafficking events *(18)*. The FYVE domain of *C. elegans* EEA-1, in a tandem repeat, specifically associates with PI3P *(19)*.

The 2xFYVE::GFP and 2xFYVE::mRFP fusion proteins, which are expressed in engulfing cells under the control of the *ced-1* promoter P*ced-1* (**Fig. 3A***) (8)*, are localized to the cytoplasmic puncta. Immediately after the closure of a phagocytic cup, these markers are enriched on the surface of nascent phagosomes and persist on phagosomal surfaces until the complete degradation of the cell corpse *(17)*. In embryos co-expressing CED-1::GFP and 2xFYVE::mRFP, the entire cell-corpse removal process can be monitored in real time (**Fig. 4B**).

In addition, the 2xFYVE::GFP or 2xFYVE::mRFP reporters are excellent tools for scoring the number of cell corpses at all stages of *C. elegans* development. We found that in wild-type embryos, the disc-like DIC morphology of a cell corpse appears when engulfment starts, and disappears ~30 min after the initiation of cell-corpse degradation (N. L. and Z. Z., unpublished observation). Compared to the DIC morphology, the 2xFYVE::GFP signal persists on the surface of a phagosome until its complete disappearance, it is thus able to detect cell corpses that partially or totally lose their distinct DIC morphology.

2. Materials

The materials and methods described here are specific for the detection of apoptotic cells in *C. elegans*. For general materials and methods for raising and handling *C. elegans*, please *see* **ref. 20**. For general introduction of using DIC microscopy in *C. elegans*, please *see* **ref. 21**.

2.1. General Materials

1. 4% agarose solution, prepared by heating 2 g agarose in 50-mL autoclaved deionized water until agarose is completely melted. After usage, the solidified solution can be stored at room temperature and melted in a microwave oven again.
2. M9 Buffer (1 L): 3 g KH_2PO_4, 6 g Na_2HPO_4, 5 g NaCl, 1 mL of 1 M $MgSO_4$ dissolved in 850 mL H_2O, add H_2O to 1 L, autoclave.
3. 30 mM sodium azide (NaN_3) in M9 buffer.
4. Microscope slides, cover slips (22 × 22 mm), Pasteur pipette and bulb, high vacuum grease (Dow Corning), DeltaVision immersion oil $N = 1.514$ (Applied Precision), a platinum wire mounted on a Pasteur pipette functioning as a worm pick.

2.2. Equipment and Software

1. Nikon SMZ645 Stereomicroscope or any stereomicroscope from other manufactures for handling of *C. elegans*.
2. An Olympus IX70-DeltaVision microscope (Applied Precision) equipped with 20×, 63×, and 100× Uplan Apo objectives, DIC

microscopy accessories, motorized stage, a Coolsnap digital camera (Photometrics), and the SoftWoRx software (for the deconvolution and processing of images) (Applied Precision).

3. A temperature control chamber mounted over the DeltaVision microscope that maintains the temperature of the stage at 20°C. Alternatively, the DeltaVision microscope can be kept in a room where the temperature is maintained at 20°C.

4. A PC computer for image processing and analysis.

5. The ImageJ software (downloaded from (http://rsb.info.nih. gov/ij/index.html) for quantitative image analyses.

3. Methods

3.1. Using DIC Microscopy to Score the Number of Cell Corpses

3.1.1. Determining Which Developmental Stages to Score

Cell corpses can be recognized as reflective, disc-like objects in living animals using DIC microscopy (**Fig. 2a**). As the execution of cell death and the clearance of apoptotic cells are dynamic processes, it is critical to score the number of cell corpses at defined developmental stages and within the defined regions of an animal for meaningful comparison of results obtained from different genetic backgrounds. To assay for the pattern of the apoptosis events of somatic cells during embryonic development, we score the number of cell corpses in the entire embryo at the following stages: bean, comma, 1.5-fold, 2-fold, 3-fold, and late 4-fold stages. Embryos at these stages, which correspond to ~320, ~380, ~420, ~460, ~520 – ~605, and ~700 – ~790 min after the first cleavage (the first cytokinesis), respectively *(6, 16)*, are easily recognized using DIC microscopy by their distinct body morphology (**Fig. 2A**). Two independent sets of data, obtained from the same wild-type strain (N2) *(20)* by two different individuals in the laboratory, are similar to each other (**Fig. 2B**) *(15, 16)*. These results are consistent with the invariable embryonic cell lineage described in *(6)*.

To study the effect of mutants defective in cell-corpse removal, the number of persistent cell corpses is often scored in the head (the area between the anterior end of the worm and the anterior boundary of the intestine) of L1 Larvae. In the head of newly hatched wild-type L1 larvae (hatched within 1 h), no cell corpses are observed (**Fig. 2B**); on the other hand, mutations that induce extra cell deaths or block cell-corpse removal result in the persistent presence of cell corpses in the head *(22, 23)*.

Germ cells that undergo apoptosis during germline development or are induced to die by DNA damaging agents can be scored in the adult hermaphrodite gonad using DIC microscopy *(7, 24)*. Again, to obtain reproducible results, it is critical to score in animals of defined age. The most commonly used samples are adult hermaphrodites that are aged 48 h post the mid-L4 larval stage.

3.1.2. Mounting Animals on an Agar Pad

1. Melt the 4% agarose solution by heating it in a microwave oven.

2. Dispense a drop of agarose solution on a glass microscope slide and flatten the drop immediately with another glass slide. Wait until agarose solidifies, then gently separate the two slides by sliding one against the other. An agarose pad provides support to the cover slip so that the living specimens are not squashed.

3. Cut the round agarose pad into an approximately 12×12 mm^2 with the edge of a glass slide. Place 3 μL of 30 mM NaN$_3$ in M9 buffer at the center of the pad (*see* **Note 1**).

4. Under the Nikon SMZ 645 Stereomicroscope, transfer animals at the stage of choice with a worm pick from a plate to the drop of 30 mM NaN$_3$ in M9 buffer, gently disperse eggs with a worm pick.

5. Gently place a cover slip over the drop of liquid. Remove any solution outside the cover slip with tissue paper.

3.1.3. Observation Under the DIC Microscope

1. Align the DIC light path carefully for optimal DIC effect according to the manufacturer's instruction (http://www.appliedprecision.com).

2. Under the 63× or 100× objective, identify cell corpses and score the number. As *C. elegans* is transparent under the light microscopy, by focusing from the bottom to the top of the animal, cell corpses in the *z*-axis of the entire desired region can be scored.

3. Alternatively, instead of scoring directly from the eyepiece, serial *z*-section DIC images could be captured (see below for *z*-sectioning) and the number of cell corpses could be scored later by replaying the serial images on the computer. This method allows a longer period of time for scoring and avoids the long-term effect of NaN$_3$ in altering the DIC appearance of cell corpses (*see* **Note 1**).

3.2. Using CED-1::GFP and 2xFYVE::mRFP1 as Reporters to Monitor the Clearance of Apoptotic Cells in Real Time in Embryos

The DeltaVision Deconvolution Microscope is a white-light microscope that relies on specially designed computer deconvolution algorithm to achieve high resolution *(25)*. Comparing with conventional confocal microscope, the DeltaVision results in less photobleaching of images and less photodamage to living specimens and offers comparable, under some conditions even superior, resolution and sensitivity. Here we described a specific protocol for image capture and time-lapse recording that we developed using the DeltaVision. For step-by-step operation of the DeltaVision microscope and the SoftWoRx software, see the manufacturer's instruction (http://www.appliedprecision.com).

3.2.1. C. elegans Strain

ZH814 is an *unc-76(e911)* mutant strain carrying reporter constructs P$_{ced-1}$*ced-1::gfp* and P$_{ced-1}$*2xfyve::mrfp* as well as pUNC-76(+), a plasmid containing the wild-type *unc-76* gene, in the same transgenic array. Transgenic animals are normal for locomotion, whereas nontransgenic animals are Unc (Uncoordinated). To cross the transgenic array to the strains of your interest, follow standard genetic operation *(20)*.

3.2.2. Mounting Embryos on a Microscope Slide	1. Follow the description of **Subheading 3.1.1** to prepare an agarose pad on a microscope slide. Spot 3 μL M9 buffer in the center of the pad, transfer eggs to the pad, disperse eggs in M9 buffer (*see* **Note 2**).

2. Gently squeeze a thin line of high vacuum grease around agarose pad and cover the pad gently with a cover slip. Avoid air bubbles. Vacuum grease prevents the drying of the agarose pad and allows air exchange. No more than 50 eggs should be loaded onto one slide, and eggs should be sufficiently dispersed in M9 solution (*see* **Note 3**).

3.2.3. Identifying Three Particular Cell Corpses C1, C2, C3 and Their Engulfing Cells

Among the 113 cells that undergo apoptosis during embryogenesis *(6)*, we choose to monitor the clearance of three apoptotic cells referred to as C1, C2, and C3 (**Fig. 3B**). These three cells are located at the ventral surface of an embryo, in approximately the same or adjacent focal planes, and are engulfed at approximately the same time, between 320–330 min post-first cleavage *(16)*. C1, C2, and C3 are each engulfed by a different ventral hypodermal cell, ABplaappppa, ABpraapppa, and ABplaappppp, respectively, while these hypodermal cells extend their cell bodies to the ventral midline (**Fig. 3b**) *(16)*. These temporal and spatial features make it easy to identify C1, C2, and C3, and their engulfing cells; furthermore, they allow the recording of the clearance of all three cell corpses in the same time-lapse series, using a z-stack containing 8–12 serial z-sections (at 0.5 μm/section) at every time point.

1. Place the prepared slide on the microscope stage, start microscope operation, prewarm the mercury light source for 10–15 min. During this time period, align the DIC light path for optimal DIC effect. Open the SoftWoRx program.

2. Using the GFP channel, identify embryos that carry the transgenic array. Under the 100× objective, identify transgenic embryos whose ventral side faces the objective and which are at ~320 min post-first cleavage (**Fig. 4A**) (*see* **Note 4**). Once an appropriate embryo is identified, its exact location on the slide should be recorded using the "point marking" function of the SoftWoRx program.

3.2.4. Time-Lapse Recording

1. Set up microscope parameters. Use the 100× objective. For capturing DIC images, exposure time is usually set at 0.1 s. For fluorescence imaging, two sets of fluorescence filters, both from Chroma, Inc., are used, including the FITC filter

(excitation wavelength 490/20 nm; emission wavelength 528/38 nm) for the GFP signal and the Rhodamine filter (excitation wavelength 555/28 nm; emission wavelength 617/73 nm) for the mRFP signal. The exposure time is 0.1 s for each channel and each *z*-section (*see* **Note 3**). If the signal is weak, 2 × 2 binning is recommended (*see* **Note 3**).

2. Set up the recording program. Serial *z*-sectioning is performed from the ventral surface of an embryo toward the center. The setting of 8–12 *z*-sections at 0.5 μm per section is sufficient to include C1, C2, C3 in one *z*-section series (cell corpses are of 2.5–3 μm in diameter). An image size of 512 × 512 pixels is sufficient for capturing the entire embryo if 2 × 2 binning is performed (*see* **Note 3**). For recording the engulfment process, which lasts for ~5 min in a wild-type strain, 30 time points at a 1-min interval is sufficient if recording starts at a time between 310 and 320 min post-first cleavage. For the degradation process, which lasts for ~50–70 min in wild-type embryos but could last much longer in degradation-defective mutants *(17)*, we record for 100–120 min at a 2-min interval. After embryos reach ~460 min post-first cleavage, rapid body movement starts, which interferes with image recording.

3. Using the "point marking" and "point visiting" functions of the software to record multiple embryos in the same program. Using the parameters described above, at least three embryos can be recorded in the same program in a time interval of 2 min.

4. Keep observing images from time to time. Adjust the starting focal plane during the interval of recording if any change of focal plane occurs. Abort recording if an embryo slows down or stops its development due to photodamage (*see* **Note 3**).

5. After recording is completed, deconvolve images using Soft-WoRx.

6. Open deconvolved files with softWoRx, save desired images as tiff or jpeg files for quantitative analysis using the ImageJ software and for further processing using Adobe Photoshop.

3.2.5. Quantitative Image Analysis

Measuring Signal Intensity on Phagosomal Surfaces

The dynamic changes of the signal intensity of CED-1 and PI3P indicate the progress of engulfment and phagosome maturation; in addition, alteration of the dynamic pattern of these signals in mutant backgrounds suggests specific defects in phagosome formation and maturation *(15, 17)*. The signal intensity of CED-1 and PI3P on phagosomal surfaces is quantified by measuring the fluorescence intensity of CED-1::GFP and 2xFYVE::mRFP, respectively. The absolute fluorescence signal intensity, however, varies from embryos to embryos due to the different expression levels of the transgene. Thus, we use the relative signal intensity represented by the ratio of the intensity on phagosomal surface to that in an adjacent area inside the cytosol to indicate the enrichment of CED-1 or PI3P on phagosomal surfaces. We use the software ImageJ to quantify fluorescence signal intensity.

1. Open an epifluorescence image file (in tiff or jpeg format) in the ImageJ program. Increase the magnification of the image until the boundary of phagosome can be clearly distinguished.

2. Use the freehand selection tool to define a donut-like and closed area with one continuous line that surrounds the surface of a phagosome.

3. Select the Measure tool from the Analyze menu to display the mean or median value of the fluorescence signal intensity measured in this area (*see* **Note 5**).

4. Use the freehand selection tool to select an area in engulfing cell cytosol adjacent to the phagosome. Repeat **step 3** to obtain the mean or median value.

5. Calculate the ratio of the values obtained from the phagosomal surface and that obtained from the cytosol.

6. Plot the ratio over time. An example of the results is shown in **Fig. 4C**.

Measuring the Volume of a Phagosome over Time

During phagosome maturation, the volume of a phagosome decreases as the content is gradually digested, and is a reliable index that reflects the progression of the degradation of apoptotic cells *(15, 17)*.

1. Among a *z*-stack of serial optical sections, identify the middle section of a phagosome in the *z*-axis, which represents the equator plane. Open this image (in tiff or jpeg format) in ImageJ.

2. Set up the μm/pixel scale (*see* **Note 6**) by selecting Set Scales in the Analyze menu and entering the scale for each pixel. As a reference, images obtained from the DeltaVision using the 100× objective and subject to 2 × 2 binning have a scale of 0.133 μm per pixel.

3. Increase the magnification of the image until the boundary of a phagosome can be clearly distinguished. Use the freehand selection tool to draw a continuous line along the phagosome surface. Always draw along the path that has the brightest signal.

4. Select the Measure tool from Analyze menu to display the area (A) of the selected shape (the phagosome) in μm².

5. Regarding a phagosome as a sphere, calculate the radius (r) of the phagosome using the formula $A = \pi r^2$. Calculate the volume of the phagosome (V) using the formula $V = (4/3)\pi r^3$.

6. Plot the phagosome volume over time. An example is shown in **Fig. 4D**.

3.3. Scoring the Number of Cell Corpses Using 2xFYVE::GFP

Using the protocols described above, mount embryos on slides and identify embryos at the stage of your choice. Capture serial z-section images of an entire embryo at 40 × 0.5 μm/s optic interval (*see* **Note 7**). Other parameters for image recording are the same as described above except a time course for recording is not necessary. Scoring the number of 2xFYVE::GFP(+) rings using deconvolved serial z-section images. The results represent the number of phagosomes, or engulfed cell corpses. As mentioned before (**Subheading 1.2**), this method is highly sensitive in identifying apoptotic cells, including those that lose their distinct DIC morphology.

4. Notes

1. NaN_3 anesthetizes and immobilizes animals. Larvae and adults are immobilized within a few minutes after incubation with the 30 mM NaN_3 solution. It takes NaN_3 a much longer time to penetrate eggshells. For scoring embryos younger than the 2-fold stage, it is not necessary to use NaN_3, since vigorous body movement of embryos does not start until that stage. Note that after 1-h incubation in the NaN_3 solution, the DIC morphology of larvae and adults starts to become abnormal, whereas that of embryos are not affected.

2. Anesthetization is not necessary since at the particular embryonic stages for recording, there is minimum embryonic body movement. NaN_3 stops embryonic development and should be avoided.

3. How to ensure that embryos develop normally during time-lapse recording? (a) To ensure normal embryonic development in a chamber with limited oxygen supply, load no more than 50 eggs onto the glass slide, disperse eggs thoroughly in the M9 solution, and carry over as little bacteria as possible. (b) To avoid photodamage of embryonic development and photobleach of fluorescence signals, use a highly sensitive CCD camera so that the light exposure time could be minimized, and restrain or avoid direct observation of fluorescent light under the eyepiece. Instead, "snapshots" with the camera should be used for finding and setting the focal plane to begin the recording. For weak fluorescence signals, use the "2 × 2 binning" function to keep the exposure time minimal. In addition, include only the necessary number of z-sections at each time point. As a rule of thumb, for recording of two channels, the exposure time of each channel should be kept below 0.2 s per z-section. (c) Signs of the photodamage of embryonic development. Data obtained from those embryos whose development is arrested

due to photodamage are not useful. We rely on a few embryonic morphology changes to determine whether the development is proceeding in the normal time course. For example, the period from the bean- to the comma-stage, lasts for 60–70 min (**Fig. 2A**)(**a, b**). During this period, an embryo rotates 90° (**Fig. 2A**)(**a, b**). In addition, a period from the comma stage to the 1.5-fold stage lasts for ~40 min. A significant elongation of any of these time intervals is a sign of developmental arrest.

4. Two distinct features that can help identify embryos at this stage are (1) the ventral surface slightly invaginates on both sides, and (2) the three soon-to-be engulfing cells are located at the lateral sides, in a edge shape, with the tip of each cell less than halfway extended toward the ventral midline (**Fig. 4A**).

5. The Median and mean values are usually very similar. An abnormally bright pixel on phagosome surface, however, is largely ignored in median value, whereas it is counted and significantly increases mean value. Therefore, the median value is more resistant to signal noise and reflects the signal intensity more accurately.

6. The µm/pixel scale designates the size of each pixel in the image, which can be obtained from the program with which the image is captured.

7. The average thickness of an embryo is 20 µm (Z. Z., unpublished observation).

Acknowledgments

Z. Z. was supported by NIH (GM067848), the Cancer Research Institute, the Rita Allen Foundation, and a Basil O' Connor Starter Scholar award from March of Dimes Foundation. X. H. was supported by NIH (GM068676).

References

1. Savill, J. and Fadok, V. (2000). Corpse clearance defines the meaning of cell death. *Nature* **407**, 784–8.

2. Jacobson, M.D., Weil, M., and Raff, M.C. (1997). Programmed cell death in animal development. *Cell* **88**, 347–54.

3. Metzstein, M.M., Stanfield, G.M., and Horvitz, H.R. (1998). Genetics of programmed cell death in *C. elegans*: past, present and future. *Trends Genet* **14**, 410–6.

4. Zhou, Z., Mangahas, P.M., and Yu, X. (2004). The genetics of hiding the corpse: engulfment and degradation of apoptotic cells in *C. elegans* and *D. melanogaster*. *Curr Top Dev Biol* **63**, 91–143.

5. Sulston, J.E. and Horvitz, H.R. (1977). Postembryonic cell lineages of the nematode, *Caenorhabditis elegans*. *Dev Biol* **56**, 110–56.

6. Sulston, J.E., Schierenberg, E., White, J.G., and Thomson, N. (1983). The embryonic

cell lineage of the nematode *Caenorhabditis elegans*. *Dev Biol* **100**, 64–119.

7. Gumienny, T.L., Lambie, E., Hartwieg, E., Horvitz, H.R., and Hengartner, M.O. (1999). Genetic control of programmed cell death in the *Caenorhabditis elegans* hermaphrodite germline. *Development* 126, 1011–22.

8. Zhou, Z., Hartwieg, E., and Horvitz, H.R. (2001b). CED-1 is a transmembrane receptor that mediates cell corpse engulfment in *C. elegans. Cell* **104**, 43–56.

9. Schwartz, H.T. (2007). A protocol describing pharynx counts and a review of other assays of apoptotic cell death in the nematode worm *Caenorhabditis elegans. Nat Protoc* **2**, 705–14.

10. Venegas, V. and Zhou, Z. (2007). Two alternative mechanisms that regulate the presentation of apoptotic cell engulfment signal in *Caenorhabditis elegans. Mol Biol Cell* **18**, 3180–92.

11. Wang, X., Wang, J., Gengyo-Ando, K., Gu, L., Sun, C.L., et al. (2007). *C. elegans* mitochondrial factor WAH-1 promotes phosphatidylserine externalization in apoptotic cells through phospholipid scramblase SCRM-1. *Nat Cell Biol* **9**, 541–9.

12. Zullig, S., Neukomm, L.J., Jovanovic, M., Charette, S.J., Lyssenko, N.N., et al. (2007). Aminophospholipid translocase TAT-1 promotes phosphatidylserine exposure during *C. elegans* apoptosis. *Curr Biol* **17**, 994–9.

13. Danial, N.N. and Korsmeyer, S.J. (2004). Cell death: critical control points. *Cell* **116**, 205–19.

14. Ooi, S.L., Priess, J.R., and Henikoff, S. (2006). Histone H3.3 variant dynamics in the germline of *Caenorhabditis elegans. PLoS Genet* **2**, e97.

15. Mangahas, P.M., Yu, X., Miller, K.G., and Zhou, Z. (2008). The small GTPase Rab2 functions in the removal of apoptotic cells in *Caenorhabditis elegans. J Cell Biol* **180**, 357–73.

16. Yu, X., Odera, S., Chuang, C.H., Lu, N., and Zhou, Z. (2006). *C. elegans* Dynamin mediates the signaling of phagocytic receptor CED-1 for the engulfment and degradation of apoptotic cells. *Dev Cell* **10**, 743–57.

17. Yu, X., Lu, N., and Zhou, Z. (2008). Phagocytic receptor CED-1 initiates a signaling pathway for degrading engulfed apoptotic cells. *PLoS Biol* **6**(3), e61.

18. Vieira, O.V., Botelho, R.J., and Grinstein, S. (2002). Phagosome maturation: aging gracefully. *Biochem J* **366**, 689–704.

19. Roggo, L., Bernard, V., Kovacs, A.L., Rose, A.M., Savoy, F., et al. (2002). Membrane transport in *Caenorhabditis elegans*: an essential role for VPS34 at the nuclear membrane. *EMBO J* **21**, 1673–83.

20. Wood, W.B. and Researchers of the *C. elegans* Community (1998). *The Nematode Caenorhabditis elegans*. Cold Spring Harbor, NY: Cold Spring Harbor Laboratory.

21. Shaham, S. (2005). Methods in cell biology, in *WormBook*, The *C. elegans* Research Community, editor.

22. Chen, F., Hersh, B.M., Conradt, B., Zhou, Z., Riemer, D., et al. (2000). Translocation of *C. elegans* CED-4 to nuclear membranes during programmed cell death. *Science* **287**, 1485–9.

23. Mangahas, P.M. and Zhou, Z. (2005). Clearance of apoptotic cells in *Caenorhabditis elegans. Semin Cell Dev Biol* **16**, 295–306.

24. Gartner, A., Milstein, S., Ahmed, S., Hodgkin, J., and Hengartner, M.O. (2000). A conserved checkpoint pathway mediates DNA damage-induced apoptosis and cell cycle arrest in *C. elegans. Mol Cell* **5**, 435–43.

25. Sibarita, J.B. (2005). Deconvolution microscopy. *Adv Biochem Eng Biotechnol* **95**, 201–43.

Chapter 26

Detection of Herpes Simplex Virus Dependent Apoptosis

Christopher R. Cotter and John A. Blaho

Summary

Subversion of the host response to virus infection is a universal theme of virology and viral immunology. Multiple mechanisms are in place to limit virus spread on behalf of the host, yet through evolution, viruses have adapted to either weaken or eliminate the effects of these host factors. Cell death or apoptosis is one such example of a host response to viral infection. As such, experimental techniques that enable analysis of viruses (and viral genes) involved in triggering, blocking, or perhaps augmenting this process represent important tools for virologists, immunologists, and cell biologists. Presented here are a series of techniques developed in our lab for the analysis of apoptosis that occurs as a consequence of herpes simplex virus type 1 infection.

Key words: Herpes simplex virus, Apoptosis, Mitochondria, Cytochrome c, Caspases

1. Introduction

A widely accepted phenomenon associated with the biology and pathogenesis of herpes simplex virus 1 (HSV-1) infection is the induction of apoptosis. Viral-induced apoptosis is subsequently blocked by the concomitant synthesis of infected cell proteins. Whether the induction of the apoptotic phenotype associated with viral infection is a host response to infection or a viral mechanism to augment the efficiency of replication (and thus a marker of viral evolutionary fitness) is both controversial and philosophical in nature. Never-the-less, the fact remains that this event is essential to understand how herpes viruses establish and maintain their conquest of the host cell. This review will summarize

Peter Erhardt and Ambrus Toth (eds.), *Apoptosis*, Methods in Molecular Biology, vol. 559
DOI 10.1007/978-1-60327-017-5_26
© Humana Press, a part of Springer Science+Business Media, LLC 2004, 2009

technology currently available for an investigator to study HSV-1-dependent apoptosis. It should be noted that essentially identical effects have been observed using HSV-2 *(1)*.

Depending on the types of tissues affected, HSV-1 infection can lead to disease as minor as a cold sore or as devastating as blinding keratitis or fatal encephalitis *(2)*. In immune-suppressed populations and neonates, HSV-1 infections commonly become disseminated to multiple organs, leading to life-threatening disease. Recent reports have linked apoptosis with the severity of herpes associated disease. Several lines of evidence support a role for apoptosis in limiting HSV replication in the eye, and thereby, protecting it from HSK. Ocular HSV-1 infection causes apoptosis in the eyes of mice *(3)*. HSV-1 infection of rabbit corneal epithelial cells induced apoptosis in the underlying keratinocytes *(4)*. Human corneal epithelial cells from patients with ocular HSV-1 infections displayed increased apoptosis as measured by Annexin V staining *(5)*. In addition, tissue sections from patients with HSV-associated acute focal encephalitis were found to contain neurons with TUNEL positive staining, active caspase 3, and cleaved PARP *(6, 7)*. HSV-1 infection also increased the levels of these apoptotic markers in rat hippocampal cultures *(7)*. Thus, results from both animal models and human infections indicate that HSV-1 infection leads to apoptosis.

Apoptosis is first triggered and later blocked in cells infected with HSV-1. The first report of this phenomenon was in 1997, when Koyama and Adachi showed that infecting the HEp-2 strain of HeLa cervical adenocarcinoma cells with HSV-1 in the presence of the protein synthesis inhibitor, cycloheximide (CHX), caused membrane blebbing, chromatin condensation, and DNA fragmentation *(8)*. Later studies have determined that other key features of apoptosis including caspase activation, cleavage of caspase substrates, mitochondrial membrane potential change, and phosphatidyl serine flipping are present during HSV-1-dependent apoptosis *(9–13)*. The finding that CHX treatment reveals apoptosis in infected cells suggested a biphasic modulation of apoptosis during HSV-1 infection in which de novo protein synthesis is required for the prevention, but not the induction of apoptosis by the virus. Wild type HSV-1 infection has been shown to confer resistance to apoptosis induced by both the extrinsic and the intrinsic signaling pathways *(14–16)*. This blocking ability has also been shown to be true in clinical HSV-1 isolates *(17, 18)*. For an accurate and up-to-date review of the history and specific details of HSV-1-dependent apoptosis, readers are encouraged to refer to a recent review on this subject *(19)*.

2. Materials

2.1. Cell Lines

1. African Green Monkey Kidney Cells (Vero Cells). Vero cells were obtained from the American Type Culture Collection (Rockville, MD) and maintained in Dulbecco's modified Eagle's medium (DMEM) supplemented with 5% fetal bovine serum. These cells are resistant to viral-induced apoptosis *(13)* and therefore represent the cell line utilized to generate viral stocks. Typical titers derived from confluent monolayers of 4×10^7 cells are in the 10^8–10^9 pfu/mL range.

2. HEp-2 Cells. HEp-2/HeLa cells are obtained from the American Type Culture Collection (Rockville, MD) and maintained in Dulbecco's modified Eagle's medium (DMEM) supplemented with 5% fetal bovine serum. Although HEp-2 cells were originally developed from a patient with laryngeal carcinoma *(20)*, it was later recognized that current isolates provided by the ATCC are HeLa cell contaminants *(21, 22)*. For this reason, they are referred to as HEp-2/HeLa cells. These cells demonstrate the apoptotic phenotype during HSV-1 infection (*see* **Note 1**).

2.2. Antibodies

Details of the origins and manufacturers of all antibodies utilized in the assays that follow are mentioned individually throughout the review.

2.3. Chemicals

1. Hoechst 33258 (Sigma).
2. Ethidium Bromide.
3. RNase/ProteinaseK.
4. Phenol/Chloroform.
5. Reagents involved in TUNEL staining.
6. Tris–HCl; EDTA; NaCl; Phenylmethylsulfonyl Fluoride (Sigma); L-I-chloro-3-(4-tosulamido)-4-phenyl-2-butanone (TPCK), L-1-chlor-3-(4-tosylanmido)-7-amino-2-heptanon-hydrochloride (TLCK).
7. Bradford Protein Assay (BioRad).
8. N,N'-diallyltartardiamide-acrylamide.
9. 5-bromo-4-chloro-3-indolyl phosphate (BCIP) and 4-nitrobluetetrazolium chloride (NBT).
10. MitoCapture Kit (BioVision).
11. HEPES pH 7.4; KCl; EGTA; $MgCl_2$; DTT; and 10% sucrose solution.
12. Dounce homogenizer and B type pestle.

2.4. Buffers

1. DNA Lysis Buffer: 10 mM Tris–HCl, pH 8.0, 10 mM EDTA, 0.6% SDS and mix with 125 µL of 5 M NaCl.
2. RIPA Buffer: 50 mM Tris–HCl, pH 7.5, 150 mM NaCl, 1% Triton X100, 1% deoxycholate, 0.1% SDS.
3. PT Lysis Buffer: 10 mM HEPES, pH 7.4, 50 mM KCl, 5 mM EGTA, 5 mM $MgCl_2$, 1 mM DTT, and 10% sucrose.

3. Methods

3.1. Apoptotic Infected Cell Morphologies

3.1.1. Biochemical Basis

Infection of cultured cells with HSV-1 ultimately results in their destruction. This is in large part due to the cytolytic nature of the virus. Alterations in the morphology of HSV-1 infected cells, referred to as cytopathic effect (CPE), occurs within the first few hours of postinfection *(23)*. This is primarily due to (1) the loss of matrix binding proteins on the cell surface; (2) modifications of membranes; (3) cytoskeletal destabilizations; and (4) a decrease in cellular macromolecular synthesis *(24–27)*. It is now recognized that there is an apoptotic component to the CPE observed with HSV-1 *(28, 29)*. Apoptotic features associated with HSV-1-dependent apoptosis are readily observable under conditions in which the infected cell apoptic prevention factors are decreased (*see* **Note 2**). The general observations of HSV-1-dependent apoptosis include gross cellular changes that present as cell shrinkage, membrane blebbing, and the production of apoptotic bodies *(10)*.

3.1.2. Procedure

1. A monolayer of HEp-2/HeLa cells is infected with virus at a predetermined multiplicity of infection (MOI) (*see* **Note 3**).
2. Infected cells are incubated at 37°C in the presence of 5% CO_2.
3. At desired time-points postinfection, cells can be removed from incubation and viewed under a phase-contrast light microscope to visualize the rounding up of cells and membrane alterations associated with infection with HSV-1 *(10)*.

3.2. Apoptotic Infected Cell Nuclei

3.2.1. Chromatin Condensation

Biochemical Basis

HSV-1 infection has additional consequences on the state of the contents of the nucleus. More specifically, viral infection alters the nucleolus and results in chromatin margination, aggregation, and damage *(23, 30)*. Nuclear alterations represent a key biochemical marker of cell death. Condensed chromatin is one aspect of this alteration that can be used to assess an apoptotic cell. Stained infected cell nuclei are readily visualized *(10)* and percentages of cell possessing condensed chromatin are quantitated *(13, 31)* using fluorescence microscopy.

Procedure

1. Virus infected cells are removed from the 37°C/5% CO_2 incubation 1 h prior to the desired harvest time.

2. Add 1 µL of a 30-ng/mL stock (in dH_2O) of Hoechst stain to each mL of media in the dish containing the infection. Allow approximately 30 min for the dye to effectively bind the cellular chromatin.

3. Cells are visualized initially by live phase-contrast light microscopy at a 40× magnification to visualize any morphological abnormalities.

4. Once a desired plane of view is found, one switches from phase-contrast to a fluorescence microscopy view to visualize the stained nuclear component of the cells.

5. Digital photographs are taken of the phase and nuclear view of the cells in the exact same plane.

6. For analysis, one utilizes desktop publishing software (e.g., Photoshop, NIH Image) to overlay the two images. Altering the opacity of the now-overlaid image allows the investigator to determine the ratio of apoptotic to healthy cells and ultimately the percent of apoptosis for this infection. Apoptotic cells are deemed as such based on the overlay. Specifically, a comparison is made of how cells with condensed chromatin match up to cells with altered morphology (membrane blebbing and cellular shrinkage).

3.2.2. DNA Laddering

Biochemical Basis

One of the chief hallmarks of apoptosis is the observed fragmentation of cellular DNA *(32)*. The cleaved DNA is roughly broken into fragment sizes that range from 180 to 200 bp in length and can be resolved by agarose gel electrophoresis, followed by visualization of fluorescence enhancement upon ethidium bromide staining *(10, 13)*.

Procedure

1. Infected cells are scrapped into media and collected by centrifugation for 5 min at $1,000 \times g$.

2. Rinse cells in ice cold PBS and recentrifuge.

3. Lyse cells in 400 µL of DNA Lysis Buffer (*see* **Subheading 2**).

4. Incubate overnight at 4°C to extract the DNA.

5. Centrifuge for 25 min at $12,000 \times g$ to precipitate the chromosomal DNA.

6. Add RNase to the supernatant at 0.1 mg/mL and incubate at 37°C for 1 h.

7. Add 1 mg/mL of proteinase K and incubate at 50°C for 2 h.

8. Phenol/chloroform extract to remove enzymes and isolate DNA.

9. Precipitate DNA using ethanol with an overnight incubation at −80°C.

10. Spin samples at $12,000 \times g$ for 20 min.

11. Precipitated DNA samples are separated in a standard 1.5% Tris–borate–EDTA agarose gel containing 0.1 µg/mL ethidium bromide and visualized using UV illumination.

3.2.3. TUNEL Staining

Biochemical Basis

Induction of apoptosis during infection can be detected by in situ end labeling of free 3′-ends of DNA. This exposed region of the DNA is generated after apoptosis has been induced either by virus or by a chemical agent. DNA strand breaks are labeled with fluorescein-conjugated dUTP using the enzyme terminal deoxynucleotide transferase. The labeled DNA is subsequently visualized by fluorescence microscopy and the ratio of TUNEL positive cells to total cells is determined as a means to quantitate the apoptotic phenotype *(33)*. However, care must be taken when interpreting TUNEL data generated during a productive HSV-1 infection (*see* **Note 4**).

Procedure

1. HEp-2 cell monolayers are grown on glass coverslips in 6-well dishes (*see* **Note 5**).

2. Infected cell are fixed with 2% (methanol free) paraformaldehyde in PBS for 20 min.

3. Cells are permeablized with 100% aceton at –20°C for 4 min.

4. DNA double-stand breaks are labeled with fluoroscein-conjugated dUTP using terminal deoxynucleotide transferase (e.g., from Boehringer Mannheim).

5. Labeled infected cell nuclei are visualized by fluorescence microscopy.

3.3. Apoptotic Infected Cell Death Factor Processing

3.3.1. Biochemical Basis

The two cellular apoptotic signaling pathways (extrinsic and intrinsic) include activation and processing of a series of aspartate-specific cysteine proteases known as caspases *(34, 35)*. Each pathway converges and culminates with the processing of the effectors caspase 3 and caspase 7. Caspase 3 drives apoptosis by its cleavage of specific target proteins necessary for the maintenance of cellular homeostasis. This cleavage event ultimately results in gross alterations of both nuclear and cytoplasmic components. Targets of caspase 3 include the nuclear DNA binding protein Poly(ADP-ribose) polymerase (PARP), Lamin B, and the DNA Fragmentation Factor (DFF) *(36, 37)*. Functionally, PARP is involved in detecting and repairing double-stranded DNA breaks. Once targeted by caspase 3, the 116-kDa form of PARP is cleaved to form a predominating 85-kDa fragment that is nonfunctional. Lamin proteins are components of a filamentous network localized at the nuclear membrane and essential for its integrity. Solubilization of Lamin B is also associated with cellular apoptosis. DFF is another DNA repair enzyme that induces DNA fragmentation upon cleavage with caspase 3 *(32)*. Western blotting to detect processing of cellular caspases, cleavage of PARP, Lamin B, and DFF can be utilized to confirm whether a cell has undergone apoptosis. Further, quantitative insight can be gained using densitometric software to assess the degree to which a population of cells has undergone apoptosis.

3.3.2. Procedure

Whole Cell Extract Preparation and Immunoblotting

1. Cells are scraped into the medium, collected by centrifugation for 5 min at 1,000 × *g*, and washed once in cold phosphate-buffered saline (PBS).

2. Cells are lysed by resuspending in RIPA buffer (*see* **Subheading 2**) supplemented with 2 mM PMSF, 1% TPCK, and 0.01 mM TLCK, and vortexed for 30 s.

3. The cell lysates are cleared by centrifuging for 10 min at 4°C at 16,000 × *g*.

4. Protein concentrations are determined using a modified Bradford protein assay (Bio-Rad).

5. Approximately 50 µg of total protein is separated on 15% *N*,*N'*-diallyltartardiamide-acrylamide gels (*see* **Note 6**) and electrically transferred to nitrocellulose using a tank apparatus (Bio-Rad).

6. Membranes are blocked for 1 h at room temperature in PBS containing 5% nonfat dry milk and incubated overnight at 4°C in primary antibody.

7. The membrane is washed four times for 10 min each in PBS at room temperature.

8. The appropriate secondary antibody (mouse or rabbit) conjugated to alkaline phosphatase[1] is reconstituted in 5% dry milk/PBS at a concentration of 1:3,000 and incubated on the membrane for 1 h at room temperature.

9. Repeat washing as in **step 7**.

10. Immunoblots are developed in buffer containing 5-bromo-4-chloro-3-indolyl phosphate and 4-nitrobluetetrazolium chloride.

Caspase Processing

Cleavage and activation of caspases is associated with HSV-1-dependent apoptosis *(33)*. It is preferred to perform direct immunoblotting for endogenous caspases as this represents a reliable method of quantitation (*see* **Note 7**). Listed below are sources and general comments regarding use of antibodies against caspases.

1. *Caspase 3.* A mouse anti-caspase 3 monoclonal may be obtained from Transduction Laboratories, Inc., which has given consistent results for over 10 years *(15)*. At least one

---

[1] Alternatively, if antibodies containing HRP are utilized during the immunoblotting procedure, traditional methods of detection such as ECL may be substituted here.

polyclonal antibody against caspase 3 has been generated *(38)* and this reagent works well in HSV-1-infected cells *(12)*.

2. *Caspase 7.* A mouse anti-procaspase 7 monoclonal obtained from BD Transduction can be used in studies of HSV-1-infected cells *(12)*.

3. *Caspase 9.* Current evidence indicates that HSV-1-dependent apoptosis results from the intrinsic cleavage of caspase 9 *(33)*. A mouse anti-caspase 9 monoclonal antibody may be obtained from Pharmingen *(33)*.

4. *Caspase 8.* Current evidence indicates that caspase 8 is not cleaved in order to initiate HSV-1-dependent apoptosis *(33)*. A mouse anti-caspase 8 monoclonal antibody may be obtained from Pharmingen *(33)*.

Caspase Inhibitors

Specific caspase inhibitors may be added to the culture medium of cells during HSV-1 infection in order to confirm which caspase is activated *(15, 33)*. These reagents may be obtained from Calbiochem. Addition of a specific caspase 3 inhibitor increases the replication of apoptotic HSV-1 *(15)*. Addition of the pan-caspase inhibitor, z-VAD-fmk, completely blocks the HSV-1-dependent apoptosis but not the release of mitochondrial cytochrome *c(33)*.

Caspase Substrates

1. *PARP cleavage.* PARP cleavage represents one of the best cellular markers of apoptotic cell death induced by HSV-1. PARP cleavage directly corresponds to the amount of condensed chromatin in an apoptotic, HSV-1-infected cell *(31)*. A high quality mouse anti-PARP monoclonal antibody may be obtained from Pharmingen *(15)*. The great advantage of PARP cleavage assessments is that both the full-length and cleaved forms are easily resolved in denaturing gels, so absolute quantitation may be performed *(15)*.

2. *DFF processing.* A goat anti-DFF polyclonal antibody may be obtained from Santa Cruz Biotechnology *(15)*. Loss of full length DFF directly corresponds to activation of caspase 3 during HSV-1 infection *(13, 15)*. It has been proposed that the loss of DFF may function in chromatin marginalization by HSV-1 *(29)*.

3. *Lamin B processing.* Lamin B cleavage during HSV-1-dependent apoptosis is dependent on caspase 3 *(12)*. An anti-lamin B antibody may be obtained from Santa Cruz Biotechnology *(12)*. It has been proposed that the loss of lamin may function in HSV-1 virion nuclear egress *(29)*.

3.3.3. Densitometric Analysis

Biochemical Basis

Once immunoblots of the above death factors have been obtained, they may be scanned and quantitated. Any appropriate imaging software can be utilized for these types of analyses. A standard method for the determination of % PARP cleavage *(31)* is given below as an example. In the absence of both the uncleaved and cleaved death factor protein band, quantitation must be done by setting the values observed on mock-infected samples to either 0 or 100%, as appropriate *(31)*. In all cases, values should be normalized to an internal loading control such as tubulin *(31)* or actin *(12)*.

Procedure

1. Developed blots are scanned and saved as tagged image file (TIF) files on a computer hard drive.

2. To quantitate the percentage of total infected cell PARP that is cleaved, NIH Image (e.g., version 1.63) may be used to measure the integrated density (ID) of the 116,000 molecular weight uncleaved and 85,000 molecular weight cleaved PARP bands.

3. These values are then used to calculated the % PARP cleavage for each lane using the following formula:

4. % cleavage = {(cleaved PARP ID)/(cleaved PARP ID plus uncleaved PARP ID)} × 100%.

3.4. Apoptotic Infected Cell Mitochondrial Changes

3.4.1. Mitochondrial Membrane Potential

Biochemical Basis

The intrinsic apoptotic pathway also commonly referred to as the mitochondrial apoptotic pathway is induced by stimuli such as growth factor withdrawal, ultraviolet (UV) irradiation, and exposure to toxic chemicals. Upon receipt of such a stimulus, a channel or a pore within the mitochondria is exposed due to the release of several mitochondrial inner and outer membrane proteins *(39)*. This newly formed structure is referred to as the mitochondrial permeability transition pore (PT). This pore allows for an equilibration of the mitochondrial membrane potential (due to unequal H^+ ion concentrations) that was previously generated by the electron transport chain *(40, 41)*. Apoptosis that results from a disruption of the function of this cellular organelle is the basis for the experimental procedure that follows. The Mito-Capture Kit commercially available from BioVision is utilized to assess changes in the membrane potential of the mitochondria during HSV-1-dependent apoptosis *(13)*.

Procedure

1. Approximately 4×10^6 cells are harvested/washed by suspending in 1 mL of diluted MitoCapture reagent and incubated at 37°C for 20 min. Special care must be taken when using the method with adherent cells (*see* **Note 8**).

2. Labeled cells are centrifuged at $500 \times g$ for 5 min.

3. The pellet is suspended in 1 mL of prewarmed MitoCapture buffer. MitoCapture-labeled cells are analyzed for green (in the FL1 channel) and red (in the FL2 channel) fluorescence using a flow cytometer, such as a Beckman Coulter Cytomics FC500. In healthy, nonapoptotic cells, the MitoCapture reagent aggregates in mitochondria and fluoresces in the red spectra. When the mitochondrial membrane potential is disrupted in apoptotic cells, the MitoCapture reagent remains in the cytoplasm as monomers. These monomers fluoresce in the green spectra.

4. The percentage of cells with disrupted mitochondrial membrane potential is determined using Cytometics RXR flow cytometry software and these fluorescence intensities (both FL1 and FL2) are plotted. The change in signal intensity is greater in the FL1 (green) channel due to the intrinsic nature of the MitoCapture reagent; the nonapoptotic and apoptotic populations are more readily discerned using this channel. Thus, it is preferred to focus the analysis on the FL1 channel. Accordingly, cells with reduced FL1 (green) fluorescence are defined as cells with disrupted mitochondrial membrane potential. A low-level FL1 fluorescence intensity peak will be observed in mock-treated cells and corresponds to background fluorescence of the MitoCapture Reagent when it is inside healthy cells. The distribution of cells with peak FL1 fluorescence intensity greater than that of this background fluorescence is defined as the distribution of cells with disrupted mitochondrial membrane potential (disrupted $\Delta\psi_m$ curve). This curve is used to calculate the percentage of cells with disrupted ψ_m using the following equation

$$\% \text{ Cell with disrupted } \Delta\psi_m = \text{area under the disrupted } \Delta\psi_m \text{ curve}/(\text{area under disrupted } \Delta\psi_m \text{ curve} + \text{area under background curve}) \times 100\%$$

3.4.2. Cytochrome c release: Immunoblotting

Biochemical Basis

The opening of the PT pore associated with the intrinsic pathway of apoptosis is in part due to the release of several proteins from the mitochondrial membranes. Many of these released proteins are known to be direct activators of the caspases that are essential for cleavage of cellular homeostatic substrates *(35)*. One such example is cytochrome *c*. Once cytosolic, this mitochondrial protein binds Apaf-1, which associates with the initiator caspase 9, to activate caspase 3 *(42)*. An assessment of released cytochrome *c* in the cytoplasm during HSV-1-dependent apoptosis is determined by immunoblotting after depletion of the mitochondria fraction *(33)*.

Procedure

1. Subconfluent cells grown in a 175-cm² flask (approximately 6×10^7 cells) are infected for desired times, with desired virus at a predetermined MOI.

2. Cells are harvested by gently scrapping directly into the medium.

3. The cells are gently pelleted by low speed centrifugation ($300 \times g$), washed once in PBS, and re-pelletted again.

4. This pellet is now resuspended in 100 µL of PT Lysis Buffer (*see* **Subheading 2**).

5. Samples are incubated on ice for 30 min to swell the cells, transferred to a Dounce homogenizer, and lysed by three gentle strokes with a B type pestle.

6. The solution is transferred to a 1.5-mL tube and centrifuged ($300 \times g$) for 10 min at 25°C.

7. The supernatant is transferred to a fresh tube and frozen on crushed dry ice for 1 min, followed by high-speed centrifugation ($14,000 \times g$) for 10 min to pellet the heavy membrane fraction and mitochondria.

8. The supernatant is removed and represents the mitochondria-depleted cytoplasmic fraction which was then immunoblotted for reactivity with anticytochrome *c* antibody (Pharmingen) *(33)*. The pellet may also be resolved on a denaturing gel and tested for the presence of cytochrome *c* (*see* **Note 8** and **9**).

3.4.3. Cytochrome c *release: Indirect Immunofluorescence*

Biochemical Basis

As a consequence of the organelle disruption that occurs due to HSV-1-mediated microtubule reorganization, the biochemical fractionation approach described above is less than ideal. An appropriate alternative is to visualize the release of cytochrome *c* from mitochondria using indirect immunofluorescence of HSV-1-infected cells *(33)*.

Procedure

1. HEp-2 cell monolayers are grown on glass coverslips in 6-well dishes (*see* **Note 5**).

2. Thirty minutes prior to performing indirect immunofluorescence, infected cells are incubated in the presence 0.5 µM MitoTracker (Molecular Probes) to stain mitochondria.

3. The cells are fixed with 2% (methanol free) paraformaldehyde in PBS for 20 min.

4. Cells are permeablized with 100% acetone at –20°C for 4 min.

5. Anti-cytochrome *c* antibody may be obtained from Pharmingen. Standard indirect immunofluorescence procedures may be performed *(43)*.

6. Infected cell mitochondria will be stained red and may be visualized by fluorescence microscopy.

3.5. Additional Methods

3.5.1. Annexin V Staining

Biochemical Basis

Annexin V is involved in the formation of the voltage-dependent Ca^{2+} channels found within the phospholipid bilayer of cells. This protein binds to phosphotidyl serines (PS) located on the underside of the cellular plasma membrane in a healthy cell. However, a cell undergoing apoptosis translocates these PS moieties to the outer side of the plasma membrane. Annexin V, which remains associated with the PS molecule, can be stained and then assessed by Flow Cytometry. To control for cell viability, the cell-impermeant, dead cell stain propidium iodide (PI) is also added to these cells. Prior to utilizing Annexin V staining, followed by flow cytometry to assess apoptosis during viral infection, a technical note must be considered. Depending on the type of cell that is used in the experiment, the ability to recover cells for FACS analysis, and most importantly, the consistency of results could vary greatly. As a general rule, adherent cells (such as HEp-2, HeLa, or Vero) are more difficult to manipulate for flow cytometry due to their intrinsic ability to stick to one another, the culture dish, and tubes utilized for FACS (*see* **Note 8**). In contrast, cells that are more amenable to flow cytometry include nonadherent cells (Jurkat T cells or dendritic cells). As such, consideration of the cell type being used in such apoptosis studies is essential before performing flow cytometry. Many commercial vendors supply reagents for use in annexin V staining (e.g., Boehringer Mannheim, Roche) and solutions referred to below are now mostly standard regardless of their source.

Procedure

1. Cells are harvested by adding trypsin to the dish and incubating at 37°C for 2 min. Care must be taken during this step (*see* **Note 8**).

2. Transfer cell suspension to a 5-mL tube and spin for 5 min at $300 \times g$.

3. Remove supernatant, resuspend in 200-µL ice cold PBS, transfer to 1.5-mL tube and respin.

4. Remove supernatant and resuspend in 200-µL ice cold Binding Buffer.

5. Transfer to a 5-mL tube suitable to use with a Flow Cytometer.

6. Add 5-µL Annexin V and 2.5-µL of PI.

7. Mix gently, and incubate on ice for 10 min in the dark.

8. Add 100-µL of Dilute Binding Buffer.

9. Analyze as per the specific recommendation of your instrument.

4. Notes

1. It is imperative that the researcher ascertain that the cells they are using are, indeed, capable of undergoing environmentally induced apoptosis. Many cell lines used in laboratories are tumor derived and have accumulated defects in their apoptotic machinery *(44)*. Other cells, like Vero cells, will actually undergo apoptotic cell death but this process takes a longer period of time and its magnitude is not great *(13)*. For this and other reasons, the HEp-2/HeLa cells system has become the prototype system on choice for investigating HSV-dependent apoptosis.

2. A delicate apoptotic balance exists in HSV-1-infected cells between proapoptotic induction and antiapoptotic prevention factors *(19)*. Certain wild-type HSV-1 isolates show a more

pronounced level of apoptosis compared to other standard isolates *(15)*. In general, all wild-type HSV-1 strains will demonstrate apoptosis upon infection if de novo protein synthesis is blocked *(45)* due to the absence of prevention factors. Infection with specific recombinant viruses possessing defined mutations in any one of the three key viral transcriptional regulatory proteins, ICP27, ICP4, and ICP22, also results in enhanced HSV-1-dependent apoptosis *(46)*, since these factors are required for optimal expression of later viral antiapoptotic factors. Finally, viruses possessing specific disruptions in certain viral genes thought to play a role in the prevention process, will also yield increased levels of apoptosis *(19, 28)*.

3. Standard virological methods are used for the infection of cultured cells by HSV-1 *(47)*. Interested readers are directed to additional appropriate methods protocols *(43, 48)*.

4. An important technical caveat worth addressing involves the labeling of DNA to assess apoptosis during infection with HSV-1 (or any DNA virus). During active viral replication (as observed during a wild-type virus infection), a slight to moderate level of TUNEL positive staining may result, indicative of the terminal transferase reaction labeling viral DNA. This could potentially lead to false positive (and thus, inaccurate) results in assessing the consequence of DNA virus infection on cellular apoptosis as measured by alterations in host cell DNA. Appropriate controls to address such experimental issues must be considered prior to utilizing this assay.

5. Our standard protocol for fixing and permeablizing cells uses glass coverslips incubated with cells in 6-well dishes. A detailed description of this method may be found elsewhere *(43)*. While this represents our preferred method, vendors of TUNEL reagents also provide detailed protocols for the specific use of their products.

6. Traditionally, SDS-polyacrylamide denaturing gels crosslinked with DATD are used for analyses of HSV polypeptides *(47)*. DATD forms a less dense matrix than BIS. This specific technique facilitates the maximum resolution of all posttranslationally modified forms of HSV proteins *(49, 50)*. In studies of apoptotic death factors in which control immunoblots for viral proteins will be included, it is recommended that DATD gels are used.

7. Direct immunoblotting of caspases is not easy. Reasons for this include the low levels of these enzymes in cells, their low molecular weight (thus, the need for high percentage acrylamide gels), and the poor quality of commercially available primary antibodies against these factors. Some investigators have resorted to using indirect methods to assay for capases, such as fluorescently labeled peptide substrates. However,

these approaches are far from being accepted in field and are now rarely reported. One problem has been the fact that caspase 3 is actually activated during HSV-1 infection, but it only cleaves a subpopulation of its substrates *(15, 45)*. Thus, as with everything else, all bets are off in infected cells. It is therefore difficult to interpret the findings of such artificial, indirect methods during productive HSV-1 infection. Another problem is the use of antibodies that are "specific" for cleaved caspases, such as caspase 3. The issue here is what the level of "active" caspase really represents (is it 1%, 10%, or 100%), since these reagents cannot assess total caspase protein levels. The recommendation is to use antibodies against full-length (pro-) caspases. In this case, activation is observed as the loss of reactivity with the procaspase. If gel systems have been optimized and good quality anti-caspase antibodies are available, the generation of the cleaved, active caspase may also be observed along with the loss of procaspase *(12)*.

8. Any time infected adherent cells are characterized by flow cytometry, extreme care must be taken when removing the cells from the dish. Fortunately, almost all HSV-1-infected cells should be easily dislodged by sharply "tapping" the dishes. Caution must be used if the only way to harvest the cells is to use a scraper or "rubber policeman." Excessive manipulations may result in elevated levels of background signals that result from the mechanical injury of the cells. If one utilizes PBS-containing EDTA (Versene) to aid in dislodging the cells, appropriate controls must be used to insure that this treatment does not cause an unwanted effect.

9. As above in **Note 8**, care must be taken when harvesting HSV-1-infected cells for determinations of cytochrome *c* release. Mitochondria and all other organelles in infected cells are extremely fragile due to the reorganization of cytoplasmic microtubules induced by HSV-1 *(51, 52)*. While the ideal situation would be to compare immunoblot levels of cytochrome *c* remaining in the mitochondrial pellet with the soluble cytoplasmic fraction, misleading results can occur due to damaged mitochondria.

Acknowledgments

We wish to thank all of the individuals in our laboratory whose hard work has set the basis for developing this interesting new research project. Individuals who played significant roles in generating the protocols and the methodologies that served as the basis of this review include Martine Aubert, Jennifer O'Toole, Renee Baranin, Lisa Pomeranz, Christine Sanfilippo, Renzo Lambardozzi, Natalie Chirimuuta, Margot Goodkin, Elise Morton, Jamie Yedowitz, Marie Nguyen, Rachel Kraft, Kristen Pena, Elisabeth Gennis, Fatima Manzoor, and Leah Kang. These studies were supported in part by grants from the United States Public Health Service (AI38873 and AI48582 to J.A.B.) and the American Cancer Society (JFRA 634 to J.A.B.). J.A. Blaho thanks the Lucille P. Markey Charitable Trust and the National Foundation for Infectious Diseases for their support. C. Cotter is a predoctoral trainee and was supported in part by a United States Public Health Service Institutional Research Training Award (AI 07647).

References

1. Yedowitz, J. C., and Blaho, J. A. (2005). Herpes simplex virus 2 modulates apoptosis and stimulates NF-kappaB nuclear translocation during infection in human epithelial HEp-2 cells, *Virology* **342**, 297–310.

2. Whitley, R. J. (2001). in *Fields Virology* (Roizman, B., and Knipe, D. M., Eds.), pp. 2462–2498, Lippincott-Raven, Philadelphia, PA.

3. Qian, H., and Atherton, S. (2003). Apoptosis and increased expression of Fas ligand after uniocular anterior chamber (AC) inoculation of HSV-1, *Curr Eye Res* **26**, 195–203.

4. Wilson, S. E., Pedroza, L., Beuerman, R., and Hill, J. M. (1997). Herpes simplex virus type-1 infection of corneal epithelial cells induces apoptosis of the underlying keratocytes, *Exp Eye Res* **64**, 775–779.

5. Miles, D., Athmanathan, S., Thakur, A., and Willcox, M. (2003). A novel apoptotic interaction between HSV-1 and human corneal epithelial cells, *Curr Eye Res* **26**, 165–174.

6. DeBiasi, R. L., Kleinschmidt-DeMasters, B. K., Richardson-Burns, S., and Tyler, K. L. (2002). Central nervous system apoptosis in human herpes simplex virus and cytomegalovirus encephalitis, *J Infect Dis* **186**, 1547–1557.

7. Perkins, D., Gyure, K. A., Pereira, E. F., and Aurelian, L. (2003). Herpes simplex virus type 1-induced encephalitis has an apoptotic component associated with activation of c-Jun N-terminal kinase, *J Neurovirol* **9**, 101–111.

8. Koyama, A. H., and Adachi, A. (1997). Induction of apoptosis by herpes simplex virus type 1, *J Gen Virol* **78**, 2909–2912.

9. Jerome, K. R., Chen, Z., Lang, R., Torres, M. R., Hofmeister, J., Smith, S., Fox, R., Froelich, C. J., and Corey, L. (2001). HSV and glycoprotein J inhibit caspase activation and apoptosis induced by granzyme B or Fas, *J Immunol* **167**, 3928–3935.

10. Aubert, M., and Blaho, J. A. (1999). The herpes simplex virus type 1 regulatory protein ICP27 is required for the prevention of apoptosis in infected human cells, *J Virol* **73**, 2803–2813.

11. Gautier, I., Coppey, J., and Durieux, C. (2003). Early apoptosis-related changes triggered by HSV-1 in individual neuronlike cells, *Exp Cell Res* **289**, 174–183.

12. Kraft, R. M., Nguyen, M. L., Yang, X. H., Thor, A. D., and Blaho, J. A. (2006). Caspase 3 activation during herpes simplex virus 1 infection, *Virus Res* **120**, 163–175.

13. Nguyen, M. L., Kraft, R. M., and Blaho, J. A. (2005). African green monkey kidney Vero cells require de novo protein synthesis for efficient herpes simplex virus 1-dependent apoptosis, *Virology* **336**, 274–290.

14. Koyama, A. H., and Miwa, Y. (1997). Suppression of apoptotic DNA fragmentation in herpes simplex virus type 1-infected cells, *J Virol* **71**, 2567–2571.

15. Aubert, M., O'Toole, J., and Blaho, J. A. (1999). Induction and prevention of apoptosis in human HEp-2 cells by herpes simplex virus type 1, *J Virol* **73**, 10359–10370.

16. Galvan, V., and Roizman, B. (1998). Herpes simplex virus 1 induces and blocks apoptosis at multiple steps during infection and protects cells from exogenous inducers in a cell- type-dependent manner, *Proc Natl Acad Sci USA* **95**, 3931–3936.

17. Jerome, K. R., Fox, R., Chen, Z., Sears, A. E., Lee, H., and Corey, L. (1999). Herpes simplex virus inhibits apoptosis through the action of two genes, Us5 and Us3, *J Virol* **73**, 8950–8957.

18. Jerome, K. R., Fox, R., Chen, Z., Sarkar, P., and Corey, L. (2001). Inhibition of apoptosis by primary isolates of herpes simplex virus, *Arch Virol* **146**, 2219–2225.

19. Nguyen, M. L., and Blaho, J. A. (2007). Apoptosis during herpes simplex virus infection, *Adv Virus Res* **69**, 67–97.

20. Moore, A. E., Sabachewsky, L., and Toolan, H. W. (1955). Culture characteristics of four permanent lines of human cancer cells, *Cancer Res.* **15**, 598–605.

21. Nelson-Rees, W. A., Zhdanov, V. M., Hawthorne, P. K., and Flandermeyer, R. R. (1974). HeLa-like marker chromosomes and type-A variant glucose-6-phosphate dehydrogenase isoenzyme in human cell cultures producing Mason-Pfizer monkey virus-like particles, *J Natl Cancer Inst* **53**, 751–757.

22. Chen, T. R. (1988). Re-evaluation of HeLa, HeLa S3, and HEp-2 karyotypes, *Cytogenet Cell Genet* **48**, 19–24.

23. Roizman, B. (1962). Polykaryocytosis induced by viruses, *Proc Natl Acad Sci USA* **48**, 228–234.

24. Avitabile, E., Di Gaeta, S., Torrisi, M. R., Ward, P. L., Roizman, B., and Campadelli-Fiume, G. (1995). Redistribution of microtubules and Golgi apparatus in herpes simplex virus-infected cells and their role in viral exocytosis, *J Virol* **69**, 7472–7482.

25. Heeg, U., Dienes, H. P., Muller, S., and Falke, D. (1986). Involvement of actin-containing microfilaments in HSV-induced cytopathology and the influence of inhibitors of glycosylation, *Arch Virol* **91**, 257–270.

26. Roizman, B., and Roanne, P. R. (1964). Multiplication of herpes simplex virus. II. The relationship between protein synthesis and the duplication of viral DNA in infected HEp-2 cells, *Virology* **22**, 262–269.

27. Roizman, B., and Furlong, D. (1974). in *Comprehensive Virology* (Fraenkel-Conrat, H., and Wagner, R. R., Eds.), pp. 229–403, Plenum, New York, NY.

28. Goodkin, M. L., Morton, E. R., and Blaho, J. A. (2004). Herpes simplex virus infection and apoptosis, *Intl Rev Immunol* **23**, 141–172.

29. Blaho, J. A. (2004). Virus infection and apoptosis (issue II) an introduction: cheating death or death as a fact of life?, *Int Rev Immunol* **23**, 1–6.v

30. Hampar, B., and S. A. Elison. (1961). Chromosomal aberrations induced by an animal virus, *Nature* **192**, 145–147.

31. Aubert, M., Rice, S. A., and Blaho, J. A. (2001). Accumulation of herpes simplex virus type 1 early and leaky-late proteins correlates with apoptosis prevention in infected human HEp-2 cells, *J Virol* **75**, 1013–1030.

32. Liu, X., Zou, H., Slaughter, C., and Wang, X. (1997). DFF, a heterodimeric protein that functions downstream of caspase-3 to trigger DNA fragmentation during apoptosis, *Cell* **89**, 175.

33. Aubert, M., Pomeranz, L. E., and Blaho, J. A. (2007). HSV blocks apoptosis by precluding mitochondrial cytochrome c release independent of caspase activation in infected human epithelial cells, *Apoptosis* **12**, 19–35.

34. Cryns, V., and Yuan, J. (1998). Proteases to die for, *Genes Dev* **12**, 1551–1570.

35. Green, D. R. (1998). Apoptotic pathways: the roads to ruin, *Cell* **94**, 695–698.

36. Salvesen, G. S., and Dixit, V. M. (1997). Caspases: intracellular signaling by proteolysis, *Cell* **91**, 443–446.

37. Vaux, D. L., and Strasser, A. (1996). The molecular biology of apoptosis, *Proc Natl Acad Sci USA* **93**, 2239–2244.

38. Wasilenko, S. T., Meyers, A. F., Vander Helm, K., and Barry, M. (2001). Vaccinia virus infection disarms the mitochondrion-mediated pathway of the apoptotic cascade by modulating the permeability transition pore, *J Virol* **75**, 11437–11448.

39. Green, D. R., and Reed, J. C. (1998). Mitochondria and apoptosis, *Science* **281**, 1309–1312.

40. Petit, P. X., Susin, S. A., Zamzami, N., Mignotte, B., and Kroemer, G. (1996). Mitochondria and programmed cell death: back to the future, *FEBS Lett* **396**, 7–13.

41. Qian, T., Nieminen, A. L., Herman, B., and Lemasters, J. J. (1997). Mitochondrial permeability transition in pH-dependent reperfusion injury to rat hepatocytes, *Am J Physiol* **273**, C1783–C1792.

42. Li, P., Nijhawan, D., Budihardjo, I., Srinivasula, S. M., Ahmad, M., Alnemri, E. S., and Wang, X. (1997). Cytochrome c and dATP-dependent formation of Apaf-1/caspase-9 complex initiates an apoptotic protease cascade, *Cell* **91**, 479–489.

43. Blaho, J. A., Morton, E. R., and Yedowitz, J. C. (2005). Herpes Simplex Virus: Propagation, Quantification, and Storage, *Curr Protoc Microbiol* **14E**, 1–23.

44. Nguyen, M. L., Kraft, R. M., and Blaho, J. A. (2007). Susceptibility of cancer cells to herpes simplex virus-dependent apoptosis, *J Gen Virol* **88**, 1866–1875.

45. Aubert, M., and Blaho, J. A. (2001). Modulation of apoptosis during herpes simplex virus infection in human cells, *Microbes Infect* **3**, 859–866.

46. Sanfilippo, C. M., and Blaho, J. A. (2006). ICP0 gene expression is a herpes simplex virus type 1 apoptotic trigger, *J Virol* **80**, 6810–6821.

47. Blaho, J. A., and Roizman, B. (1998). in *Methods in Molecular Medicine: Herpes Simplex Virus Protocols* (Brown, S. M., and Maclean, A. R., Eds.), pp. 237–256, Human Press, Totowa, NJ.

48. Brown, S. M., and MacLean, A. R., Eds. (1998). *Herpes Simplex Virus Protocols: Methods in Molecular Medicine*, Vol. 10, Human Press, Totowa, NJ.

49. Blaho, J. A., Mitchell, C., and Roizman, B. (1993). Guanylylation and adenylylation of the alpha regulatory proteins of herpes simplex virus require a viral beta or gamma function, *J Virol* **67**, 3891–3900.

50. Blaho, J. A., Zong, C. S., and Mortimer, K. A. (1997). Tyrosine phosphorylation of the herpes simplex virus type 1 regulatory protein ICP22 and a cellular protein which shares antigenic determinants with ICP22, *J Virol* **71**, 9828–9832.

51. Yedowitz, J. C., Kotsakis, A., Schlegel, E. F., and Blaho, J. A. (2005). Nuclear localizations of the herpes simplex virus type 1 tegument proteins VP13/14, vhs, and VP16 precede VP22-dependent microtubule reorganization and VP22 nuclear import, *J Virol* **79**, 4730–4743.

52. Kotsakis, A., Pomeranz, L. E., Blouin, A., and Blaho, J. A. (2001). Microtubule reorganization during herpes simplex virus type 1 infection facilitates the nuclear localization of VP22, a major virion tegument protein, *J Virol* **75**, 8697–8711.

C

Printed by Books on Demand, Germany